# Mathematical Olympiads
# 2000–2001

## Problems and Solutions
## From Around the World

© *2003 by*
*The Mathematical Association of America (Incorporated)*
*Library of Congress Catalog Card Number 2003103274*
ISBN 0-88385-810-X
*Printed in the United States of America*
Current Printing (last digit):
10 9 8 7 6 5 4 3 2 1

# Mathematical Olympiads
# 2000–2001

## Problems and Solutions
## From Around the World

Edited by
Titu Andreescu,
Zuming Feng,
and
George Lee, Jr.

*Published and distributed by*
The Mathematical Association of America

# MAA PROBLEM BOOKS SERIES

Problem Books is a series of the Mathematical Association of America consisting of collections of problems and solutions from annual mathematical competitions; compilations of problems (including unsolved problems) specific to particular branches of mathematics; books on the art and practice of problem solving, etc.

*A Friendly Mathematics Competition: 35 Years of Teamwork in Indiana,* edited by Rick Gillman

*The Inquisitive Problem Solver,* Paul Vaderlind, Richard K. Guy, and Loren C. Larson

*Mathematical Olympiads 1998–1999: Problems and Solutions From Around the World,* edited by Titu Andreescu and Zuming Feng

*Mathematical Olympiads 1999–2000: Problems and Solutions From Around the World,* edited by Titu Andreescu and Zuming Feng

*Mathematical Olympiads 2000–2001: Problems and Solutions From Around the World,* edited by Titu Andreescu, Zuming Feng, and George Lee, Jr.

*The William Lowell Putnam Mathematical Competition 1985–2000: Problems, Solutions, and Commentary,* Kiran S. Kedlaya, Bjorn Poonen, Ravi Vakil

*USA and International Mathematical Olympiads 2000,* edited by Titu Andreescu and Zuming Feng

*USA and International Mathematical Olympiads 2001,* edited by Titu Andreescu and Zuming Feng

MAA Service Center
P. O. Box 91112
Washington, DC 20090-1112
1-800-331-1622     fax: 1-301-206-9789
www.maa.org

# Contents

**Preface**      **vii**

**Acknowledgments**      **ix**

**1   2000 National Contests: Problems and Solutions**      **1**

    1.1   Belarus . . . . . . . . . . . . . . . . . . . . . . . . . . .   1

    1.2   Bulgaria . . . . . . . . . . . . . . . . . . . . . . . . . .   12

    1.3   Canada . . . . . . . . . . . . . . . . . . . . . . . . . . .   24

    1.4   China . . . . . . . . . . . . . . . . . . . . . . . . . . . .   27

    1.5   Czech and Slovak Republics . . . . . . . . . . . . . . .   35

    1.6   Estonia . . . . . . . . . . . . . . . . . . . . . . . . . . .   40

    1.7   Hungary . . . . . . . . . . . . . . . . . . . . . . . . . .   46

    1.8   India . . . . . . . . . . . . . . . . . . . . . . . . . . . .   52

    1.9   Iran . . . . . . . . . . . . . . . . . . . . . . . . . . . .   55

    1.10   Israel . . . . . . . . . . . . . . . . . . . . . . . . . . .   65

    1.11   Italy . . . . . . . . . . . . . . . . . . . . . . . . . . . .   68

    1.12   Japan . . . . . . . . . . . . . . . . . . . . . . . . . . .   72

    1.13   Korea . . . . . . . . . . . . . . . . . . . . . . . . . . .   77

    1.14   Mongolia . . . . . . . . . . . . . . . . . . . . . . . . .   81

    1.15   Poland . . . . . . . . . . . . . . . . . . . . . . . . . . .   88

    1.16   Romania . . . . . . . . . . . . . . . . . . . . . . . . . .   93

    1.17   Russia . . . . . . . . . . . . . . . . . . . . . . . . . . .   101

    1.18   Taiwan . . . . . . . . . . . . . . . . . . . . . . . . . . .   137

    1.19   Turkey . . . . . . . . . . . . . . . . . . . . . . . . . . .   142

    1.20   United Kingdom . . . . . . . . . . . . . . . . . . . . .   151

    1.21   United States of America . . . . . . . . . . . . . . . . .   153

    1.22   Vietnam . . . . . . . . . . . . . . . . . . . . . . . . . .   161

**2  2000 Regional Contests: Problems and Solutions**          **167**
2.1  Asian Pacific Mathematical Olympiad . . . . . . . . . .  167
2.2  Austrian-Polish Mathematics Competition . . . . . . . .  172
2.3  Balkan Mathematical Olympiad . . . . . . . . . . . . .  177
2.4  Mediterranean Mathematical Competition . . . . . . . .  181
2.5  St. Petersburg City Mathematical Olympiad (Russia) . . .  185

**3  2001 National Contests: Problems**                          **209**
3.1  Belarus . . . . . . . . . . . . . . . . . . . . . . . . .  209
3.2  Bulgaria . . . . . . . . . . . . . . . . . . . . . . . . .  212
3.3  Canada . . . . . . . . . . . . . . . . . . . . . . . . .  215
3.4  China . . . . . . . . . . . . . . . . . . . . . . . . . .  216
3.5  Czech and Slovak Republics . . . . . . . . . . . . . . .  219
3.6  Hungary . . . . . . . . . . . . . . . . . . . . . . . . .  220
3.7  India . . . . . . . . . . . . . . . . . . . . . . . . . . .  222
3.8  Iran . . . . . . . . . . . . . . . . . . . . . . . . . . .  223
3.9  Japan . . . . . . . . . . . . . . . . . . . . . . . . . .  226
3.10 Korea . . . . . . . . . . . . . . . . . . . . . . . . . .  227
3.11 Poland . . . . . . . . . . . . . . . . . . . . . . . . . .  229
3.12 Romania . . . . . . . . . . . . . . . . . . . . . . . . .  230
3.13 Russia . . . . . . . . . . . . . . . . . . . . . . . . . .  233
3.14 Taiwan . . . . . . . . . . . . . . . . . . . . . . . . . .  241
3.15 United States of America . . . . . . . . . . . . . . . . .  243
3.16 Vietnam . . . . . . . . . . . . . . . . . . . . . . . . .  245

**4  2001 Regional Contests: Problems**                          **247**
4.1  Asian Pacific Mathematical Olympiad . . . . . . . . . .  247
4.2  Austrian-Polish Mathematics Competition . . . . . . . .  249
4.3  Balkan Mathematical Olympiad . . . . . . . . . . . . .  251
4.4  Baltic Mathematics Competition . . . . . . . . . . . . .  252
4.5  Czech-Slovak-Polish Match . . . . . . . . . . . . . . .  253
4.6  St. Petersburg City Mathematical Olympiad (Russia) . . .  255

**Glossary**                                                     **259**

**Classification of Problems**                                   **277**

# *Preface*

This book is a continuation of *Mathematical Olympiads 1999–2000: Problems and Solutions From Around the World*, published by the Mathematical Association of America. It contains solutions to the problems from 27 national and regional contests featured in the earlier book, together with selected problems (without solutions) from national and regional contests given during 2001. In many cases multiple solutions are provided in order to encourage students to compare different problem-solving strategies.

This collection is intended as practice for the serious student who wishes to improve his or her performance on the USA Math Olympiad (USAMO) and Team Selection Test (TST). Some of the problems are comparable to the USAMO in that they came from national contests. Others are harder, as some countries first have a national Olympiad, and later one or more exams to select a team for the IMO. Some problems come from regional international contests ("mini-IMOs").

Different nations have different mathematical cultures, so you will find some of these problems extremely hard and some rather easy. We have tried to present a wide variety of problems, especially from those countries that have often done well at the IMO.

Each contest has its own time limit. We have not furnished this information, because we have not always included complete exams. As a rule of thumb, most contests allow time ranging between one-half to one full hour per problem.

The problems themselves should provide much enjoyment for all those fascinated by solving challenging mathematics questions.

# Acknowledgments

Thank you to the following participants of the Mathematical Olympiad Summer Program who helped in preparing and proofreading solutions: Reid Barton, Steve Byrnes, Gabriel Carroll, Kamaldeep Gandhi, Stephen Guo, Luke Gustafson, Michael Hamburg, Daniel Jerison, Daniel Kane, Ian Le, Tiankai Liu, Po-Ru Loh, Sean Markan, Alison Miller, Christopher Moore, Gregory Price, Michael Rothenberg, Inna Zakharevich, Tony Zhang, and Yan Zhang. A special thank you to Kiran Kedlaya for all his help.

Without their efforts this work would not have been possible.

Titu Andreescu    Zuming Feng    George Lee, Jr.

# I

# 2000 National Contests: Problems and Solutions

## 1.1 Belarus

### Problem 1

Let $M$ be the intersection point of the diagonals $AC$ and $BD$ of a convex quadrilateral $ABCD$. The bisector of angle $ACD$ hits ray $BA$ at $K$. If $MA \cdot MC + MA \cdot CD = MB \cdot MD$, prove that $\angle BKC = \angle CDB$.

**Solution.** Let $N$ be the intersection of lines $CK$ and $BD$. By the Angle Bisector Theorem applied to triangle $MCD$, $\frac{CD}{DN} = \frac{MC}{MN}$, or $CD = \frac{MC \cdot DN}{MN}$. We then have

$$MB \cdot MD = MA \cdot MC + MA \cdot \frac{MC \cdot ND}{MN} = (MA \cdot MC) \cdot \frac{MD}{MN},$$

or $MA \cdot MC = MB \cdot MN$. Because $M$ lies inside quadrilateral $ABCN$, the Power of a Point Theorem implies that $A$, $B$, $C$, and $N$ are concyclic. Hence, $\angle KBD = \angle ABN = \angle ACN = \angle NCD = \angle KCD$, implying that $K, B, C,$ and $D$ are concyclic. Thus, $\angle BKC = \angle CDB$, as desired.

### Problem 2

In an equilateral triangle of $\frac{n(n+1)}{2}$ pennies, with $n$ pennies along each side of the triangle, all but one penny shows heads. A *move* consists of choosing two adjacent pennies with centers $A$ and $B$ and flipping every penny on line $AB$. Determine all initial arrangements — the value of $n$ and the position of the coin initially showing tails — from which one can make all the coins show tails after finitely many moves.

**Solution.** Every move flips 0 or 2 of the coins in the corners, so the parity of the number of heads in the three corners is preserved. If the coin showing tails is not in a corner, all three coins in the corners initially show heads, so there will always be an odd number of heads in the corners. Hence, the three corners will never simultaneously show tails. Conversely, if the coin showing tails is in a corner, we prove that we can make all the coins show tails. Orient the triangle to make the side opposite that corner horizontal. In each of the $n - 1$ horizontal rows of two or more coins, choose two adjacent pennies and flip all the coins in that row; all the coins will then show tails. Therefore, the desired initial arrangements are those in which the coin showing tails is in the corner.

## Problem 3

We are given triangle $ABC$ with $\angle C = \pi/2$. Let $M$ be the midpoint of the hypotenuse $\overline{AB}$, $H$ be the foot of the altitude $\overline{CH}$, and $P$ be a point inside the triangle such that $AP = AC$. Prove that $\overline{PM}$ bisects angle $BPH$ if and only if $\angle A = \pi/3$.

**First Solution.** Point $P$ lies on the circle $\omega$ centered at $A$ with radius $AC$. Let $\omega$ intersect lines $CH$ and $PH$ at $D$ and $Q$, respectively, and let $\omega$ intersect ray $AB$ at $N$. Because $MA = MC$, $\angle A = \pi/3$ if and only if triangle $ACM$ is equilateral, i.e., if and only if $M = N$. Thus, it suffices to show that $\overline{PM}$ bisects angle $HPB$ if and only if $M = N$.

Because $\overline{AH}$ is the altitude to the base of isosceles triangle $ACD$, $H$ is the midpoint of $\overline{CD}$ and hence lies in $\omega$. By the Power of a Point Theorem, $PH \cdot HQ = CH \cdot HD = CH^2$. Because $\overline{CH}$ is the altitude to the hypotenuse of right triangle $ABC$, $CH^2 = AH \cdot HB$. Hence, $PH \cdot HQ = AH \cdot HB$, and because $H$ lies on segments $\overline{AB}$ and $\overline{PQ}$, quadrilateral $APBQ$ must be cyclic in that order. Note also that in circle $\omega$, $\angle QAB = \angle QAN = 2\angle QPN = 2\angle HPN$. Thus, $\angle HPB = \angle QPB = \angle QAB = 2\angle HPN$, and because $N$ lies on $\overline{HB}$ it follows that segment $\overline{PN}$ bisects angle $HPB$. Therefore, segment $\overline{PM}$ bisects angle $HPB$ if and only if $M = N$, as desired.

**Second Solution.** Without loss of generality, assume that $AC = 1$. Introduce coordinate axes such that $C$ is the origin, $A$ has coordinates $(0,1)$, and $B$ has coordinates $(n,0)$ where $n > 0$. If $n = 1$, then $M = H$ and then $\overline{PM}$ cannot bisect angle $BPH$. In this case, $\angle A = \pi/4 \neq \pi/3$, consistent with the desired result. Thus, we can disregard this case and assume that $n \neq 1$. Using the distance formula, we find that $AP = AC$

if and only if $P$ has coordinates of the form $(\pm\sqrt{(m)(2-m)}, m)$ for some $m$ between 0 and 2. It is clear that $M$ has coordinates $(n/2, 1/2)$, and, because $CH$ has slope $n$ and $H$ lies on $AB$, we find that $H$ has coordinates $(n/(n^2+1), n^2/(n^2+1))$. Using the distance formula twice and simplifying with some calculations yields $BP/HP = \sqrt{n^2+1}$.

Also, comparing ratios in similar right triangles $AHC$ and $ACB$ shows that $AH = b^2/c$, where $b = CA$ and $c = AB$. Therefore,

$$\frac{MB}{MH} = \frac{c/2}{c/2 - b^2/c} = \frac{c^2}{c^2 - 2b^2} = \frac{n^2+1}{n^2-1}.$$

By the Angle Bisector Theorem, $\overline{PM}$ bisects angle $BPH$ if and only if $BP/HP = MB/MH$. Equating the expressions found above, we find that this is true if and only if $n^2(n^2-3) = 0$. Because $n > 0$, it follows that $\overline{PM}$ bisects angle $BPH$ if and only if $n = \sqrt{3}$, i.e., if and only if $\angle A = \pi/3$.

## Problem 4

Does there exist a function $f : \mathbb{N} \to \mathbb{N}$ such that

$$f(f(n-1)) = f(n+1) - f(n)$$

for all $n \geq 2$?

**Solution.** The answer is "no". For the sake of contradiction, assume that such a function exists. From the given equation, $f(n+1) - f(n) > 0$ for $n \geq 2$, implying that $f$ is strictly increasing for $n \geq 2$. Thus, $f(n) \geq f(2) + (n-2) \geq n-1$ for all $n \geq 2$.

We can also bound $f(n)$ from above: the given equation implies that $f(f(n-1)) < f(n+1)$ for $n \geq 2$, or equivalently that

$$f(f(n)) < f(n+2)$$

for $n \geq 1$. Because $f$ is increasing on values greater than 1, this inequality implies that either $f(n) = 1$ or $f(n) < n+2$ for all $n \geq 1$. In either case, $f(n) < n+2$.

Hence, $n-1 \leq f(n) \leq n+1$ for all $n \geq 2$. Let $n$ be an arbitrary integer greater than 4. On the one hand, $f(n-1) \geq 2$ and $n-1 \geq 2$ so that applying our lower bound twice yields

$$f(f(n-1)) \geq f(n-1) - 1 \geq n-3.$$

On the other hand, from the given equation we have

$$f(f(n-1)) = f(n+1) - f(n) \le (n+2) - (n-1) = 3.$$

Thus, $n - 3 \le 3$ for arbitrary $n > 4$, which is impossible. Therefore, our original assumption was incorrect, and no such function exists.

## Problem 5

In a convex polyhedron with $m$ triangular faces (and possibly faces of other shapes), exactly four edges meet at each vertex. Find the minimum possible value of $m$.

**Solution.** Consider a polyhedron with $m$ triangular faces and four edges meeting at each vertex. Let $F$, $E$, and $V$ be the number of faces, edges, and vertices, respectively, of the polyhedron. For each edges, count the 2 vertices at its endpoints; because each vertex is the endpoint of exactly 4 edges, we count each vertex 2 times in this fashion. Hence, $2E = 4V$. Also, counting the number of edges on each face and summing the $F$ tallies yields a total of at least $3m + 4(F - m)$. Every edge is counted twice in this manner, implying that $2E \ge 3m + 4(F - m)$.

By Euler's formula for planar graphs, $F + V - E = 2$. Combined with $2E = 4V$, this equation yields $2E = 4F - 8$. Thus,

$$4F - 8 = 2E \ge 3m + 4(F - m),$$

or $m \ge 8$. Equality occurs if and only if every face of the polyhedron is triangular or quadrilateral. A regular octahedron provides an example of such a polyhedron, implying that $m = 8$ is indeed attainable.

## Problem 6

(a) Prove that $\{n\sqrt{3}\} > \frac{1}{n\sqrt{3}}$ for every positive integer $n$, where $\{x\}$ denotes the fractional part of $x$.

(b) Does there exist a constant $c > 1$ such that $\{n\sqrt{3}\} > \frac{c}{n\sqrt{3}}$ for every positive integer $n$?

**Solution.** The condition $\{n\sqrt{3}\} > c/n\sqrt{3}$ can hold for $n = 1$ only if $\sqrt{3} - 1 > c/\sqrt{3}$, i.e., only if $3 - \sqrt{3} > c$. Let $c \in [1, 3 - \sqrt{3})$ be such a constant.

For each $n$, $\{n\sqrt{3}\} = n\sqrt{3} - \lfloor n\sqrt{3} \rfloor$ is greater than $c/n\sqrt{3}$ if and only if $n\sqrt{3} - c/n\sqrt{3} > \lfloor n\sqrt{3} \rfloor$. Because $c < 3 - \sqrt{3} < \sqrt{3} < 3n^2$, both sides

of this inequality are positive, and we may square each side to obtain the
equivalent inequality

$$3n^2 - 2c + \frac{c^2}{3n^2} > \lfloor n\sqrt{3} \rfloor^2. \qquad (*)$$

For each $n$, $3n^2 - 1$ is not a perfect square because no perfect square is
congruent to 2 modulo 3, and $3n^2$ is also not a perfect square. Therefore,
$\lfloor n\sqrt{3} \rfloor = \lfloor \sqrt{3n^2} \rfloor$ — the largest integer whose square is less than or equal
to $3n^2$ — is at most $\sqrt{3n^2 - 2}$, with equality if and only if $3n^2 - 2$ is a
perfect square. We claim that equality indeed holds for arbitrarily large $n$.
Define $(m_0, n_0) = (1, 1)$ and $(m_{k+1}, n_{k+1}) = (2m_k + 3n_k, m_k + 2n_k)$
for $k \geq 1$. It is easily verified that $m_{k+1}^2 - 3n_{k+1}^2 = m_k^2 - 3n_k^2$. Thus,
because the equation $3n_k^2 - 2 = m_k^2$ holds for $k = 0$, it holds for all $k \geq 1$.
Because $n_1, n_2, \ldots$ is an increasing sequence, it follows that $3n^2 - 2$ is a
perfect square for arbitrarily large $n$, as needed.

If $c = 1$, then

$$3n^2 - 2c + \frac{c^2}{3n^2} > 3n^2 - 2c = 3n^2 - 2 \geq \lfloor n\sqrt{3} \rfloor^2$$

for all $n$. Thus, $(*)$ and hence the inequality in (a) holds for all $n$.

However, if $c > 1$, then

$$3n^2 - 2c + \frac{c^2}{3n^2} \leq 3n^2 - 2$$

for all sufficiently large $n$. Thus, there exists such an $n$ with the additional
property that $3n^2 - 2$ is a perfect square. For this $n$, $(*)$ and hence the
inequality in (b) fails. Therefore, the answer to the question in part (b) is
"no."

## Problem 7

Let $M = \{1, 2, \ldots, 40\}$. Find the smallest positive integer $n$ for which it
is possible to partition $M$ into $n$ disjoint subsets such that whenever $a$, $b$,
and $c$ (not necessarily distinct) are in the same subset, $a \neq b + c$.

**Solution.** The answer is 4. Assume, for the sake of contradiction, that it
is possible to partition $M$ into 3 such sets $X$, $Y$, and $Z$. One of the sets —
say, $X$ — has at least 6 elements $x_1, x_2, \ldots, x_6$ with $x_6 > x_1, x_2, \ldots, x_5$.
The differences $x_6 - x_1, x_6 - x_2, \ldots, x_6 - x_5$ lie in $M$. However, none
of these differences $x_6 - x_k$ can lie in $X$ because then setting $a = x_6$,
$b = x_6 - x_k$, $c = x_k$ shows that $X$ does not satisfy the given condition.

By the Pigeonhole Principle, one of the sets $Y$ and $Z$ — say, $Y$ — contains some three of the differences $x_6 - x_k$ ($1 \leq k \leq 5$). Without loss of generality, assume that $y_1 = x_6 - x_1$, $y_2 = x_6 - x_2$, and $y_3 = x_6 - x_3$ lie in $Y$, where $x_1 < x_2 < x_3$ and $y_1 > y_2 > y_3$.

For $1 \leq j < k \leq 3$, the value $x_k - x_j$ is in $M$ but cannot be in $X$, because otherwise $(x_j) + (x_k - x_j) = x_k$. Similarly, $y_j - y_k \notin Y$ for $1 \leq j < k \leq 3$. Therefore, the three common differences $x_2 - x_1 = y_1 - y_2$, $x_3 - x_2 = y_2 - y_3$, and $x_3 - x_1 = y_1 - y_3$ are in $M \setminus (X \cup Y) = Z$. However, setting $a = x_3 - x_1$, $b = x_3 - x_2$, and $c = x_2 - x_1$, we have $a = b + c$ and $a, b, c \in Z$, a contradiction. Therefore, our original assumption was incorrect, and it is impossible to partition $M$ into three sets with the desired property.

We now prove that $M$ is the union of 4 sets $B_1, B_2, B_3, B_4$ (not necessarily disjoint) such that $a \neq b + c$ whenever $a, b, c$ lie in the same $B_k$. (Removing duplicate elements from the $B_k$ will yield a partition of $M$ with the required property.)

For each positive integer $k$, let $A_k$ consist of those integers congruent to an integer in $(\frac{1}{2} \cdot 3^{k-1}, 3^{k-1}]$ modulo $3^k$. We first claim that there do not exist $a, b, c$ in the same $A_k$ such that $a = b + c$. Indeed, if $b, c \in A_k$, then $b + c$ is congruent to an integer in $(3^{k-1}, 2 \cdot 3^{k-1}]$ modulo $3^k$, so that $b + c \notin A_k$.

We next prove by induction on $k \geq 1$ that an integer $s$ lies in $\bigcup_{i=1}^{k} A_i$ if $s \equiv t \pmod{3^k}$ for some $t \in [1, \frac{1}{2}(3^k - 1)]$. For $k = 1$, the claim follows from the definition of $A_1$. Now assuming that the claim is true for $k = n$, we prove that it is true for $k = n + 1$. Suppose that $s \equiv t \pmod{3^{n+1}}$, where $t \in [1, \frac{1}{2}(3^{n+1} - 1)]$. Then one of the following conditions holds:

- $t \in [1, \frac{1}{2}(3^n - 1)]$. Note that $s \equiv t \pmod{3^n}$. By the induction hypothesis, $s \in \bigcup_{i=1}^{n} A_i \subseteq \bigcup_{i=1}^{n+1} A_i$.

- $t \in [\frac{1}{2}(3^n + 1), 3^n]$. Then $s \in A_{n+1} \subseteq \bigcup_{i=1}^{n+1} A_i$ by the definition of $A_{n+1}$.

- $t \in [3^n + 1, \frac{1}{2}(3^{n+1} - 1)]$. Then $s \equiv t - 3^n \pmod{3^n}$, and $t - 3^n \in [1, \frac{1}{2}(3^n - 1)]$. By the induction hypothesis, $s \in \bigcup_{i=1}^{n} A_i \subseteq \bigcup_{i=1}^{n+1} A_i$.

In all three cases, $s \in \bigcup_{i=1}^{n+1} A_i$, as desired. This completes the induction.

Therefore, $S = \{1, 2, \ldots, \frac{1}{2}(3^k - 1)\}$ is the union of the $k$ sets $B_1 = A_1 \cap S$, $B_2 = A_2 \cap S$, $\ldots$, $B_k = A_k \cap S$. As we showed before, $a \neq b + c$ whenever $a, b, c$ lie in the same set $A_i$, and hence whenever they lie in the same set $B_i$. Setting $k = 4$ yields the desired union $M = B_1 \cup B_2 \cup B_3 \cup B_4$.

**Note.** For $n, k \in \mathbb{Z}^+$ and a partition of $\{1, 2, \ldots, k\}$ into $n$ sets, a triple $(a, b, c)$ such that $a + b = c$ and $a, b, c$ are in the same set is called a *Schur triple*. For each $n \in \mathbb{Z}^+$, there exists a maximal integer $k$ such that there are no Schur triples for some partition $\{1, 2, \ldots, k\}$ into $n$ sets; this integer is denoted by $S(n)$ and is called the *nth Schur number*. (Sometimes, $S(n) + 1$ is called the $n$th Schur number.) Although lower and upper bounds exist for all $S(n)$, no general formula is known. The lower bound found in this solution is sharp for $n = 1, 2, 3$, but $S(4) = 44$.

## Problem 8

A positive integer is called *monotonic* if its digits in base 10, read from left to right, are in nondecreasing order. Prove that for each $n \in \mathbb{N}$, there exists an $n$-digit monotonic number which is a perfect square.

**Solution.** Any 1-digit perfect square (namely, 1, 4, or 9) is monotonic, proving the claim for $n = 1$. We now assume $n > 1$.

If $n$ is odd, write $n = 2k - 1$ for an integer $k \geq 2$, and let

$$x_k = (10^k + 2)/6 = 1\underbrace{66\ldots6}_{k-2}7.$$

Then

$$x_k^2 = \frac{10^{2k} + 4 \cdot 10^k + 4}{36} = \frac{10^{2k}}{36} + \frac{10^k}{9} + \frac{1}{9}. \qquad (*)$$

Observe that

$$\frac{10^{2k}}{36} = 10^{2k-2}\left(\frac{72}{36} + \frac{28}{36}\right)$$

$$= 2 \cdot 10^{2k-2} + 10^{2k-2} \cdot \frac{7}{9}$$

$$= 2\underbrace{77\ldots7}_{2k-2} + \frac{7}{9}.$$

Thus, the right-hand side of (*) equals

$$\left(2\underbrace{77\ldots7}_{2k-2} + \frac{7}{9}\right) + \left(\underbrace{11\ldots1}_{k} + \frac{1}{9}\right) + \frac{1}{9} = 2\underbrace{77\cdots7}_{k-2}8\underbrace{8\cdots8}_{k-1}9,$$

an $n$-digit monotonic perfect square.

If $n$ is even, write $n = 2k$ for an integer $k \geq 1$, and let

$$y_k = \frac{10^k + 2}{3} = \underbrace{33\ldots34}_{k-1}.$$

Then

$$y_k^2 = (10^{2k} + 4 \cdot 10^k + 4)/9$$
$$= \frac{10^{2k}}{9} + 4 \cdot \frac{10^k}{9} + \frac{4}{9}$$
$$= \left(\underbrace{11\ldots1}_{2k} + \frac{1}{9}\right) + \left(\underbrace{44\ldots4}_{k} + \frac{4}{9}\right) + \frac{4}{9}$$
$$= \underbrace{11\ldots1}_{k}\underbrace{55\ldots56}_{k-1},$$

an $n$-digit monotonic perfect square. This completes the proof.

## Problem 9

Given a pair $(\vec{r}, \vec{s})$ of vectors in the plane, a *move* consists of choosing a nonzero integer $k$ and then changing $(\vec{r}, \vec{s})$ to either (i) $(\vec{r} + 2k\vec{s}, \vec{s})$ or (ii) $(\vec{r}, \vec{s} + 2k\vec{r})$. A *game* consists of applying a finite sequence of moves, alternating between moves of types (i) and (ii), to some initial pair of vectors.

(a) Is it possible to obtain the pair $((1,0),(2,1))$ during a game with initial pair $((1,0),(0,1))$, if the first move is of type (i)?

(b) Find all pairs $((a,b),(c,d))$ that can be obtained during a game with initial pair $((1,0),(0,1))$, where the first move can be of either type.

**Solution.** Let $\|\vec{z}\|$ denote the length of vector $\vec{z}$, and let $|z|$ denote the absolute value of the real number $z$.

(a) Let $(\vec{r}, \vec{s})$ be the pair of vectors, where $\vec{r}$ and $\vec{s}$ change throughout the game. Observe that if $\vec{x}, \vec{y}$ are vectors such that $\|\vec{x}\| < \|\vec{y}\|$, then

$$\|\vec{x} + 2k\vec{y}\| \geq \|2k\vec{y}\| - \|\vec{x}\| > 2\|\vec{y}\| - \|\vec{y}\| = \|\vec{y}\|.$$

After the first move of type (i), we have $\vec{r} = (1, 2k)$ and $\vec{s} = (0, 1)$ for some nonzero $k$ so that $\|\vec{r}\| > \|\vec{s}\|$. Applying the above result with $\vec{x} = \vec{s}$ and $\vec{y} = \vec{r}$, we see that after the next move (of type (ii)), the magnitude of $\vec{r}$ does not change while that of $\vec{s}$ increases to over $\|\vec{r}\|$. Applying the above result again with $\vec{x} = \vec{r}$ and $\vec{y} = \vec{s}$, we see that after the next move (of type (i)), the magnitude of $\vec{s}$ remains the same while that of $\vec{r}$ increases to over $\|\vec{s}\|$. Continuing in this fashion, we find that $\|\vec{r}\|$ and $\|\vec{s}\|$ never decrease as a result of a move. Because after the very first move, the first vector has magnitude greater than 1, we can never obtain $((1,0),(2,1))$.

(b) We modify the game slightly by not requiring that moves alternate between types (i) and (ii) and by allowing the choice $k = 0$. Of course, any pair that can be obtained under the original rules can be obtained under these new rules as well. The converse is true as well: by repeatedly discarding any moves under the new rules with $k = 0$ and combining any adjacent moves of the same type into one move, we obtain a sequence of moves valid under the original rules that yields the same pair.

Let $((w,x),(y,z))$ represent the pair of vectors, where $w$, $x$, $y$, and $z$ change throughout the game. It is easy to verify that the value of $wz - xy$, and the parity of $x$ and $y$, are invariant under any move in the game. In a game that starts with $((w,x),(y,z)) = ((1,0),(0,1))$, we must always have $wz - xy = 1$ and $x \equiv y \equiv 0 \pmod 2$. Because $x$ and $y$ are always even, $w$ and $z$ remain constant modulo 4 as well; specifically, we must have $w \equiv z \equiv 1 \pmod 4$ throughout the game.

Call a pair $((a,b),(c,d))$ *desirable* when $ad - bc = 1$, $a \equiv d \equiv 1 \pmod 4$, and $b \equiv c \equiv 0 \pmod 2$. Above we showed that any pair obtainable during a game with initial pair $((1,0),(0,1)$ must be desirable; we now prove the converse. Assume, for the sake of contradiction, that there are desirable pairs $((a,b),(c,d))$ that are *not* obtainable; let $((e,f),(g,h))$ be such a pair for which $|ac|$ is minimal.

If $g = 0$, then $eh = 1 + fg = 1$; because $e \equiv h \equiv 1 \pmod 4$, $e = h = 1$. If $f = 0$, the pair is clearly obtainable. Otherwise, by performing a move of type (i) with $k = f/2$, we can transform $((1,0),(0,1))$ into the pair $((e,f),(g,h))$, a contradiction.

Thus, $g \neq 0$. Because $g$ is even and $e$ is odd, either $|e| > |g|$ or $|g| > |e|$. In the former case, $e - 2k_0 g$ is in the interval $(-|e|, |e|)$ for some $k_0 \in \{1, -1\}$. Performing a type-(i) move on $((e,f),(g,h))$ with $k = -k_0$ yields another desirable pair $((e',f'),(g,h))$. Because $|e'| < |e|$ and $g \neq 0$, we have $|e'g| < |eg|$. Therefore, because $|ac| = |ag|$ is minimal among unobtainable desirable pairs, the new desirable pair $((e',f'),(g,h))$ can be obtained from $((1,0),(0,1))$ through some sequence of moves $\mathcal{S}$. We can then obtain $((e,f),(g,h))$ from $((1,0),(0,1))$ as well, by first applying the moves in $\mathcal{S}$ to $((1,0),(0,1))$, then applying one additional move of type (i) with $k = k_0$. Thus, our minimal pair is obtainable — a contradiction.

A similar proof holds if $|e| < |g|$, where we instead choose $k_0$ such that $g - 2k_0 e \in (-|g|, |g|)$ and perform type-(ii) moves. Thus, in all cases, we get a contradiction. Therefore, we can conclude that every obtainable pair of vectors is indeed desirable. This completes the proof.

## Problem 10

Prove that
$$\frac{a^3}{x} + \frac{b^3}{y} + \frac{c^3}{z} \geq \frac{(a+b+c)^3}{3(x+y+z)}$$

for all positive real numbers $a, b, c, x, y, z$.

**Solution.** By Hölder's inequality,

$$\prod_{i=1}^{3} \left(p_i^3 + q_i^3 + r_i^3\right)^{1/3} \geq p_1 p_2 p_3 + q_1 q_2 q_3 + r_1 r_2 r_3$$

for all positive reals $p_i, q_i, r_i$. Hence,

$$\left(\frac{a^3}{x} + \frac{b^3}{y} + \frac{c^3}{z}\right)^{1/3} (1+1+1)^{1/3}(x+y+z)^{1/3} \geq a+b+c.$$

Cubing both sides and then dividing both sides by $3(x+y+z)$ gives the desired result.

## Problem 11

Let $P$ be the intersection point of the diagonals $\overline{AC}$ and $\overline{BD}$ of the convex quadrilateral $ABCD$ in which $AB = AC = BD$. Let $O$ and $I$ be the circumcenter and incenter, respectively, of triangle $ABP$. Prove that if $O \neq I$, then lines $OI$ and $CD$ are perpendicular.

**Solution.** We first prove a fact that is very helpful in proving that two segments are perpendicular: Given two segments $\overline{XY}$ and $\overline{UV}$, $\overline{XY} \perp \overline{UV}$ if and only if $UX^2 - XV^2 = UY^2 - YV^2$. To prove this result, let $X'$ and $Y'$ be the feet of the perpendiculars of $X$ and $Y$, respectively, to line $UV$. Then $\overline{XY} \perp \overline{UV}$ if and only if $X' = Y'$, that is, if and only if

$$UX' - X'V = UY' - Y'V. \tag{$*$}$$

where all distances are directed distances. Because $UX' + X'V = UV = UY' + Y'V$, equation $(*)$ holds if and only if $UX'^2 - X'V^2 = UY'^2 - Y'V^2$, or equivalently $UX^2 - XV^2 = UY^2 - YV^2$.

Thus, it suffices to show that $DO^2 - CO^2 = DI^2 - CI^2$. Let $AB = AC = BD = p$, $PC = a$, and $PD = b$. Then $AP = p - a$ and $BP = p - b$. Let $R$ be the circumradius of triangle $ABP$. By the Power of a Point Theorem, $pb = DP \cdot DB = DO^2 - R^2$. Likewise, $pa = CO^2 - R^2$. Hence, $DO^2 - CO^2 = p(b - a)$.

Because triangle $ABD$ is isosceles with $BA = BD$, and $I$ lies on the bisector of angle $ABD$, $ID = IA$. Likewise, $IB = IC$. Let $T$ be the point of tangency of the incircle of triangle $ABC$ to side $\overline{AB}$. Then $AT = (AB + AP - BP)/2 = (p + b - a)/2$ and $BT = (p + a - b)/2$. Because $IT \perp AB$, $AI^2 - BI^2 = AT^2 - BT^2$. Putting the above arguments together, we find that

$$DI^2 - CI^2 = AI^2 - BI^2 = AT^2 - BT^2 = (AT + BT)(AT - BT)$$
$$= p(b - a) = DO^2 - CO^2,$$

as desired.

## 1.2  Bulgaria

### Problem 1

A line $\ell$ is drawn through the orthocenter of acute triangle $ABC$. Prove that the reflections of $\ell$ across the sides of the triangle are concurrent.

**Solution.**  Because triangle $ABC$ is acute, its orthocenter $H$ is inside the triangle. Without loss of generality, we may assume that $\ell$ intersects sides $\overline{AC}$ and $\overline{BC}$ at $Q$ and $P$, respectively. If $\ell \parallel \overline{AB}$, let $R$ be any point on the reflection of $\ell$ across line $AB$. Otherwise, let $R$ be the intersection of $\ell$ and line $AB$, and assume without loss of generality that $R$ lies on ray $BA$. Let $A_1, B_1, C_1$ be the reflections of $H$ across lines $BC, CA, AB$, respectively. It is well known that $A_1, B_1, C_1$ lie on the circumcircle $\omega$ of triangle $ABC$. (Note that $\angle A_1 CB = \angle BCH = \angle HAB = \angle A_1 AB$.) It suffices to prove that lines $A_1 P, B_1 Q, C_1 R$ are concurrent.

Because lines $AC$ and $BC$ are not parallel or perpendicular, lines $B_1 Q$ and $A_1 P$ are not parallel. Let $S$ be the intersection of lines $A_1 P$ and $B_1 Q$. Because

$$\angle SA_1 C + \angle SB_1 C = \angle PA_1 C + \angle QB_1 C = \angle PHC + \angle QHC = \pi,$$

quadrilateral $SA_1 CB_1$ is cyclic. Hence, $S$ is the intersection of line $B_1 Q$ and circle $\omega$.

Likewise, lines $B_1 Q$ and $C_1 R$ are not parallel, and their intersection is also the intersection of line $B_1 Q$ and circle $\omega$. Hence, lines $A_1 P, B_1 Q, C_1 R$ are concurrent at a point on the circumcircle of triangle $ABC$.

### Problem 2

There are 2000 white balls in a box. There are also unlimited supplies of white, green, and red balls, initially outside the box. During each turn, we can replace two balls in the box with one or two balls as follows: two whites with a green, two reds with a green, two greens with a white and red, a white and green with a red, or a green and red with a white.

(a) After finitely many of the above operations there are three balls left in the box. Prove that at least one of them is a green ball.

(b) Is it possible after finitely many operations to have only one ball left in the box?

**Solution.** Assign the value $i$ to each white ball, $-i$ to each red ball, and $-1$ to each green ball. A quick check verifies that the given operations preserve the product of the values of the balls in the box. This product is initially $i^{2000} = 1$. If three balls were left in the box, none of them green, then the product of their values would be $\pm i$, a contradiction. Hence, if three balls remain, at least one is green, proving the claim in part (a). Furthermore, because no ball has value 1, the box must contain at least two balls at any time. Therefore, the answer to the question in part (b) is "no."

(To prove the claim in part (a), we could also assign the value 1 to each green ball and $-1$ to each red ball and white ball.)

## Problem 3

The incircle of the isosceles triangle $ABC$ touches the legs $\overline{AC}$ and $\overline{BC}$ at points $M$ and $N$, respectively. A line $t$ is drawn tangent to minor arc $MN$, intersecting $\overline{NC}$ and $\overline{MC}$ at points $P$ and $Q$, respectively. Let $T$ be the intersection point of lines $AP$ and $BQ$.

(a) Prove that $T$ lies on $\overline{MN}$;

(b) Prove that the sum of the areas of triangles $ATQ$ and $BTP$ is smallest when $t$ is parallel to line $AB$.

**Solution.** (a) The degenerate hexagon $AMQPNB$ is circumscribed about the incircle of triangle $ABC$. By Brianchon's Theorem, its diagonals $\overline{AP}$, $\overline{MN}$, and $\overline{QB}$ concur. Therefore, $T$ lies on $\overline{MN}$.

One can also use a more elementary approach. Let $R$ and $S$ be the points of tangency of the incircle with sides $\overline{AB}$ and $\overline{PQ}$, respectively. Let $\overline{BQ}$ intersect $\overline{MN}$ and $\overline{SR}$ at $T_1$ and $T_2$, respectively. Because $\angle QMN = \angle PNM = \frac{MN}{2}$, we have $\sin \angle QMN = \sin \angle PNM = \sin \angle BNM$. Applying the Law of Sines to triangles $MQT_1$ and $NBT_1$ yields

$$\frac{QT_1}{QM} = \frac{\sin \angle QMN}{\sin \angle QT_1 M} = \frac{\sin \angle BNM}{\sin \angle BT_1 N} = \frac{BT_1}{BN},$$

or

$$\frac{QT_1}{BT_1} = \frac{MQ}{BN}.$$

Likewise,

$$\frac{QT_2}{BT_2} = \frac{SQ}{BR}.$$

By equal tangents, $BN = BR$ and $QM = QS$. Hence ,

$$\frac{QT_1}{BT_1} = \frac{QT_2}{BT_2}.$$

Because $T_1$ and $T_2$ both lie on $\overline{BQ}$, we must have $T_1 = T_2$. Hence, $\overline{BQ}, \overline{MN}, \overline{SR}$ are concurrent. In exactly the same manner, we can prove that $\overline{AP}, \overline{MN}, \overline{SR}$ are concurrent. It follows that $T$ lies on $\overline{MN}$.

(b) Let $\alpha = \angle CAB = \angle CBA$ and $\beta = \angle ACB$. Let $f = [AQT] + [BPT] = [ABQ] + [ABP] - 2[ABT]$. Because triangle $ABC$ is isosceles, $\overline{MN} \parallel \overline{AB}$, implying that $[ABT]$ is constant. Hence, minimizing $f$ is equivalent to minimizing $g = [ABQ] + [ABP]$. Note that

$$2g = AB(AQ + PB) \sin \alpha = AB(AB + PQ) \sin \alpha,$$

where $AQ + PB = AB + QP$ because quadrilateral $ABCD$ has an inscribed circle. Thus, it suffices to minimize $PQ$.

Let $I$ be the incenter of triangle $ABC$, so that $I$ is the excenter of triangle $CPQ$ opposite $C$. Hence, $PC + CQ + QP = 2CM$ is constant. Let $\angle CPQ = p$ and $\angle CQP = q$. Then $p + q = \pi - \beta$ is constant as well. Applying the Law of Sines to triangle $CPQ$ yields

$$\frac{CM}{PQ} = 1 + \frac{CP}{PQ} + \frac{CQ}{PQ} = 1 + \frac{\sin p + \sin q}{\sin \beta}$$
$$= 1 + \frac{2 \sin \frac{p+q}{2} \cos \frac{p-q}{2}}{\sin \beta}.$$

Hence, it suffices to maximize $\cos \frac{p-q}{2}$. It follows that $[ATQ] + [BTP]$ is minimized when $p = q$, that is, when $\overline{PQ} \parallel \overline{AB}$.

## Problem 4

We are given $n \geq 4$ points in the plane such that the distance between any two of them is an integer. Prove that at least $\frac{1}{6}$ of these distances are divisible by 3.

**Solution.**   In this solution, all congruences are taken modulo 3.

We first show that if $n = 4$, then at least two points are separated by a distance divisible by 3. Denote the points by $A, B, C, D$. We show that at least one of the six distances $AB$, $BC$, $CD$, $DA$, $AC$, $BD$ is divisible by 3. We approach indirectly by assuming that all those distances are not divisible by 3.

Without loss of generality, we assume that $\angle BAD = \angle BAC + \angle CAD$. Let $\angle BAC = x$ and $\angle CAD = y$. Also, let $\alpha = 2AB \cdot AC \cdot \cos x$, $\beta = 2AD \cdot AC \cos y$, and $\gamma = 2AB \cdot AD \cdot \cos(x + y)$. Applying the Law of Cosines in triangles $ABC, ACD, ABD$ gives

$$BC^2 = AB^2 + AC^2 - \alpha,$$
$$CD^2 = AD^2 + AC^2 - \beta,$$
$$BD^2 = AB^2 + AD^2 - \gamma.$$

Because the square of each distance is an integer congruent to 1, it follows from the above equations that $\alpha$, $\beta$, and $\gamma$ are also integers congruent to 1. Also,

$$2AC^2\gamma = 4AC^2 \cdot AB \cdot AD \cdot \cos(x + y)$$
$$= 4AC^2 \cdot AB \cdot AD \cdot (\cos x \cos y - \sin x \sin y)$$
$$= \alpha\beta - 4AC^2 \cdot AB \cdot AD \cdot \sin x \sin y,$$

implying that $4AC^2 \cdot AB \cdot AD \cdot \sin x \sin y$ is an integer congruent to 2. Thus, $\sin x \sin y = \sqrt{(1 - \cos^2 x)(1 - \cos^2 y)}$ is a rational number which, when written in lowest terms, has a numerator that is not divisible by 3. Let $p = 2AB \cdot AC$ and $q = 2AD \cdot AC$, so that $\cos x = \frac{\alpha}{p}$ and $\cos y = \frac{\beta}{q}$. Because

$$\sin x \sin y = \frac{\sqrt{(p^2 - \alpha^2)(q^2 - \beta^2)}}{pq}$$

is rational, the numerator on the right-hand side must be an integer. This numerator is divisible by 3 because $p^2 \equiv \alpha^2 \equiv 1$, but the denominator is not divisible by 3. Therefore, when $\sin x \sin y$ is written in lowest terms, its numerator *is* divisible by 3, a contradiction. Therefore, our assumption was wrong and there is at least one distance divisible by 3 for $n = 4$.

Now assume that $n \geq 4$. From the set of $n$ given points, there exist $\binom{n}{4}$ four-element subsets $\{A, B, C, D\}$. At least two points in each subset are separated by a distance divisible by 3, and each such distance is counted in at most $\binom{n-2}{2}$ subsets. Hence, at least $\binom{n}{4}/\binom{n-2}{2} = \binom{n}{2}/6$ distances are divisible by 3.

## Problem 5

In triangle $ABC$, $\overline{CH}$ is an altitude, and cevians $\overline{CM}$ and $\overline{CN}$ bisect angles $ACH$ and $BCH$, respectively. The circumcenter of triangle $CMN$

coincides with the incenter of triangle $ABC$. Prove that $[ABC] = \frac{AN \cdot BM}{2}$.

**Solution.** Let $I$ be the incenter of triangle $ABC$, and let the incircle of triangle $ABC$ intersect sides $\overline{AC}$ and $\overline{AB}$ at $E$ and $F$, respectively. Because $IM = IN$ and $\overline{IF} \perp \overline{MN}$, we have $\angle FIN = \frac{1}{2}\angle MIN$. Furthermore, because $I$ is the circumcenter of triangle $CMN$, $\frac{1}{2}\angle MIN = \angle MCN = \frac{1}{2}\angle ACB = \angle ECI$. Thus, $\angle FIN = \angle ECI$.

In addition, $\angle NFI = \pi/2 = \angle IEC$. Hence, triangles $NFI$ $IEC$ are similar. Because $NI = IC$, these two triangles are actually congruent, and $NF = IE = IF$. Right triangle $NFI$ is thus isosceles, $\angle FIN = \pi/4$, and $\angle ACB = 2\angle FIN = \pi/2$.

Thus, $\angle HCB = \pi/2 - \angle CBH = \angle BAC$ and

$$\angle ACN = \angle ACB - \frac{1}{2}\angle HCB = \frac{\pi}{2} - \angle BAC/2.$$

Therefore,

$$\angle CNA = \pi - (\angle ACN + \angle NAC) = \frac{\pi}{2} - \angle BAC/2 = \angle ACN,$$

and $AN = AC$. Similarly, $BM = BC$. It follows that $\frac{1}{2}AN \cdot BM = \frac{1}{2}AC \cdot BC = [ABC]$, as desired.

## Problem 6

Let $a_1, a_2, \ldots$ be a sequence such that $a_1 = 43$, $a_2 = 142$, and $a_{n+1} = 3a_n + a_{n-1}$ for all $n \geq 2$. Prove that

(a) $a_n$ and $a_{n+1}$ are relatively prime for all $n \geq 1$;

(b) for every natural number $m$, there exist infinitely many natural numbers $n$ such that $a_n - 1$ and $a_{n+1} - 1$ are both divisible by $m$.

**Solution.** (a) Suppose there exist $n$, $g > 1$ such that $g \mid a_n$ and $g \mid a_{n+1}$. Then $g$ would divide $a_{n-1} = a_{n+1} - 3a_n$ as well. If $n - 1 > 1$ then $g$ would also divide $a_{n-2} = a_n - 3a_{n-1}$. Continuing, it follows that, $g$ must divide each of $a_{n+1}, a_n, \ldots, a_1$. However, this is impossible because $\gcd(a_1, a_2) = 1$. Therefore, $a_n$ and $a_{n+1}$ are relatively prime for all $n \geq 1$.

(b) Define the sequence $a'_1, a'_2, \ldots$ recursively by setting $a'_1 = 1$, $a'_2 = 1$, and $a'_{n+1} = 3a'_n + a'_{n-1}$ for all $n \geq 2$. Observe that $(a'_3, a'_4, a'_5, a'_6) = (4, 13, 43, 142)$, and hence $(a'_5, a'_6) = (a_1, a_2)$. Because the two sequences satisfy the same recursive relation, $a_n = a'_{n+4}$ for all $n \geq 1$.

Let $b_n$ be the remainder of $a'_n$ when divided by $m$, and consider the pairs $(b_n, b_{n+1})$ for $n \geq 1$. Because there are infinitely many such pairs but only $m^2$ ordered pairs of integers $(r, s)$ with $0 \leq r, s < m$, two of these pairs must be equal: say, $(b_i, b_{i+1}) = (b_{i+t}, b_{j+t})$ where $t > 0$. By applying the recursive relation, it follows easily by induction on $|n|$ that $b_{i+n} = b_{i+n+t}$ for all integers $n$ such that $i + n \geq 1$. Therefore, $(b_{1+kt}, b_{2+kt}) = (b_1, b_2) = (1, 1)$ for all $k \geq 1$. Hence, $a_{kt-3} - 1$ and $a_{kt-2} - 1$ are both divisible by $m$ for all $k \geq 4$.

## Problem 7

In convex quadrilateral $ABCD$, $\angle BCD = \angle CDA$. The bisector of angle $ABC$ intersects $\overline{CD}$ at point $E$. Prove that $\angle AEB = \pi/2$ if and only if $AB = AD + BC$.

**Solution.** If $\angle AEB = \pi/2$, then $\angle CEB < \pi/2$. It follows that there is a point $F$ on side $\overline{AB}$ such that $\angle BEF = \angle BEC$. Then triangles $BEC$ and $BEF$ are congruent, implying that $BC = BF$ and $\angle BFE = \angle BCE = \angle EDA$ from the given. Thus, quadrilateral $ADEF$ is cyclic. Because $\angle AEB = \pi/2$ and $\angle CEB = \angle BEF$, we have $\angle FEA = \angle AED$. It follows that $\angle FDA = \angle FEA = \angle AED = \angle AFD$. Hence, $AF = AD$, and $AB = AF + BF = AD + BC$.

If $AB = BC + AD$, then there is a point $F$ on $\overline{AB}$ such that $BF = BC$ and $AF = AD$. Then triangles $BCE$ and $BFE$ are congruent, and again we see that $ADEF$ is cyclic. Also, $\angle FDA = \angle AFD$. Hence, $\angle FEA = \angle FDA = \angle AFD = \angle AED$, so line $AE$ bisects angle $FED$. Because triangles $BCE$ and $BFE$ are congruent, line $BE$ bisects angle $CEF$. Hence, $\overline{AE} \perp \overline{BE}$, and $\angle AEB = \pi/2$.

## Problem 8

In the coordinate plane, a set of 2000 points $\{(x_1, y_1), (x_2, y_2), \ldots, (x_{2000}, y_{2000})\}$ is called *good* if $0 \leq x_i \leq 83, 0 \leq y_i \leq 1$ for $i = 1, 2, \ldots, 2000$ and $x_i \neq x_j$ when $i \neq j$. Find the largest positive integer $n$ such that, for any good set, the interior and boundary of some unit square contains exactly $n$ of the points in the set on its interior or its boundary.

**Solution.** We first prove that for any good set, some unit square contains exactly 25 of the points in the set. We call a unit square *proper* if two of its sides lie on the lines $y = 0$ and $y = 1$. Each of the given points lies in the region $\mathcal{R} = \{(x, y) \mid 0 \leq x \leq 83, 0 \leq y \leq 1\}$, which can be divided into proper unit squares whose left sides lie on a line of the form $x = i$ for $i = 0, 1, \ldots, 82$. Because $83 \cdot 24 < 2000$, one of these squares contains more than 24 points. Because $83 \cdot 26 - 82 > 2000$, one of these squares contains less than 26 points.

In addition to these 83 unit squares, consider the proper unit squares whose left sides lie on lines of the form $x = x_i$ or $x = x_i - 1$. Order all these unit squares $S_1, \ldots, S_k$ from left to right, where the left side of $S_i$ lies on the line $x = z_i$. For $i = 1, 2, \ldots, k - 1$, at most one of the given points lies in the region determined by $z_i \le x < z_{i+1}$, and at most one of the given points lies in the region determined by $z_i + 1 < x \le z_{i+1} + 1$. Hence, for all such $i$, the number of points in $S_i$ differs from the number of points in $S_{i+1}$ by either $-1$, $0$, or $1$. Because there exists an $S_{i_1}$ containing at least 25 points and an $S_{i_2}$ containing at most 25 points, it follows that some $S_{i_3}$ (with $i_3$ between $i_1$ and $i_2$, inclusive) contains exactly 25 points.

We now prove that no $n > 25$ has the required property. Let $d = 2 \cdot \frac{83}{1999}$, $x_i = (i - 1) \cdot \frac{1}{2}d$ for $i = 1, 2, \ldots, 2000$, and $y_{2k-1} = 0$, $y_{2k} = 1$ for $k = 1, 2, \ldots, 1000$. Any two distinct points $(x_i, y_i)$ that lie on the same horizontal line (either $y = 0$ or $y = 1$) are separated by distance at least $d > \frac{2}{25}$. Let $XYZW$ be any unit square. For $j = 0, 1$, the region $\mathcal{R}_0$ bounded by this square intersects each line $y = j$ in a closed interval (possibly consisting of zero points or one point) of length $r_j$. If at least one of $r_0, r_1$ is zero, then the corresponding interval contains at most 1 of the points $(x_i, y_i)$. The other interval has length at most $\sqrt{2}$, and hence can contain at most $\lfloor \sqrt{2}/d \rfloor + 1 \le 18$ of the required points, for a total of no more than 19. Also, if $XYZW$ has a pair of horizontal sides, then $\mathcal{R}_0$ contains at most $\lfloor \frac{1}{d/2} \rfloor + 1 \le 25$ of the required points. Otherwise, $\mathcal{R}_0$ intersects the lines $y = 0$ and $y = 1$ at some points $P$, $Q$ and $R$, $S$, respectively, where $P$ and $R$ lie to the left of $Q$ and $S$. Also, $\overline{PQ}$ and $\overline{RS}$ contain at most $\lfloor PQ/d \rfloor + 1$ and $\lfloor RS/d \rfloor + 1$ of the chosen points, respectively.

Translate $\mathcal{R}_0$ in a direction parallel to either of its pairs of sides until its center is on the line $y = \frac{1}{2}$. Let $\mathcal{R}_1$ be the image of $\mathcal{R}_0$ under the translation, and let $P'$, $Q'$, $R'$, and $S'$ be its intersections with $y = 0$ and $y = 1$, defined analogously as before. Then $P'Q' + R'S' = PQ + RS$. Also, $P'Q' = R'S'$ by symmetry. Let $\mathcal{R}_2$ be the region formed by rotating $\mathcal{R}_1$ about its center so that two of its sides are on $y = 0$ and $y = 1$. Then the region $\mathcal{R}_1 \cup \mathcal{R}_2 - \mathcal{R}_1 \cap \mathcal{R}_2$ is the union of eight congruent triangular regions. Let $T$ and $U$ be the left and right vertices of $\mathcal{R}_2$ on $y = 1$, and let $V$ be the vertex of $\mathcal{R}_1$ above the line $y = 1$. Finally, let $K$ and $L$ be the uppermost points on the left and right vertical sides of $\mathcal{R}_2$, respectively, that also belong to the boundary of $\mathcal{R}_1$. We have

$$\triangle KTR' \cong \triangle S'VR' \cong \triangle S'UL.$$

Also,

$$TR' + R'S' + S'U = TU = 1.$$

On the other hand, by the triangle inequality,

$$TR' + S'U = R'V + S'V > R'S'.$$

It follows that $R'S' < \frac{1}{2}$. Because $P'Q' = R'S'$, the number of points $(x_i, y_i)$ in $XYZW$ is at most

$$\left\lfloor \frac{PQ}{d} \right\rfloor + \left\lfloor \frac{RS}{d} \right\rfloor + 2 \leq \frac{P'Q' + R'S'}{d} + 2 < \frac{1}{d} + 2 < 15,$$

which completes the proof.

## Problem 9

We are given the acute triangle $ABC$.

(a) Prove that there exist unique points $A_1$, $B_1$, and $C_1$ on $\overline{BC}$, $\overline{CA}$, and $\overline{AB}$, respectively, with the following property: If we project any two of the points onto the corresponding side, the midpoint of the projected segment is the third point.

(b) Prove that triangle $A_1B_1C_1$ is similar to the triangle formed by the medians of triangle $ABC$.

**Solution.** (a) We work backward by first assuming such a triangle exists. Let $T$ be the midpoint of $\overline{A_1B_1}$. By definition, $\overline{C_1T} \perp \overline{AB}$. Let $P$ be the centroid of triangle $A_1B_1C_1$. Because $\overline{PA_1} \perp \overline{BC}$, $\overline{PB_1} \perp \overline{CA}$, and $\overline{PC_1} \perp \overline{AB}$, $P$ uniquely determines triangle $A_1B_1C_1$.

It is clear that quadrilaterals $AB_1PC_1$, $BC_1PA_1$, $CA_1PB_1$ are cyclic. Let $\alpha = \angle CAB$, $\beta = \angle ABC$, $x = \angle A_1B_1P$, and $y = \angle B_1A_1P$. Because quadrilaterals $AB_1PC_1$ and $CA_1PB_1$ are cyclic,

$$\angle TPB_1 = \alpha, \quad \angle TPA_1 = \beta, \quad \angle A_1CP = x, \quad \angle B_1CP = y.$$

Applying the Law of Sines to triangles $A_1TP$ and $B_1TP$ yields

$$\frac{\sin y}{\sin \beta} = \frac{TP}{TA_1} = \frac{TP}{TB_1} = \frac{\sin x}{\sin \alpha},$$

or

$$\frac{\sin x}{\sin y} = \frac{\sin \alpha}{\sin \beta}.$$

In exactly the same way, we can show that

$$\frac{\sin \angle ACF}{\sin \angle BCF} = \frac{\sin \alpha}{\sin \beta},$$

where $F$ is the midpoint of side $\overline{AB}$. Because triangle $ABC$ is acute, we conclude that $\angle A_1 CP = x = \angle ACF$ and $\angle B_1 CP = y = \angle BCF$. Hence, lines $CP$ and $CF$ are symmetric with respect to the angle bisector of angle $ACB$. Analogous results hold for lines $AP$ and $AD$, $BP$ and $BE$, where $D$ and $E$ are the midpoints of sides $\overline{BC}$ and $\overline{CA}$, respectively. It follows that $P$ is the isogonal conjugate of $G$, where $G$ is the centroid of triangle $ABC$. Thus $P$ is unique, and reversing our steps shows that the $P$ we found generates a unique triangle $A_1 B_1 C_1$ satisfying the conditions of the problem.

(b) Let $D, E, F$ be midpoints of $\overline{BC}, \overline{CA}, \overline{AB}$, respectively. Extend $\overline{AG}$ through $G$ to $K$ such that $GD = DK$. Then $BGCK$ is a parallelogram and $CK = BG = \frac{2}{3}BE$, $CG = \frac{2}{3}CF$, $GK = AG = \frac{2}{3}AD$. Hence, triangle $CGK$ is similar to the triangle formed by the medians of triangle $ABC$. It suffices to prove that triangles $A_1 B_1 C_1$ and $CGK$ are similar. But this is true because

$$\angle B_1 C_1 A_1 = \angle B_1 C_1 P + \angle A_1 C_1 P = \angle B_1 AP + \angle A_1 BP$$
$$= \angle BAG + \angle GBA = \angle KGB = \angle GKC,$$

and (analogously) $\angle C_1 A_1 B_1 = \angle KCG$.

## Problem 10

Let $p \geq 3$ be a prime number and $a_1, a_2, \ldots, a_{p-2}$ be a sequence of positive integers such that $p$ does not divide either $a_k$ or $a_k^k - 1$ for all $k = 1, 2, \ldots, p-2$. Prove that the product of some terms of the sequence is congruent to 2 modulo $p$.

**Solution.** We prove by induction on $k = 2, \ldots, p-1$ that there exists a set of integers $\{b_{k,1}, b_{k,2}, \ldots, b_{k,k}\}$ such that (i) each $b_{k,i}$ either equals 1 or is the product of some terms of the sequence $a_1, a_2, \ldots, a_{p-2}$, and (ii) $b_{k,i} \not\equiv b_{k,j} \pmod p$ for $1 \leq i < j \leq k$.

For the base case $k = 2$, we may choose $b_{1,1} = 1$ and $b_{1,2} = a_1 \not\equiv 1 \pmod{p}$.

Suppose that we have chosen the set $\{b_{k,1}, b_{k,2}, \ldots, b_{k,k}\}$ for some integer $2 \leq k \leq p - 2$. Because $a_k \not\equiv 0 \pmod{p}$, no two of the numbers $a_k b_{k,1}, \ldots, a_k b_{k,k}$ are congruent modulo $p$. Also, because $a_k^k \not\equiv 1 \pmod{p}$, we have

$$(a_k b_{k,1})(a_k b_{k,2}) \cdots (a_k b_{k,k}) \not\equiv b_{k,1} b_{k,2} \cdots b_{k,k} \pmod{p}$$

Hence, we cannot permute $(a_k b_{k,1}, \ldots, a_k b_{k,k})$ so that each term is congruent modulo $p$ to the corresponding term in $(b_{k,1}, \ldots, b_{k,k})$. Because the $a_k b_{k,i}$ are distinct modulo $p$, there must exist $j$ such that no two elements of the set $\{b_{k,1}, \ldots, b_{k,k}, a_k b_{k,j}\}$ are congruent modulo $p$. We relabel this set of numbers as $\{b_{k+1,1}, b_{k+1,2}, \ldots, b_{k+1,k+1}\}$. Each of these $k + 1$ numbers equals 1 or is the product of some terms of the sequence $a_1, \ldots, a_{p-2}$, and the induction is complete.

Consider the resulting list $b_{p-1,1}, \ldots, b_{p-1,p-1}$. Exactly one of these numbers is congruent to 2 modulo $p$; because this number is not equal to 1, it is congruent to the product of some of the $a_k$, as desired.

## Problem 11

Let $D$ be the midpoint of base $\overline{AB}$ of the isosceles acute triangle $ABC$. Choose a point $E$ on $\overline{AB}$, and let $O$ be the circumcenter of triangle $ACE$. Prove that the line through $D$ perpendicular to $\overline{DO}$, the line through $E$ perpendicular to $\overline{BC}$, and the line through $B$ parallel to $\overline{AC}$ are concurrent.

**Solution.** Let $\ell$ denote the line passing through $B$ and parallel to line $AC$, and let $F_1$ and $F_2$ be points on line $\ell$ such that $\overline{OD} \perp \overline{DF_1}$ and $\overline{BC} \perp \overline{EF_2}$. Let $H_1$ and $H_2$ be the feet of the perpendiculars from $F_1$ and $F_2$ to line $AB$, respectively. Because $\angle CAB$ is acute, $O$ is an interior point of $\angle ADC$. It follows that $F_1$ is between rays $AB$ and $AC$. Because angle $ABC$ is acute, $F_2$ is also between rays $AB$ and $AC$.

It suffices to prove that $F_1 H_1 = F_2 H_2$. Let $G$ be the circumcenter of triangle $ABC$, and let $O_1$ and $G_1$ be the feet of the perpendiculars from $O$ to line $AB$ and $G$ to line $OO_1$, respectively. Because $\overline{OD} \perp \overline{DF_1}$, triangles $OO_1 D$ and $DH_1 F_1$ are similar. Hence,

$$\frac{DH_1}{F_1 H_1} = \frac{OO_1}{O_1 D}.$$

Let $\angle BAC = \angle CBA = x$. Because $AG = GC$ and $AO = OC$, line $GO$ bisects angle $AGC$. Hence, $\angle CGO = x$. Because $\overline{CG} \parallel \overline{OO_1}$,

$\angle G_1OG = \angle CGO = x$. Therefore, right triangles $GOG_1$ and $F_1BH_1$ are similar. Hence

$$\frac{BH_1}{F_1H_1} = \frac{OG_1}{GG_1}.$$

Combining the last two equalities yields

$$F_1H_1 = \frac{BH_1 \cdot O_1D}{OG_1} = \frac{DH_1 \cdot O_1D}{OO_1}$$

$$= \frac{DH_1 \cdot O_1D - BH_1 \cdot O_1D}{OO_1 - OG_1}$$

$$= \frac{BD \cdot O_1D}{G_1O_1} = \frac{BD \cdot O_1D}{GD}.$$

Because $\angle DGB = \angle ACB = \pi - 2x$, we obtain

$$F_1H_1 = -\tan 2x \cdot O_1D.$$

Let $I$ be the intersection of $\overline{BC}$ and $\overline{EF_2}$. Because $\overline{BF_2} \parallel \overline{AC}$, $\angle F_2BI = \angle ACB = \pi - 2x$ and $\angle H_2BF_2 = x$. Note that $BE = AB - AE = 2(AD - AO_1) = 2O_1D$. It follows that

$$F_2H_2 = BF_2 \cdot \sin \angle H_2BF_2 = BF_2 \cdot \sin x$$

$$= \frac{BI}{\cos \angle F_2BI} \cdot \sin x = \frac{BI \cdot \sin x}{-\cos 2x}$$

$$= -\frac{BE \cos x \sin x}{\cos 2x} = -O_1D \tan 2x = F_1H_1,$$

as desired.

## Problem 12

Let $n$ be a positive integer. A *binary sequence* is a sequence of integers, all equal to 0 or 1. Let $\mathcal{A}$ be the set of all binary sequences with $n$ terms, and let $\mathbf{0} \in \mathcal{A}$ be the sequence of all zeroes. The sequence $c = c_1, c_2, \ldots, c_n$ is called the sum $a + b$ of $a = a_1, a_2, \ldots, a_n$ and $b = b_1, b_2, \ldots, b_n$ if $c_i = 0$ when $a_i = b_i$ and $c_i = 1$ when $a_i \neq b_i$. Let $f : \mathcal{A} \to \mathcal{A}$ be a function with $f(\mathbf{0}) = \mathbf{0}$ such that whenever the sequences $a$ and $b$ differ in exactly $k$ terms, the sequences $f(a)$ and $f(b)$ also differ in exactly $k$ terms. Prove that if $a$, $b$, and $c$ are sequences from $\mathcal{A}$ such that $a + b + c = \mathbf{0}$, then $f(a) + f(b) + f(c) = \mathbf{0}$.

**Solution.** Consider the sequences $e_1 = 1, 0, 0, \ldots, 0$, $e_2 = 0, 1, 0, \ldots, 0$, $\ldots$, $e_n = 0, 0, \ldots, 0, 1$. For each $i$, $\mathbf{0}$ and $e_i$ differ in 1 term, so $f(\mathbf{0}) = \mathbf{0}$ and $f(e_i)$ do as well — that is, $f(e_i) = e_j$ for some $j$. Also, because $e_i$

and $e_j$ differ in two terms for any $i \neq j$, so do $f(e_i)$ and $f(e_j)$, implying that $f(e_i) \neq f(e_j)$. Therefore,

$$\{f(e_1), f(e_2), \ldots, f(e_n)\} = \{e_1, e_2, \ldots, e_n\}.$$

Consider an arbitrary sequence $x = (x_1, x_2, \ldots, x_n)$ with $f(x) = (y_1, y_2, \ldots, y_n)$. If $x$ has $t$ 1's, then so does $f(x)$. If $f(e_i) = e_j$ and $x_i = 1$, then $e_i$ and $x$ differ in $t - 1$ terms, implying that $f(e_i) = e_j$ and $f(x)$ do as well. This is only possible if $y_j = 1$, because otherwise $e_j$ and $f(x)$ would differ in $t + 1$ terms. Likewise, if $x_i = 0$ then $y_j = 0$.

If $a = (a_1, a_2, \ldots, a_n)$, $b = (b_1, b_2, \ldots, b_n)$, $c = (c_1, c_2, \ldots, c_n)$, and $a + b + c = 0$, then $a_i + b_i + c_i$ is even for $i = 1, 2, \ldots, n$. For each $e_j$, we can choose $e_i$ such that $f(e_i) = e_j$. The $j$th terms of $f(a), f(b), f(c)$ equal $a_i, b_i, c_i$, respectively, implying that these three terms have even sum. Therefore, $f(a) + f(b) + f(c)$ has $j$th term 0 for all $j$, and $f(a) + f(b) + f(c) = 0$.

## 1.3  Canada

### Problem 1

Let $a_1, a_2, \ldots, a_{2000}$ be a sequence of integers each lying in the interval $[-1000, 1000]$. Suppose that $\sum_{i=1}^{2000} a_i = 1$. Show that the terms in some nonempty subsequence of $a_1, a_2, \ldots, a_{2000}$ sum to zero.

**Solution.** First we show that we can rearrange $a_1, a_2, \ldots, a_{2000}$ into a sequence $b_1, b_2, \ldots, b_{2000}$ such that $\sum_{i=1}^{n} b_i \in [-999, 1000]$ for $n = 1, 2, \ldots, 2000$. We construct the $b_i$ term by term. Not all the $a_i$ equal $-1000$, so we may set $b_1$ equal to such an $a_i \in [-999, 1000]$. Call that index $i$ *assigned*.

Suppose that we have constructed $b_1, b_2, \ldots, b_k$ (with $1 \le k < 2000$) and that $k$ corresponding indices are assigned. If $\sum_{i=1}^{k} b_i$ is in $[-999, 0]$ (resp. in $[1, 1000]$), then the sum of the $a_i$ at the unassigned indices is positive (resp. nonpositive); hence, at least one such $a_i$ is positive (resp. nonpositive). Let $b_{k+1}$ equal that value $a_i$ and call the corresponding index assigned. Then $b_{k+1}$ is in $[1, 1000]$ (resp. in $[-1000, 0]$), implying that $\sum_{i=1}^{k+1} b_i$ is in $[-999, 1000]$. Repeating this construction, we can construct all 2000 terms $b_1, b_2, \ldots, b_{2000}$.

By construction, each of the 2000 partial sums $\sigma_n = \sum_{i=1}^{n} b_i$ (for $1 \le n \le 2000$) equals one of the 2000 integers in $[-999, 1000]$. Therefore, either $\sigma_i = \sigma_j$ for some $i < j$ or else $\sigma_i = 0$ for some $i$. In the first case, the terms in the subsequence $b_{i+1}, b_{i+2}, \ldots, b_j$ sum to zero; in the second, the terms in the subsequence $b_1, b_2, \ldots, b_i$ sum to zero. It follows that the terms in a corresponding subsequence of $a_1, a_2, \ldots, a_{2000}$ sum to zero as well, as desired.

### Problem 2

Let $ABCD$ be a quadrilateral with $\angle CBD = 2\angle ADB$, $\angle ABD = 2\angle CDB$, and $AB = CB$. Prove that $AD = CD$.

**Note.** By "quadrilateral" it is implied that no three of the vertices of $ABCD$ are collinear.

**First Solution.** Let $x = \angle ADB$ and $y = \angle CDB$ so that $\angle CBD = 2x$ and $\angle ABD = 2y$. Applying the Law of Sines in triangles $ABD$ and $CBD$, we find that

$$\frac{\sin(\pi - (2y + x))}{\sin x} = \frac{BD}{BA} = \frac{BD}{BC} = \frac{\sin(\pi - (2x + y))}{\sin y}.$$

Cross-multiplying and applying a product-to-difference trigonometric formula, we find that

$$\sin(2y + x)\sin y = \sin(2x + y)\sin x$$

$$\frac{1}{2}(\cos(y + x) - \cos(3y + x)) = \frac{1}{2}(\cos(x + y) - \cos(3x + y))$$

$$\cos(3y + x) = \cos(3x + y).$$

The last equation implies that either $(3y+x)+(3x+y)$ or $(3y+x)-(3x+y)$ is a multiple of $2\pi$. That is, either $x + y$ is a multiple of $\pi/2$ or $x - y$ is a multiple of $\pi$. Now, as $\angle ABC + \angle CDA = 3(x + y) < 2\pi$, we have $0 < x, y < x + y < \pi$, so either $x = y$ or $x + y = \pi/2$. In the case that $x + y = \pi/2$, we have $\angle ABC = 2(x + y) = \pi$, which is impossible, as that would imply that $A, B, C$ are collinear. Therefore, $x = y$. It follows that quadrilateral $ABCD$ is symmetric about $\overline{BD}$ and that $AD = CD$.

**Second Solution.** Let $E, F$ be the intersections of the internal angle bisectors of $\angle ABD$ and $\angle CBD$ with $\overline{AD}$ and $\overline{CD}$, respectively. Note that $\angle EBD = \frac{1}{2}\angle ABD = \angle FDB$ and $\angle FBD = \frac{1}{2}\angle CBD = \angle EDB$, and it follows that $BEDF$ is a parallelogram. Let $M$ and $N$ be the intersections of line $BD$ with $\overline{EF}$ and $\overline{AC}$, respectively. By the angle bisector theorem,

$$\frac{AE}{ED} = \frac{AB}{BD} = \frac{CB}{BD} = \frac{CF}{FD},$$

so lines $AC$ and $EF$ are parallel. Because $EM = MF$, we have $AN = NC$. Then $AB/AN = BC/CN$. Note that $N$ does not coincide with $B$ because $N$ lies on line $AC$ and $B$ does not. Thus, $BN$ bisects $\angle ABC$, and $\angle ABD = \angle DBC$. Thus, triangles $ABD$ and $CBD$ are similar, so $AD = CD$.

## Problem 3

Suppose that the real numbers $a_1, a_2, \ldots, a_{100}$ satisfy (i) $a_1 \geq a_2 \geq \cdots \geq a_{100} \geq 0$, (ii) $a_1 + a_2 \leq 100$, and (iii) $a_3 + a_4 + \cdots + a_{100} \leq 100$. Determine the maximum possible value of $a_1^2 + a_2^2 + \cdots + a_{100}^2$, and find all possible sequences $a_1, a_2, \ldots, a_{100}$ for which this maximum is achieved.

**Solution.** For $i \geq 3$, we have $0 \leq a_i \leq a_2$, and hence $a_i(a_i - a_2) \leq 0$, with equality only if $a_i \in \{0, a_2\}$. Adding these 98 inequalities together yields

$$\sum_{i=3}^{100} a_i^2 \leq a_2 \cdot \sum_{i=3}^{100} a_i.$$

By (iii), this is at most $100a_2$, with equality only if $\sum_{i=3}^{100} a_i = 100$ or $a_2 = 0$.

Also, (i) and (ii) imply that $0 \le a_1 \le 100 - a_2$. Thus, $a_1^2 \le (100 - a_2)^2$, with equality only if $a_1 = 100 - a_2$.

Conditions (i) and (ii) further imply that $0 \le a_2 \le 100 - a_1 \le 100 - a_2$, or $0 \le a_2 \le 50$. Hence, $2a_2(a_2 - 50) \le 0$ with equality only if $a_2$ equals 0 or 50.

Therefore,

$$\sum_{i=1}^{100} a_i^2 = a_1^2 + a_2^2 + \sum_{i=3}^{100} a_i^2 \le (100 - a_2)^2 + a_2^2 + 100a_2$$

$$= 10000 + 2a_2(a_2 - 50) \le 10000.$$

For equality to hold, equality must hold in each inequality found above — that is, we must have: (a) $\{a_3, a_4, \ldots, a_{100}\} \subseteq \{0, a_2\}$; (b) $\sum_{i=3}^{100} a_i = 100$ or $a_2 = 0$; (c) $a_1 = 100 - a_2$; and (d) $a_2 \in \{0, 50\}$. These conditions hold only when the sequence $a_1, a_2, \ldots, a_{100}$ equals

$$100, 0, 0, \ldots, 0 \qquad \text{or} \qquad 50, 50, 50, 50, 0, 0, \ldots, 0.$$

Indeed, these sequences satisfy conditions (i)-(iii), and $\sum_{i=1}^{100} a_i^2 = 10000$ for each sequence. Therefore, 10000 is the maximum sum of squares, and this maximum is achieved only for the two sequences above.

**Note.** Although the claim $\sum_{i=3}^{100} a_i^2 \le a_2 \sum_{i=3}^{100} a_i$ may seem to appear from nowhere, it actually arises quite naturally. In general, suppose that $x_1, x_2, \ldots, x_n \in [a, b]$ have a fixed sum $\sigma$ and that $f$ is a convex function on $[a, b]$. Then $\sum_{i=1}^{n} f(x_i)$ is maximized when the $x_i$ are "spread out" as much as possible, i.e. at most one value is not in $\{a, b\}$. With $[a, b] = [0, a_2]$ and $f(x) = x^2$, the maximum sum would occur when as many values as possible were equal to $a_2$. If $\sigma/a_2$ were an integer and $\sigma/a_2$ values did equal $a_2$, the sum of squares would be $a_2\sigma$. This suggests that we should attempt to prove that

$$\sum_{i=3}^{100} a_i^2 \le a_2 \sum_{i=3}^{100} a_i.$$

## 1.4  China

### Problem 1

In triangle $ABC$, $BC \leq CA \leq AB$. Let $R$ and $r$ be the circumradius and inradius, respectively, of triangle $ABC$. As a function of $\angle C$, determine whether $BC + CA - 2R - 2r$ is positive, negative, or zero.

**Solution.**  Set $AB = c$, $BC = a$, $CA = b$, $\angle A = 2x$, $\angle B = 2y$, $\angle C = 2z$. Then $0 < x \leq y \leq z$ and $x + y + z = \pi/2$. Let $s$ denote the given quantity $BC + CA - 2R - 2r = a + b - 2R - 2r$. Using the well-known formulas

$$2R = \frac{a}{\sin \angle A} = \frac{b}{\sin \angle B} = \frac{c}{\sin \angle C} = \frac{a}{\sin 2x} = \frac{b}{\sin 2y} = \frac{c}{\sin 2z},$$

$$r = 4R \sin \frac{\angle A}{2} \sin \frac{\angle B}{2} \sin \frac{\angle C}{2} = 4R \sin x \sin y \sin z,$$

we find that $s = 2R(\sin 2x + \sin 2y - 1 - 4 \sin x \sin y \sin z)$. Note that in a right triangle $ABC$ with $\angle C = \pi/2$, we have $2R = c$ and $2r = a + b - c$, implying that $s = 0$. Hence, we try to factor $\cos 2z$ from our expression for $s$:

$$\begin{aligned}
\frac{s}{2R} &= 2 \sin (x + y) \cos (x - y) - 1 + 2(\cos (x + y) - \cos (x - y)) \sin z \\
&= 2 \cos z \cos (x - y) - 1 + 2(\sin z - \cos (x - y)) \sin z \\
&= 2 \cos (x - y)(\cos z - \sin z) - \cos 2z \\
&= 2 \cos (y - x) \cdot \frac{\cos^2 z - \sin^2 z}{\cos z + \sin z} - \cos 2z \\
&= \left[ \frac{2 \cos (y - x)}{\cos z + \sin z} - 1 \right] \cos 2z,
\end{aligned}$$

where we may safely introduce the quantity $\cos z + \sin z$ because it is positive when $0 < z < \pi/2$.

Observe that $0 \leq y - x < \min\{y, x + y\} \leq \min\{z, \pi/2 - z\}$. Because $z \leq \pi/2$ and $\pi/2 - z \leq \pi/2$, we have $\cos (y - x) > \max\{\cos z, \cos (\pi/2 - z)\} = \max\{\cos z, \sin z\}$. Hence,

$$\frac{2 \cos (x - y)}{\cos z + \sin z} - 1 > 0.$$

Thus, $s = p \cos 2z$ for some $p > 0$. It follows that $s = BC + CA - 2R - 2r$ is positive, zero, or negative if and only if $\angle C$ is acute, right, or obtuse, respectively.

## Problem 2

Define the infinite sequence $a_1, a_2, \ldots$ recursively as follows: $a_1 = 0$, $a_2 = 1$, and

$$a_n = \frac{1}{2}na_{n-1} + \frac{1}{2}n(n-1)a_{n-2} + (-1)^n \left(1 - \frac{n}{2}\right)$$

for all $n \geq 3$. Find an explicit formula for

$$f_n = a_n + 2\binom{n}{1}a_{n-1} + 3\binom{n}{2}a_{n-2} + \cdots + n\binom{n}{n-1}a_1.$$

**First Solution.** Rewrite the recursive relation as

$$a_n = (-1)^n + \frac{1}{2}na_{n-1} + \frac{1}{2}n\left((-1)^{n-1} + (n-1)a_{n-2}\right).$$

If $(-1)^{n-1} + (n-1)a_{n-2} = a_{n-1}$, then we have that

$$a_n = (-1)^n + \frac{1}{2}na_{n-1} + \frac{1}{2}na_{n-1} = (-1)^n + na_{n-1}.$$

Then, it is straightforward to show by induction that $a_n = (-1)^n + na_{n-1}$, which implies that

$$a_n = n! - \frac{n!}{1!} + \frac{n!}{2!} - \frac{n!}{3!} + \cdots + (-1)^n \frac{n!}{n!}.$$

Therefore, by a famous result of Euler's, $a_n$ is the number of derangements of $(1, 2, \ldots, n)$, i.e. the number of permutations of this $n$-tuple with no fixed points.

To each pair $(\pi, j)$ of a permutation $\pi$ distinct from the identity and an integer $j$ in $\{1, 2, \ldots, n\}$, assign one mark if $j$ is a fixed point of $\pi$. For a fixed $k = 1, 2, \ldots, n$, there are $\binom{n}{n-k}a_k$ permutations $\pi$ with exactly $n-k$ fixed points: there are $\binom{n}{n-k}$ ways to choose which points are fixed, and $a_k$ derangements of the remaining $k$ points. For each such permutation $\pi$, exactly $n - k$ pairs $(\pi, j)$ are assigned one mark. Summing over all permutations, we find that the total number of marks assigned is

$$\sum_{k=1}^{n}(n - k)\binom{n}{n-k}a_k = f_n - \sum_{k=1}^{n}\binom{n}{n-k}a_k = f_n - (n! - 1),$$

where the sum $\sum_{k=1}^{n}\binom{n}{n-k}a_k$ counts all the $n! - 1$ permutations with fewer than $n$ fixed points.

On the other hand, for each $j \in \{1, 2, \ldots, n\}$, there are exactly $(n - 1)! - 1$ permutations distinct from the identity fix $j$. Thus, summing over all $j$, we find that the total number of marks assigned is

$$\sum_{j=1}^{n} ((n-1)! - 1) = n(n-1)! - n.$$

Setting the two calculated sums equal to each other, we find that $f_n = 2 \cdot n! - n - 1$.

**Note.** Alternatively, after discovering that $f_n = 2 \cdot n! - n - 1$ for small values of $n$, one could use the given recursive relation and combinatorial identities to prove that the formula is true for all $n$.

**Second Solution.** We present another method proving that $a_n$ is the number of derangements of $(1, 2, \ldots, n)$. For $n \geq 3$, we have

$$\begin{aligned}
a_n &= na_{n-1} + (-1)^n = a_{n-1} + (n-1)a_{n-1} + (-1)^n \\
&= [(n-1)a_{n-2} + (-1)^{n-1}] + (n-1)a_{n-1} + (-1)^n \\
&= (n-1)(a_{n-1} + a_{n-2}).
\end{aligned}$$

Now, let $b_n$ be the number of derangements of $(1, 2, \ldots, n)$. Each derangement is of exactly one of the following types:

(a) For some $k \neq 1$, 1 maps to $k$ and $k$ maps to 1. Then there are $n - 1$ possible values for $k$, and for each $k$ there are $b_{n-2}$ derangements for the other $n - 2$ elements. Hence, there are $(n-1)b_{n-2}$ such derangements.

(b) 1 maps to $k$ and $k$ does *not* map to 1. Fix $k$. Then there is a bijection between the set of all such derangements $\pi$ and the set of permutations which fix only 1, via the map $\pi \mapsto \tau\pi$, where $\tau$ is the transposition that swaps 1 and $k$. Because there are $b_{n-1}$ maps which fix only 1, there are $b_{n-1}$ such permutations $\pi$. Letting $k$ vary from 2 to $n$, we find that there are $(n-1)b_{n-2}$ derangements of type (b).

Therefore, $b_n = (n-1)(b_{n-1} + b_{n-2})$. Because $a_1 = b_1 = 0$ and $a_2 = b_2 = 1$, $a_n = b_n$ for all $n \geq 1$, as claimed.

## Problem 3

A table tennis club wishes to organize a doubles tournament, a series of matches where in each match one pair of players competes against a pair of two different players. Let a player's *match number* for a tournament be the number of matches he or she participates in. We are given a set

$A = \{a_1, a_2, \ldots, a_k\}$ of distinct positive integers all divisible by 6. Find with proof the minimal number of players among whom we can schedule a doubles tournament such that

(i) each participant belongs to at most 2 pairs;

(ii) any two different pairs have at most 1 match against each other;

(iii) if two participants belong to the same pair, they never compete against each other; and

(iv) the set of the participants' match numbers is exactly $A$.

**Solution.**

**Lemma.** *Suppose that $k \geq 1$ and $1 \leq b_1 < b_2 < \cdots < b_k$. Then there exists a graph of $b_k + 1$ vertices such that the set of degrees of the $b_k + 1$ vertices is the set $\{b_1, b_2, \ldots, b_k\}$ consists of the degrees of the $b_k + 1$ vertices.*

*Proof:* We prove the lemma by strong induction on $k$. If $k = 1$, the complete graph on $b_1 + 1$ vertices suffices. If $k = 2$, then take $b_2 + 1$ vertices, distinguish $b_1$ of these vertices, and connect two vertices by an edge if and only if one of the vertices is distinguished.

We now prove that the claim is true when $k = i \geq 3$ assuming that it is true when $k < i$. We construct a graph $G$ of $b_i + 1$ vertices, forming the edges in two steps and thus "changing" the degrees of the vertices at each step. Take $b_i + 1$ vertices, and partition them into three sets $S_1, S_2, S_3$ with $|S_1| = b_1$, $|S_2| = b_{i-1} - b_1 + 1$, and $|S_3| = b_i - (b_{i-1} + 1)$. By the induction hypothesis, we can construct edges between the vertices in $S_2$ such that the degrees of those vertices form the set $\{b_2 - b_1, \ldots, b_{i-1} - b_1\}$. Further construct every edge which has some vertex in $S_1$ as an endpoint. Each vertex in $S_1$ now has degree $b_i$, each vertex in $S_3$ has degree $b_1$, and the degrees of the vertices in $S_2$ form the set $\{b_2, \ldots, b_{i-1}\}$. Hence, altogether, the degrees of the $b_i + 1$ vertices in $G$ form the set $\{b_1, b_2, \ldots, b_i\}$. This completes the inductive step and the proof. ∎

Suppose that we have a doubles tournament among $n$ players satisfying the given conditions. At least one player, say X, has match number $\max(A)$. Let $m$ be the number of different pairs she has played against. Each of these pairs contains two players for a count of $2m$. Any player is counted at most twice in this fashion since any player belongs to at most two pairs. Hence, player X must have played against at least $m$ players. If X is in $j$ pairs (where $j$ equals 1 or 2), then there are at least $m + j + 1$

players in total. Also, X played in at most $jm$ matches, implying that $jm \geq \max(A)$. Hence,

$$n \geq m + j + 1 \geq \max(A)/j + j + 1$$
$$\geq \min\{\max(A) + 2, \max(A)/2 + 3\}.$$

Because $\max(A) \geq 6$, we have $\max(A) + 2 > \max(A)/2 + 3$, implying that $n \geq \max(A)/2 + 3$.

We now prove that $n = \max(A)/2 + 3$ is attainable. From the lemma, we can construct a graph $G$ of $\max(A)/6 + 1$ vertices whose degrees form the set $\{a_1/6, a_2/6, \ldots, a_k/6\}$. Partition the $n$ players into $\max(A)/6 + 1$ triples, and let two players be in a pair if and only if they are in the same triple. Assign each triple (and, at the same time, the three pairs formed by the corresponding players) to a vertex in $G$, and let two pairs compete if and only if their corresponding vertices are adjacent. Suppose that we have a pair assigned to a vertex $v$ of degree $a_i/6$. For each of the $a_i/6$ vertices $w$ adjacent to $v$, that pair competes against the three pairs assigned to $w$, for a total of $a_i/2$ matches. Each player assigned to $v$ is in two pairs and hence has match number $2(a_i/2) = a_i$. Therefore, the set of the participants' match numbers is $\{a_1, a_2, \ldots, a_k\}$, as desired.

## Problem 4

We are given an integer $n \geq 2$. For any ordered $n$-tuple of real numbers $A = (a_1, a_2, \ldots, a_n)$, let $A$'s *domination score* be the number of values $k \in \{1, 2, \ldots, n\}$ such that $a_k > a_j$ for all $1 \leq j < k$. Consider all permutations $A = (a_1, a_2, \ldots, a_n)$ of $(1, 2, \ldots, n)$ with domination score 2. Find with proof the arithmetic mean of the first elements $a_1$ of these permutations.

**Solution.** For any ordered $n$-tuple of real numbers $A = (a_1, a_2, \ldots, a_n)$, if $a_k > a_j$ for all $1 \leq j < k$, then we call $a_k$ a *dominator*. If a permutation $A = (a_1, a_2, \ldots, a_n)$ of $(1, 2, \ldots, n)$ has domination score 2, then the two dominators must be $a_1$ and $n$, where $n = a_k$ for some $2 \leq k \leq n$.

Fix $m$ in $\{1, 2, \ldots, n-1\}$. We call the numbers $m+1, m+2, \ldots n$ *big* and the numbers $1, 2, \ldots, m-1$ *small*. In a permutation with 2 dominators and $a_1 = m$, $n$ must appear in the permutation before all the other big numbers. Thus, to form all such permutations, we first choose the $n - m$ positions occupied by big numbers, placing $n$ at the first chosen position and then arranging the other $n - m - 1$ big numbers into the rest of the chosen places. We then arrange all the small numbers in the remaining

$m - 1$ places. Hence, there are

$$x_m = \binom{n-1}{n-m}(n-m-1)!(m-1)! = \frac{(n-1)!}{n-m}$$

such permutations.

Therefore, the desired average is equal to

$$\frac{\sum_{m=1}^{n-1} m x_m}{\sum_{m=1}^{n-1} x_m} = \frac{(n-1)! \sum_{m=1}^{n-1} \frac{m}{n-m}}{(n-1)! \sum_{m=1}^{n-1} \frac{1}{n-m}} = \frac{\sum_{m=1}^{n-1} \frac{m}{n-m}}{\sum_{m=1}^{n-1} \frac{1}{n-m}}$$

$$= \frac{\sum_{m=1}^{n-1} \frac{n}{m} - \sum_{m=1}^{n-1} \frac{m}{m}}{\sum_{m=1}^{n-1} \frac{1}{m}}$$

$$= n - \frac{n-1}{1 + \frac{1}{2} + \cdots + \frac{1}{n}}.$$

## Problem 5

Find all positive integers $n$ such that there exist integers $n_1, n_2, \ldots, n_k > 3$ with

$$n = n_1 n_2 \cdots n_k = 2^{\frac{1}{2^k}(n_1-1)(n_2-1)\cdots(n_k-1)} - 1.$$

**Solution.** If $n$ satisfies the given conditions, then $n = 2^m - 1$ for some positive integer $m$. It is easy to check that 3 is the only integer $m < 10$ such that $n = 2^m - 1$ satisfies the given condition.

Given $m \geq 10$, we now prove that $2^m - 1$ does not satisfy the given condition. Suppose, for the sake of contradiction, that the equation holds for some $k$ and $n_1, n_2, \ldots, n_k$ such that

$$m = \frac{1}{2^k}(n_1 - 1)(n_2 - 1) \cdots (n_k - 1) \geq 10.$$

For $\ell \geq 10$, we have

$$\left(\frac{\ell+1}{\ell}\right)^3 < \left(\frac{5}{4}\right)^3 < 2.$$

Using this fact, it is easy to prove by induction that $2^\ell - 1 > \ell^3$ for integers $\ell \geq 10$. Hence,

$$2^m - 1 > m^3 = \left(\frac{n_1-1}{2}\right)^3 \left(\frac{n_2-1}{2}\right)^3 \cdots \left(\frac{n_k-1}{2}\right)^3. \qquad (1)$$

Because $n = 2^m - 1$ is odd, the $n_i$ are all odd; because each $n_i > 3$, each $n_i$ is at least 5. Hence,

$$\left(\frac{n_i - 1}{2}\right)^3 \geq 4 \cdot \frac{n_i - 1}{2} > n_i \qquad (2)$$

for $i = 1, 2, \ldots, k$. Putting (1) and (2) together, we obtain

$$n = 2^m - 1 > n_1 n_2 \cdots n_k = n,$$

a contradiction. Hence, our assumption was wrong, and $n = 2^3 - 1 = 7$ is the only solution.

## Problem 6

An exam paper consists of 5 multiple-choice questions, each with 4 different choices; 2000 students take the test, and each student chooses exactly one answer per question. Find the smallest value of $n$ for which it is possible for the students' answer sheets to have the following property: among any $n$ of the students' answer sheets, there exist 4 of them among which any two have at most 3 common answers.

**Solution.**   First we prove that $n \geq 25$. Let $1, 2, 3, 4$ denote the four different choices of each problem. Represent each student's answer sheet by an ordered 5-tuple $(a_1, a_2, a_3, a_4, a_5)$, $a_i \in \{1, 2, 3, 4\}$, where the student's answer to problem $i$ is $a_i$. We say that two answer sheets are of the same *type* if their corresponding 5-tuples belong to a set of the form

$$\{ (k, a_2, a_3, a_4, a_5) \mid k \in \{1, 2, 3, 4\} \},$$

where $a_2, a_3, a_4, a_5 \in \{1, 2, 3, 4\}$. Since there are 256 such sets, and $2000 = 256 \times 7 + 208$, at least eight answer sheets are of the same type by the pigeonhole principle. Among the 1992 remaining answer sheets, again some eight are of the same type. Finally, among the 1984 remaining answer sheets, another eight are of the same type. Consider the set $A$ of these 24 answer sheets. Given any four answer sheets in $A$, two of them must be of the same type, that is, their solutions for the last 4 problems are identical. This violates the assumption that there are 4 answer sheets in $A$, among which any two have at most 3 common answers. Hence, $n \geq 25$.

Now we show that $n = 25$ is indeed attainable. Define the set

$$S = \{(a_1, a_2, a_3, a_4, a_5) \mid \textstyle\sum_{i=1}^{5} a_i \equiv 0 \,(\text{mod } 4), \ a_i \in \{1, 2, 3, 4\}\}.$$

Then $|S| = 4^4 = 256$, and any two answer sheets have at most 3 common answers if their corresponding 5-tuples are distinct elements of $S$. Pick any 250 elements of $S$, and assume that exactly eight students turn in answer sheets that correspond to each of these 250 5-tuples. Among any $25 > 3 \cdot 8$ answer sheets, there are four whose corresponding 5-tuples are distinct elements in $S$, and they satisfy the given conditions of the problem.

Therefore, the answer is $n = 25$.

## 1.5  Czech and Slovak Republics

### Problem 1
Show that

$$\sqrt[3]{\frac{a}{b}} + \sqrt[3]{\frac{b}{a}} \leq \sqrt[3]{2(a+b)\left(\frac{1}{a}+\frac{1}{b}\right)}$$

for all positive real numbers $a$ and $b$, and determine when equality occurs.

**First Solution.**  Multiplying both sides of the desired inequality by $\sqrt[3]{ab}$ gives the equivalent inequality

$$\sqrt[3]{a^2} + \sqrt[3]{b^2} \leq \sqrt[3]{2(a+b)^2}.$$

Setting $\sqrt[3]{a} = x$ and $\sqrt[3]{b} = y$, we see that it suffices to prove that

$$x^2 + y^2 \leq \sqrt[3]{2(x^3+y^3)^2} \qquad (*)$$

for $x, y > 0$.

By the arithmetic mean-geometric mean inequality,

$$3x^4y^2 \leq x^6 + x^3y^3 + x^3y^3$$

and

$$3x^2y^4 \leq y^6 + x^3y^3 + x^3y^3,$$

with equality if and only if $x^6 = x^3y^3 = y^6$, or equivalently if and only if $x = y$. Adding these two inequalities and adding $x^6 + y^6$ to both sides yields

$$x^6 + y^6 + 3x^2y^2(x^2+y^2) \leq 2(x^6 + y^6 + 2x^3y^3).$$

Taking the cube root of both sides yields $(*)$, as desired. Equality occurs when $x = y$, or equivalently when $a = b$.

**Second Solution.**  By the power mean inequality, we have

$$\left(\frac{\sqrt[3]{\frac{a}{b}} + \sqrt[3]{\frac{b}{a}}}{2}\right)^3 \leq \left(\frac{\sqrt{\frac{a}{b}} + \sqrt{\frac{b}{a}}}{2}\right)^2, \qquad (\dagger)$$

with equality if and only if $a/b = b/a$, or equivalently $a = b$.

The desired result follows from $(\dagger)$ and the identity

$$\left(\sqrt{\frac{a}{b}} + \sqrt{\frac{b}{a}}\right)^2 = (a+b)\left(\frac{1}{a}+\frac{1}{b}\right).$$

## Problem 2

Find all convex quadrilaterals $ABCD$ for which there exists a point $E$ inside the quadrilateral with the following property: Any line which passes through $E$ and intersects sides $\overline{AB}$ and $\overline{CD}$ divides the quadrilateral $ABCD$ into two parts of equal area.

**Solution.** Quadrilateral $ABCD$ has the desired property if and only if $\overline{AB} \parallel \overline{CD}$.

Suppose that convex quadrilateral $ABCD$ has the desired property. Let $X_1$, $X_2$, and $X_3$ be three points on side $\overline{AB}$ with $AX_1 < AX_2 < AX_3$, such that line $X_k E$ intersects side $\overline{CD}$ at $Y_k$ for $k = 1, 2, 3$. Because quadrilateral $ABCD$ is convex, $CY_1 < CY_2 < CY_3$. We have

$$
\begin{aligned}
0 &= \frac{1}{2}[ABCD] - \frac{1}{2}[ABCD] \\
&= [AX_1Y_1D] - [AX_2Y_2D] \\
&= [EY_1Y_2] - [EX_1X_2] \\
&= \frac{1}{2}\sin \angle Y_1 E Y_2 \left( EY_1 \cdot EY_2 - EX_1 \cdot EX_2 \right),
\end{aligned}
$$

implying that $EX_1 \cdot EX_2 = EY_1 \cdot EY_2$. Similarly, $EX_2 \cdot EX_3 = EY_2 \cdot EY_3$. Hence, $EX_1/EY_1 = EX_3/EY_3$ and $\triangle Y_1 E Y_3 \sim \triangle X_1 E X_3$. Therefore, $\overline{X_1 X_3} \parallel \overline{Y_1 Y_3}$, that is, $\overline{AB} \parallel \overline{CD}$.

On the other hand, for any convex quadrilateral $ABCD$ with $\overline{AB} \parallel \overline{CD}$, let $E$ be the midpoint of segment $\overline{M_1 M_2}$, where $M_1$ and $M_2$ are the midpoints of sides $\overline{AB}$ and $\overline{CD}$, respectively. Suppose a line passes through $E$ and intersects sides $\overline{AB}$ at $X$ and $\overline{CD}$ at $Y$. Reflecting the figure across $E$ sends line $AB$ to line $CD$ and hence $\overline{XM_1}$ to $\overline{YM_2}$. It follows that $XM_1 = YM_2$ and $AX + DY = BX + CY$. Thus, quadrilaterals $AXYD$ and $BXYC$ — where each quadrilateral is a trapezoid or possibly a parallelogram — have the same heights and the same sums of base lengths. Therefore they have equal areas, as desired.

## Problem 3

An isosceles triangle $ABC$ is given with base $\overline{AB}$ and altitude $\overline{CD}$. Point $P$ lies on $\overline{CD}$. Let $E$ be the intersection of line $AP$ with side $\overline{BC}$, and let $F$ be the intersection of line $BP$ with side $\overline{AC}$. Suppose that the incircles of triangle $ABP$ and quadrilateral $PECF$ are congruent. Show that the incircles of the triangles $ADP$ and $BCP$ are also congruent.

**First Solution.** Let $\omega_1$ and $\omega_2$ be the incircles of quadrilateral $CEPF$ and triangle $ABP$, respectively, and let $I_1$ and $I_2$ are the centers of circles $\omega_1$ and $\omega_2$, respectively. Because the figure is symmetric about $\overline{CD}$, $I_1$ and $I_2$ lie on segment $\overline{CD}$ with $P$ between these two points. Because $\omega_1$ and $\omega_2$ are congruent and inscribed in vertical angles, they are reflections of each other across $P$. Therefore, $PI_1 = PI_2$.

Because triangles $ADP$ and $BDP$ are congruent, we only need to prove that the inradius $r_1$ of triangle $BCP$ equals the inradius $r_2$ of triangle $BDP$. Let $X$ and $Y$ be the incenters of triangles $BCP$ and $BDP$, respectively. Observe that $I_1$ is also the incenter of triangle $CBF$, so that $I_1$ lies on the bisector of angle $CBF$, that is, of angle $CBP$. Hence, $X$ is on segment $\overline{BI_1}$, and likewise, $Y$ is on segment $\overline{BI_2}$.

Because $PI_1 = PI_2$, $[BI_1P] = [BI_2P]$. Therefore,

$$r_1(PI_1 + BP) = 2([I_1PX] + [XPB]) = 2[I_1PB]$$
$$= 2[PI_2B] = 2([PI_2Y] + [PYB])$$
$$= r_2(PI_2 + BP).$$

Therefore, $r_1 = r_2$, as desired.

**Second Solution.** As in the first solution, let $\omega_1$ and $\omega_2$ be the incircles of quadrilateral $CEPF$ and triangle $ABP$, respectively. Of the common external tangents of these circles, let the tangent closer to $A$ intersect lines $BC$ and $BD$ at $C'$ and $D'$, respectively. Then $\overline{C'D'} \parallel \overline{CD}$. Let segments $\overline{C'D'}$ and $\overline{BF}$ intersect at $P'$. Observe that $\omega_1$ and $\omega_2$ are the incircles of triangle $BC'P'$ and $BD'P'$, respectively.

Consider the homothety $\mathbf{H}$ centered at $B$ with ratio $CD/C'D'$. Then $\mathbf{H}$ sends triangles $BC'P'$ and $BD'P'$ to triangles $BCP$ and $BDP$, respectively. Hence, $\mathbf{H}$ sends circles $\omega_1$ and $\omega_2$ to the incircles of triangles $BCP$ and $BDP$, respectively. Because $\omega_1$ and $\omega_2$ are congruent, the incircles of triangles $BCP$ and $BDP$ are as well. Because triangles $BDP$ and $ADP$ have congruent incircles as well, the desired result follows.

## Problem 4

In the plane are given 2000 congruent triangles of area 1, which are images of a single triangle under different translations. Each of these triangles contains the centroids of all the others. Show that the area of the union of these triangles is less than $\frac{22}{9}$.

**Solution.** Orient the figure in the problem such that each of the 2000 given triangles has one horizontal side, with the opposite vertex above that side. Let triangle $ABC$ be one of the given triangles, with $\overline{AB}$ horizontal and $A$ to the left of $B$, such that none of the other 1999 triangles' horizontal sides lie below line $AB$.

We begin by defining notions of distance different from usual Euclidean distance, and using these definitions to formally describe some relations between the 2000 triangles. Define an $\alpha$-*object* to be a point, or a line, or segment parallel to $\overline{BC}$. Let the $\alpha$-*distance* $d_\alpha$ from one $\alpha$-object to another be the signed distance between the two lines parallel to $\overline{BC}$ passing through the objects, where the signs are chosen such that $\tilde{a} = d_\alpha(\overline{BC}, A)$ is positive. Similarly define $\beta$- and $\gamma$-objects and distances with respect to $\overline{CA}$ and $\overline{AB}$, with $\tilde{b} = d_\beta(\overline{CA}, B) > 0$ and $\tilde{c} = d_\gamma(\overline{AB}, C) > 0$. Notice that if a translation maps an $\alpha$-object through some $\alpha$-distance, it maps *any* $\alpha$-object through that $\alpha$-distance; analogous results hold for the $\beta$- and $\gamma$-distances.

Suppose that triangles $XYZ$ and $X'Y'Z'$ are any two triangles from among the 2000, with $\overline{XY} \parallel \overline{X'Y'} \parallel \overline{AB}$ and $\overline{YZ} \parallel \overline{Y'Z'} \parallel \overline{BC}$. Let $\mathbf{T}$ be the translation which maps triangle $XYZ$ to triangle $X'Y'Z'$. Because the centroid of triangle $XYZ$ lies on or to the left of $Y'Z' = \mathbf{T}(YZ)$, $d_\alpha(\overline{YZ}, \mathbf{T}(\overline{YZ})) \le \frac{1}{3}\tilde{a}$. Therefore, $d_\alpha(X, X') = d_\alpha(X, \mathbf{T}(X)) \le \frac{1}{3}\tilde{a}$ and hence

$$d_\alpha(\overline{YZ}, X') \le \frac{4}{3}\tilde{a}.$$

Similarly, $d_\beta(\overline{ZX}, Y') \le \frac{4}{3}\tilde{b}$ and $d_\gamma(\overline{XY}, Z') \le \frac{4}{3}\tilde{c}$.

Now, let $\mathcal{T}$ be the image of triangle $ABC$ under a dilation about $C$ with ratio $\frac{1}{3}$, so that $[\mathcal{T}] = \frac{1}{9}$. Let $\mathcal{U}_2$ and $\mathcal{V}_1$ be translations of $\mathcal{T}$ whose bottom-right vertices are at $C$ and $A$, respectively, and let $\mathcal{U}_1$ and $\mathcal{V}_2$ be translations of $\mathcal{T}$ whose bottom-left vertices are at $C$ and $B$, respectively. Let $\mathbf{T}_1$ and $\mathbf{T}_2$ be the translations such that $\mathbf{T}_1(\mathcal{U}_1) = \mathcal{V}_1$ and $\mathbf{T}_2(\mathcal{U}_2) = \mathcal{V}_2$. Let the convex hull of these four triangles be $\mathcal{F}$, bounded by line $\ell_1$ on the right, line $\ell_2$ on the left, line $\ell_3$ above, and line $AB$ below. Observe that $\mathcal{F}$ is a trapezoid with area $\frac{24}{9}$.

Because $d_\alpha(\overline{AB}, \ell_3) = \frac{4}{3}\tilde{c}$, the points of any triangle containing the centroid of triangle $ABC$ lie on or below $\ell_3$. Similarly, because $d_\beta(\ell_2, B) = \frac{4}{3}\tilde{b}$ and $d_\gamma(\ell_3, C) = \frac{4}{3}\tilde{c}$, in order for triangle $ABC$ to contain the centroid of a triangle, the points of that triangle must lie on or to the right of $\ell_2$ and on or to the left of $\ell_1$. Combined with the extremal definition of triangle $ABC$, these results imply that the region $\mathcal{R}$ covered by the 2000 triangles lies within the trapezoid $\mathcal{F}$ defined earlier.

Of the lines passing through the 2000 triangle sides parallel to $\ell_1$, let $k$ be the line closest to $\ell_1$. Because $d_\alpha(\mathcal{U}_1, \mathcal{V}_1) = \frac{4}{3}\tilde{a}$, $\mathbf{T}_1$ moves $\alpha$-objects through $\alpha$-distance $\frac{4}{3}\tilde{a}$. Thus, each of the 2000 triangles lies in the region $\mathcal{R}'$ bounded by $k$ and $\mathbf{T}_1(k)$. Observe that $\mathcal{V}_1 = \mathbf{T}_1(\mathcal{U}_1)$, so the regions $\mathcal{R}' \cap \mathcal{V}_1$ and $\mathcal{R}' \cap \mathcal{U}_1$ fit together to form a triangle congruent to $\mathcal{T}$. In other words, the area in $\mathcal{V}_1 \cup \mathcal{U}_1$ also in $\mathcal{R}'$, and hence in $\mathcal{R}$, is at most $\frac{1}{9}$. Said differently, the area in $\mathcal{V}_1 \cup \mathcal{U}_1$ *not* in $\mathcal{R}$ is at least $\frac{1}{9}$.

Similarly, the area in $\mathcal{U}_2 \cup \mathcal{V}_2$ not in $\mathcal{R}$ is at least $\frac{1}{9}$. Furthermore, consider the segment $m_1 \cap \mathcal{F}$. Any point on this segment inside $\mathcal{R}$ must be the top vertex of one of the 2000 triangles. Because there can only be finitely many such points, there must be some segment along $m_1 \cap \mathcal{F}$ not containing any such points. Thus, this segment borders some uncovered triangular region in $\mathcal{F}$ with positive area, such that this triangle, the region $\mathcal{U}_1 \cup \mathcal{V}_1$, and the region $\mathcal{U}_2 \cup \mathcal{V}_2$ are pairwise disjoint. Therefore,

$$[\mathcal{R}] \le [\mathcal{F}] - 1/9 - 1/9 - \kappa = 22/9 - \kappa < 22/9,$$

as desired.

## 1.6  Estonia

### Problem 1

Five real numbers are given such that, no matter which three of them we choose, the difference between the sum of these three numbers and the sum of the remaining two numbers is positive. Prove that the product of all these 10 differences (corresponding to all the possible triples of chosen numbers) is less than or equal to the product of the squares of these five numbers.

**Solution.**  Let the five numbers be $x_1$, $x_2$, $x_3$, $x_4$, $x_5$, where indices are taken modulo 5. The 10 given differences are $a_1, a_2, \ldots, a_5$ and $b_1, b_2, \ldots, b_5$, where

$$a_i = -x_{i-2} + x_{i-1} + x_i + x_{i+1} - x_{i+2};$$

$$b_i = x_{i-2} - x_{i-1} + x_i - x_{i+1} + x_{i+2}$$

for $i = 1, 2, 3, 4, 5$. For each such $i$, we have

$$x_i^2 - a_i b_i = \left(\frac{a_i + b_i}{2}\right)^2 - a_i b_i = \left(\frac{a_i - b_i}{2}\right)^2 \geq 0,$$

or $x_i^2 \geq a_i b_i$. Because $a_i b_i \geq 0$ for each $i$, we may multiply the five inequalities $a_i b_i \leq x_i^2$ for $1 \leq i \leq 5$ to obtain

$$\prod_{i=1}^{5} x_i^2 \geq \prod_{i=1}^{5} a_i b_i,$$

as desired.

### Problem 2

Prove that it is not possible to divide any set of 18 consecutive positive integers into two disjoint sets $A$ and $B$, such that the product of the elements in $A$ equals the product of the elements in $B$.

**Solution.**  Suppose, for the sake of contradiction, that we could partition a set $S = \{n, n+1, \ldots, n+17\}$ of 18 consecutive positive integers into two disjoint sets $A$ and $B$ such that $\prod_{a \in A} a = \prod_{b \in B} b$. Because the product of the elements of $A$ equals the product of elements of $B$, if one set contains a multiple of 19, then the other must as well. Thus, $S$ contains either no multiples of 19 or at least 2 multiples of 19. Because only one of any

18 consecutive integers can be a multiple of 19, $S$ must contain no such multiples. Therefore, $n, n + 1, \ldots, n + 17$ are congruent to $1, 2, \ldots, 18$, respectively, modulo 19. Thus,

$$\prod_{a \in A} a \cdot \prod_{b \in B} b = n(n + 1) \cdots (n + 17) \equiv 18! \equiv -1 \pmod{19},$$

by Wilson's Theorem. However, the two products on the left hand side are equal, which is impossible because $-1$ is not a square modulo 19 (or of any prime congruent to 3 modulo 4, a well known result in number theory). Therefore, no such sets $A$ and $B$ exist.

## Problem 3

Let $M$, $N$, and $K$ be the points of tangency of the incircle of triangle $ABC$ with the sides of the triangle, and let $Q$ be the center of the circle drawn through the midpoints of $\overline{MN}$, $\overline{NK}$, and $\overline{KM}$. Prove that the incenter and circumcenter of triangle $ABC$ are collinear with $Q$.

**First Solution.** Assume that $M$, $N$, and $K$ lie on sides $\overline{BC}$, $\overline{CA}$, $\overline{AB}$, respectively, and define $X$, $Y$, $Z$ as the midpoints of $\overline{NK}$, $\overline{KM}$, $\overline{MN}$, respectively. According to the given information, $Q$ is the circumcenter of triangle $XYZ$. Line $AX$ is the median to base $\overline{KN}$ of isosceles triangle $AKN$, implying that it is also an angle bisector and an altitude in this triangle. Thus, $A$, $X$, and $I$ are collinear, and $\angle AXN = \pi/2$. Hence, right triangles $AXK$ and $AKI$ are similar, and $IA \cdot IX = IK^2$. Therefore, $X$ is the image of $A$ under the inversion through the incircle of triangle $ABC$. Similarly, $Y$ is the image of $B$ and $Z$ is the image of $C$ under this same inversion. It follows that this inversion maps the circumcircle of triangle $ABC$ to the circumcircle of triangle $XYZ$, so the centers of these circles are collinear with $I$. In other words, $Q$, $I$, and the circumcenter of $ABC$ are collinear.

**Second Solution.** For triangle $XYZ$, let $H_{XYZ}, I_{XYZ}, O_{XYZ}$ denote its orthocenter, incenter, and circumcenter, respectively. By definition, $Q$ is the center of the nine-point circle of triangle $MNK$. It is well known that $Q$ is the on line $\ell$ passing through $H_{MNK}$ and $O_{MNK} = I_{ABC}$.

Let $\omega$ and $\omega'$ be the circumcircle and incircle of triangle $ABC$, respectively. Let $A'$ be the midpoint of arc $BC$ not including $A$. Define $B'$ and $C'$ analogously. Tangent lines to circle $\omega$ at $A', B', C'$ form a triangle $A_1 B_1 C_1$ with its sides parallel to the sides of triangle $ABC$. Hence there is homothety $\mathbf{H}$, centered at $P$, that sends triangle $ABC$ to triangle $A_1 B_1 C_1$. With correct labelling, $\mathbf{H}(MNK) = A'B'C'$. Hence $\mathbf{H}(H_{MNK}) = H_{A'B'C'}$, implying that $P, H_{MNK}, H_{A'B'C'}$ are collinear. Similarly, $P, O_{MNK} = I_{ABC}, O_{A'B'C'} = O_{ABC}$ are collinear.

But it is also well known, and not difficult to prove, that $H_{A'B'C'} = I_{ABC}$. Hence $P, Q, H_{MNK}, H_{A'B'C'} = I_{ABC} = O_{MNK}, O_{A'B'C'} = O_{ABC}$ all lie on line $\ell$, as desired.

## Problem 4

Find all functions $f : \mathbb{N} \to \mathbb{N}$ such that

$$f\big(f(f(n))\big) + f\big(f(n)\big) + f(n) = 3n$$

for all $n \in \mathbb{N}$.

**Solution.** Observe that if $f(a) = f(b)$, then setting $n = a$ and $n = b$ into the given equation yields $3a = 3b$, or $a = b$. Therefore, $f$ is injective. We now prove by induction on $n \in \mathbb{Z}^+$ that $f(n) = n$. Suppose that for all $n < n_0$, $f(n) = n$; we prove that $f(n_0) = n_0$. (This proof applies to $n_0 = 1$ as well.) Because $f$ is injective, if $n \geq n_0 > k$, then $f(n) \neq f(k) = k$. Thus,

$$f(n) \geq n_0. \tag{$*$}$$

for all $n \geq n_0$. Specifically, $(*)$ holds for $n = n_0$, i.e., $f(n_0) \geq n_0$. Then $(*)$ holds for $n = f(n_0)$ and similarly for $f\big(f(n_0)\big)$ as well. Substituting $n = n_0$ in the given equation, we find that

$$3n_0 = f\big(f(f(n_0))\big) + f\big(f(n_0)\big) + f(n_0) \geq n_0 + n_0 + n_0.$$

Equality must occur, so $f(n_0) = n_0$, as desired.

## Problem 5

In a triangle $ABC$ we have $AC \neq BC$. Take a point $X$ in the interior of this triangle and let $\alpha = \angle A$, $\beta = \angle B$, $\phi = \angle ACX$, and $\psi = \angle BCX$. Prove that

$$\frac{\sin \alpha \sin \beta}{\sin(\alpha - \beta)} = \frac{\sin \phi \sin \psi}{\sin(\phi - \psi)}$$

if and only if $X$ lies on the median of triangle $ABC$ drawn from the vertex $C$.

**Solution.** Let $M$ be the midpoint of $\overline{AB}$, and let $\phi' = \angle ACM$ and $\psi' = \angle MCB$. Without loss of generality, assume that $\alpha > \beta$ and

$BC > AC$ (otherwise, swapping the labels of $B$ and $C$ negates each side of the desired equation without affecting whether the equation is true). By the Angle Bisector Theorem, the bisector of angle $ACB$ meets $\overline{AB}$ at a point closer to $A$ than to $B$, i.e., at a point on $\overline{AM}$. Thus, $\psi' > \phi'$.

Reflect $B$ across line $CX$ to produce a new point $B'$, and construct $D$ on line $CB'$, on the opposite side of line $AB$ as $C$, such that $\angle BAD = \psi'$. We now use heavy angle-chasing in order to apply the trigonometric form of Ceva's Theorem in triangle $ABC$ to the concurrent lines $AD$, $XD$, and $B'D$.

Triangle $BCB'$ is isosceles with $BC = B'C$ and $\angle B'CB = 2\psi'$, implying that $\angle CBB' = \pi/2 - \psi'$. Hence, $\angle ABB' = \angle CBB' - \angle CBA = \pi/2 - \psi' - \beta$.

Because $MA = MB = MB'$, $\angle AB'B = \pi/2$ and triangle $MAB'$ is isosceles. Thus, the angles of this triangle are

$$\angle B'AM = \angle B'AB = \pi/2 - \angle ABB' = \psi' + \beta,$$

$$\angle MB'A = \psi' + \beta,$$

$$\angle AMB' = \pi - 2(\psi' + \beta) = (\alpha - \beta) + (\phi' - \psi'),$$

where the last equality is true because $\alpha + \beta + \phi' + \psi' = \pi$.

Because $\angle MAB' > \psi' = \angle MAD$, $D$ lies between rays $AM$ and $AB'$ and hence inside triangle $AXB'$. Also, $\angle MAD = \psi' = \angle BCM = \angle MCB' = \angle MCD$, implying that quadrilateral $MCAD$ is cyclic.

We now calculate the angles lines $AD$, $MD$, and $B'D$ make with the sides of triangle $AMB'$. We first have $\angle MAD = \psi'$, $\angle DMA = \angle DCA = \phi' - \psi'$, and $\angle DB'M = \angle CB'M = \angle MBC = \alpha$.

Combined with the expressions found earlier for the angles of triangle $AM'B$, these equations also yield $\angle DAB' = \angle MAB' - \angle MAD = \beta$, $\angle B'MD = \angle B'MA - \angle DMA = \alpha - \beta$, and $\angle AB'D = \angle AB'M - \angle DB'M = \psi'$.

Applying Ceva's Theorem and the Law of Sines, we find that

$$1 = \frac{\sin \angle MAD}{\sin \angle DAB'} \frac{\sin \angle AB'D}{\sin \angle DB'M} \frac{\sin \angle B'MD}{\sin \angle DMA}$$

$$= \frac{\sin \psi'}{\sin \beta} \frac{\sin \psi'}{\sin \beta} \frac{\sin(\alpha - \beta)}{\sin(\phi' - \psi')} = \frac{MB}{MC} \frac{\sin \psi'}{\sin \beta} \frac{\sin(\alpha - \beta)}{\sin(\phi' - \psi')}$$

$$= \frac{MA}{MC} \frac{\sin \psi'}{\sin \beta} \frac{\sin(\alpha - \beta)}{\sin(\phi' - \psi')} = \frac{\sin \phi'}{\sin \alpha} \frac{\sin \psi'}{\sin \beta} \frac{\sin(\alpha - \beta)}{\sin(\phi' - \psi')}.$$

If $X$ lies on line $CM$, then $(\phi, \psi) = (\phi', \psi')$, and the above equation implies that the given equation holds. Conversely, suppose that the given equation holds. Let $\gamma = \angle BCA$ and let

$$f(\theta) = \frac{\sin\theta\sin(\gamma-\theta)}{\sin(\theta-(\gamma-\theta))}.$$

We are given that $f(\phi) = f(\phi')$; this common value is nonzero because $0 < \phi, \phi' < \gamma$. Thus, $1/f(\theta)$ is defined and takes on equal values at $\theta = \phi$ and $\theta = \phi'$. However,

$$\frac{1}{f(\theta)} = \frac{\sin\theta\cos(\gamma-\theta) - \cos\theta\sin(\gamma-\theta)}{\sin\theta\sin(\gamma-\theta)} = \cot\theta - \cot(\gamma-\theta),$$

which is a strictly decreasing function for $\theta \in (0,\gamma)$. Therefore, $\phi = \phi'$, and $X$ must lie on line $CM$. This completes the proof.

## Problem 6

We call an infinite sequence of positive integers an *F-sequence* if every term of this sequence (starting from the third term) equals the sum of the two preceding terms. Is it possible to decompose the set of all positive integers into

(a) a finite;

(b) an infinite

number of $F$-sequences having no common members?

**Solution.**  (a) We prove that $\mathbb{N}$ cannot be partitioned into finitely many $F$-sequence. Suppose, for the sake of contradiction, that there exist $m$ $F$-sequences that partition the positive integers. Let the $i$th sequence be $F_1^{(i)}, F_2^{(i)}, \ldots$. Because $F_{n+2}^{(i)} - F_{n+1}^{(i)} = F_n^{(i)}$ is increasing for $n \geq 2$, there exists $N_i$ such that $F_{n+2}^{(i)} - F_{n+1}^{(i)} > m$ for all $n > N_i$. Let $N = \max\{N_1, N_2, \ldots, N_i\}$, and choose a positive integer $k$ that exceeds the first $N$ terms of each sequence. By the Pigeonhole Principle, two of the numbers $k, k+1, \ldots, k+m$ appear in the same $F$-sequence. These two integers will differ by at most $m$, a contradiction. Therefore, the answer to part (a) is "no."

(b) The answer is "yes". Define the Fibonacci sequence $\{F_n\}$ by $F_0 = F_1 = 1$ and the recursive relation $F_{n+1} = F_n + F_{n-1}$ for $n > 1$. It can be shown by induction on $j$ that each positive integer $j$ has a unique *Zeckendorf representation* (or "base Fibonacci" representation)

$\overline{a_k a_{k-1} \ldots a_1}$ with the following properties: $a_k = 1$; each $a_i$ equals 0 or 1; no two consecutive digits equal 1; and $j = \sum_{i=1}^{k} a_k F_k$.

There are infinitely many positive integers $m$ whose Zeckendorf representation $\overline{a_k a_{k-1} \ldots a_1}$ with $a_1 = 1$. For each such $m$, define a sequence $\mathcal{F}_m$ as follows: let the $n$th term be the number $\sum_{k=1}^{n} a_k F_{k+n-1}$ whose Zeckendorf representation is $\overline{a_k a_{k-1} \ldots a_1}$ followed by $n-1$ zeroes. Then the sum of the $n$th and $(n+1)$th terms is

$$\sum_{k=1}^{n} a_k F_{k+n-1} + \sum_{k=1}^{n} a_k F_{k+n} = \sum_{k=1}^{n} a_k (F_{k+n-1} + F_{k+n})$$

$$= \sum_{k=1}^{n} a_k F_{k+n+1},$$

the $(n+2)$th term. Hence, $\mathcal{F}_m$ is an $F$-sequence. Any positive integer $j$ appears in $\mathcal{F}_m$ for exactly one positive integer $m$ — namely, the one whose Zeckendorf representation is the same as $j$'s, except without any trailing zeroes. Hence, these sequences partition the positive integers, proving that the answer to part (b) is "yes."

# 1.7 Hungary

## Problem 1

Find all positive primes $p$ for which there exist positive integers $n, x, y$ such that $p^n = x^3 + y^3$.

**Solution.** The answers are $p = 2$ and $p = 3$. To see that these numbers work note $2^1 = 1^3 + 1^3$ and $3^2 = 1^3 + 2^3$. Now suppose that $p > 3$, and assume, for the sake of contradiction, that the equation holds for some triples of positive integers $(n, x, y)$. Choose $n$, $x$, and $y$ such that $n$ is minimal.

Because $p \neq 2$, we have $(x, y) \neq (1, 1)$. Therefore, $x^2 - xy + y^2 = (x - y)^2 + xy$ is greater than 1, as is $x + y$. Because both these quantities divide $x^3 + y^3$, they must be multiples of $p$. Hence, $p$ divides $(x + y)^2 - (x^2 - xy + y^2) = 3xy$. Because $p \nmid 3$, $p$ divides at least one of $x$ or $y$. Furthermore, $p$ cannot divide only one of $x$ and $y$ because $p \mid (x + y)$. It follows that $n > 3$ and that $p^{n'} = x'^3 + y'^3$, where $(n', x', y') = (n - 3, x/3, y/3)$. But $n' < n$, contradicting the minimal definition of $n$.

Therefore, $p = 2$ and $p = 3$ are the only primes that work.

## Problem 2

Is there a polynomial $f$ of degree 1999 with integer coefficients, such that $f(n), f(f(n)), f(f(f(n))), \ldots$ are pairwise relatively prime for any integer $n$?

**Solution.** The answer is "yes". Let $g(x)$ be any polynomial of degree 1997 with integer coefficients, and let $f(x) = x(x-1)g(x)+1$. We prove that $f$ satisfies the condition in the problem. It suffices to show that if $n$ is any integer and $p$ is a prime that divides $f(n)$, then $p \nmid f^k(n)$ for any positive integer $k > 1$. More specifically, we prove that $f^k(n) \equiv 1 \pmod{p}$ for all $k > 1$.

We induct on $k$ to prove our assertion. It is well known that for a polynomial $h$ with integer coefficients, $a \equiv b \pmod{c}$ implies $h(a) \equiv h(b) \pmod{c}$. For the base case $k = 2$, $f(n) \equiv 0 \pmod{p} \implies f(f(n)) \equiv f(0) \equiv 1 \pmod{p}$. For the inductive step, assume that $f^k(n) \equiv 1 \pmod{p}$. Then $f(f^k(n)) \equiv f(1) \equiv 1 \pmod{p}$. This completes the proof.

## Problem 3

The feet of the angle bisectors of triangle $ABC$ are $X$, $Y$, and $Z$. The circumcircle of triangle $XYZ$ cuts off three segments from lines $AB$, $BC$, and $CA$. Prove that two of these segments' lengths add up to the third segment's length.

**Solution.** We use signed distances throughout the problem, where $a = BC$, $b = CA$, and $c = AB$ are positive. Also, let the specified circle intersect $BC$ at $X$ and $P$, $CA$ at $Y$ and $Q$, and $AB$ at $Z$ and $R$, and let $x = PX$, $y = QY$, $z = RZ$. By the Angle Bisector Theorem, $YA = bc/(c+a)$ and $AZ = bc/(a+b)$. Hence, $QA = bc/(c+a) + y$ and $AR = bc/(a+b) - z$. Applying the Power of a Point Theorem to point $A$ and the specified circle gives

$$\frac{bc}{c+a}\left(\frac{bc}{c+a} + y\right) = \frac{bc}{a+b}\left(\frac{bc}{a+b} - z\right).$$

After multiplying through by $a/bc$ and rearranging, we obtain

$$\frac{a}{c+a}y + \frac{a}{a+b}z = \frac{abc}{(a+b)^2} - \frac{abc}{(c+a)^2}.$$

Similarly, we have

$$\frac{b}{a+b}z + \frac{b}{b+c}x = \frac{abc}{(b+c)^2} - \frac{abc}{(a+b)^2},$$

$$\frac{c}{b+c}x + \frac{c}{c+a}y = \frac{abc}{(c+a)^2} - \frac{abc}{(b+c)^2}.$$

When we add these three equations, we get $x + y + z = 0$. It follows that two of the quantities $x, y, z$ are of the same sign and that the third is of the other sign. Hence, the sum of the absolute values of the former two quantities equals the absolute value of the latter, which is what we wished to prove.

## Problem 4

Let $k$ and $t$ be relatively prime integers greater than 1. Starting from the permutation $(1, 2, \ldots, n)$ of the numbers $1, 2, \ldots, n$, we may swap two numbers if their difference is either $k$ or $t$. Prove that we can get any permutation of $1, 2, \ldots, n$ with such steps if and only if $n \geq k + t - 1$.

**Solution.** Construct a graph $G$ whose vertices are the integers $1, 2, \ldots, n$, with an edge between $a$ and $b$ if and only if $|a - b| \in \{k, t\}$. We show that the following conditions are equivalent: (i) every permutation is obtainable; (ii) $G$ is connected; (iii) $n \geq k + t - 1$.

(i) $\Rightarrow$ (ii): Because every step swaps two numbers in the same component of $G$, it follows that no number can ever be sent to a position formerly occupied by a number in a different component. Consequently, we cannot obtain every permutation unless all numbers are in the same component.

(ii) $\Rightarrow$ (i): More generally, we demonstrate by induction on $m$ that given a connected graph on $m$ integers, any permutation $\pi$ of these integers maybe be obtained from any other by successive swaps of the form $(a \; b)$, where $a$ and $b$ are adjacent vertices of the graph. The claim is obvious when $m = 1$. Otherwise, choose a vertex $a$ such that the graph remains connected when $a$ is deleted — for instance, we may let $a$ be a leaf of a spanning tree of the vertices. Some path of distinct vertices $a_0 a_1 \ldots a_r$ connects $a_0 = \pi^{-1}(a)$ and $a_r = a$. By successively performing transpositions $(a_r \; a_{r-1}), (a_{r-1} \; a_{r-2}), \ldots, (a_1 \; a_1)$, we can bring $a$ to the position initially occupied by $\pi^{-1}(a)$. By the induction hypothesis, the numbers other than $a$ can then be permuted as needed, so that the permutation $\pi$ is obtained. Applying this claim with $m = n$ and our graph $G$ proves that (ii) $\Rightarrow$ (i).

(ii) $\Rightarrow$ (iii): If $k$ were at least $n$, then every edge would connect two numbers congruent modulo $t$. Then there would be no path between 1 and 2, a contradiction. Thus, we must have $k < n$; likewise, $t < n$. Then, there are $n - k$ edges of the form $\{a, a+k\}$ and $n - t$ of the form $\{a, a+t\}$. Connectedness requires at least $n - 1$ edges, so $(n-k) + (n-t) \geq n - 1 \Rightarrow n \geq k + t - 1$.

(iii) $\Rightarrow$ (ii): Certainly $k, t < n$ in this case. Notice that any two numbers which are congruent modulo $k$ are connected to each other (via edges of the form $\{a, a+k\}$), so it suffices to show that all the numbers $1, 2, \ldots, k$ are mutually connected. Because $t$ is relatively prime to $k$, the numbers $t, 2t, 3t, \ldots, kt$ represent all the congruence classes modulo $k$. Thus, we may rearrange $1, 2, \ldots, k$ in the order $b_1, b_2, \ldots, b_k$, where $b_i \equiv it \pmod{k}$. Notice that $k \equiv 0 \equiv kt \pmod{k}$, so $b_k = k$. Thus, when $1 \leq i \leq k - 1$, we have $b_i \leq k - 1$ and hence $b_i + t \leq k + t - 1 \leq n$. Thus, vertex $b_i + t$ exists and is connected by an edge to $b_i$. Furthermore, $b_i + t \equiv b_{i+1} \pmod{k}$, so $b_i + t$ is connected to $b_{i+1}$. Thus, $b_i$ is connected to $b_{i+1}$ in $G$ for each $i = 1, \ldots, k - 1$. Hence, these numbers are all mutually connected, and the proof is complete.

## Problem 5

For any positive integer $k$, let $e(k)$ denote the number of positive even divisors of $k$, and let $o(k)$ denote the number of positive odd divisors of $k$. For all $n \geq 1$, prove that $\sum_{k=1}^{n} e(k)$ and $\sum_{k=1}^{n} o(k)$ differ by at most $n$.

**Solution.** The number of integers divisible by $d$ among $1, 2, \ldots, n$ is $\lfloor n/d \rfloor$. Thus, the sum of $o(k)$ (resp., $e(k)$) over $k = 1, 2, \ldots, n$ equals the sum of $\lfloor n/d \rfloor$ over all positive odd (resp. even) integers $d$.

Because $\lfloor n/d \rfloor \geq \lfloor n/(d+1) \rfloor$ for positive integers $a$ and $n$, we have

$$\sum_{k=1}^{n} o(k) - \sum_{k=1}^{n} e(k) = \sum_{i=1}^{\infty} \left( \left\lfloor \frac{n}{2i-1} \right\rfloor - \left\lfloor \frac{n}{2i} \right\rfloor \right) \geq 0,$$

where the infinite sum is well-defined because the summands equal zero for $i > \lceil n/2 \rceil$. Similarly,

$$\sum_{k=1}^{n} o(k) - \sum_{k=1}^{n} e(k) = \left\lfloor \frac{n}{1} \right\rfloor - \sum_{i=1}^{\infty} \left( \left\lfloor \frac{n}{2i} \right\rfloor - \left\lfloor \frac{n}{2i+1} \right\rfloor \right) \leq n.$$

## Problem 6

Given a triangle in the plane, show how to construct a point $P$ inside the triangle which satisfies the following condition: if we drop perpendiculars from $P$ to the sides of the triangle, the feet of the perpendiculars determine a triangle whose centroid is $P$.

**Solution.** Let the triangle be $ABC$, with side lengths $a = BC$, $b = CA$, $c = AB$. Let $P$ be a point inside the triangle, to be determined later. Let $X, Y, Z$ be the feet of the perpendiculars to lines $BC, CA, AB$, respectively, and $x = PX$, $y = PY$, $z = PZ$. Notice that $\sin \angle YPZ = \sin(\pi - \angle BAC) = \sin \angle BAC$. Similarly, $\sin \angle ZPX = \sin \angle CBA$ and $\sin \angle XPY = \sin \angle ACB$. The following are readily seen to be equivalent:

1. $P$ is the centroid of triangle $XYZ$;

2. triangles $YPZ, ZPX, XPY$ have equal areas;

3. $yz \sin \angle YPZ = zx \sin \angle ZPX = xy \sin \angle XPY$;

4. $\sin \angle BAC / x = \sin \angle CBA / y = \sin \angle ACB / z$;

5. $a/x = b/y = c/z$. (by the Law of Sines)

Construct a line parallel to line $BC$, at distance $a$ from it, on the same side of line $BC$ as $A$. Likewise, construct a line parallel to line $CA$, at a distance of $b$, and on the same side of line $CA$ as $B$. Let $Q$ be their intersection; and notice that ray $CQ$ passes through the interior of the triangle. Given any point $P'$ on $CQ$, consider the ratio of its distance from line $BC$ to its distance from line $AB$. If $P' = Q$, this ratio is $a/b$; because all such points $P'$ are homothetic images of each other about $C$, the ratio is independent of $P'$ and must *always* equal $a/b$. Likewise, we can construct a ray from $A$, directed into the triangle, such that for any point $P'$ on the ray, the ratio of its distance from line $AB$ to its distance from line $CA$ equals $c/b$. These two rays intersect at some point $P$ inside the triangle. If we let $P$ be their intersection, we obtain $a/x = b/y$ and $b/y = c/z$, and the problem is solved.

## Problem 7

Given a natural number $k$ and more than $2^k$ different integers, prove that a set $S$ of $k + 2$ of these numbers can be selected such that for any positive integer $m \le k + 2$, all the $m$-element subsets of $S$ have different sums of elements.

**Solution.** Given a positive integer $m$, we call a set *weakly m-efficient* if its $m$-element subsets have different sums of elements, and we call a set *strongly m-efficient* if it is weakly $i$-efficient for $1 \le i \le m$. Also, given any set $T$ of integers, let $\sigma(T)$ equal the sum of the elements of $T$.

We prove the desired claim by induction on $k$. For $k = 1$, it is easy to check that we may let $S$ consist of any three of the given integers. Now assuming that the claim is true for $k = n$, we prove that it is true for $k = n + 1$.

Given more than $2^{n+1}$ different integers $a_1, a_2, \ldots, a_t$, let $2^\alpha$ be the largest power of 2 such that $a_1 \equiv a_i \pmod{2^\alpha}$ for each $i = 1, 2, \ldots, t$. Write $b_i = (a_i - a_1)/2^\alpha$ for $1 \le i \le t$, yielding $t$ distinct integers $b_1, b_2, \ldots, b_t$.

By the Pigeonhole Principle, among the $b_i$ there exist more than $2^n$ different integers of the same parity. By the induction hypothesis, from among these integers we may choose an $(n+2)$-element, strongly $(n+2)$-efficient set $S_1$. Furthermore, there exists a $b_{i_0}$ of the opposite parity because the $b_i$ are not all of the same parity: $b_1 = 0$ is even, and by the maximal definition of $2^\alpha$, at least one of the $b_i$ is odd.

We claim that the $(n + 3)$-element set $S_2 = S_1 \cup \{b_{i_0}\}$ is strongly $(n + 3)$-efficient. Suppose, for the sake of contradiction, that $X$ and $Y$ are two distinct $m$-element subsets of $S$ with the same sums of elements, where $1 \le m \le n+3$. Because $X \ne Y$, $m > 1$. $X$ and $Y$ cannot both be subsets of $S_1$ because $S_1$ is weakly $m$-efficient. Nor can they both contain $b_{i_0}$ because then $X \setminus \{b_{i_0}\}$ and $Y \setminus \{b_{i_0}\}$ would be two distinct $(m-1)$-element subsets of $S_1$ with the same sums of elements, which is impossible because $S_1$ is weakly $(m-1)$-efficient. Therefore, one of $X$ and $Y$ contains $b_{i_0}$ and the other does not. This in turn implies that $\sigma(X)$ and $\sigma(Y)$ are of opposite parity, a contradiction.

Let $\Phi$ be the map which sends any set $A$ of reals to

$$\left\{ \frac{a - a_1}{2^\alpha} \mid a \in A \right\}.$$

There exists an $(n+3)$-element subset $S \subseteq \{a_1, a_2, \ldots, a_t\}$ such that $\Phi(S) = S_2$. Suppose that there existed $X, Y \subseteq S$ such that $X \ne Y$, $|X| = |Y| = m$, and $\sigma(X) = \sigma(Y)$. Then we would also have

$$\Phi(X), \quad \Phi(Y) \subseteq \Phi(S) = S_2,$$

$$\Phi(X) \ne \Phi(Y), \quad |\Phi(X)| = m = |\Phi(Y)|,$$

and

$$\sigma\big(\Phi(X)\big) = \frac{\sigma(X) - ma_1}{2^\alpha} = \frac{\sigma(Y) - ma_1}{2^\alpha} = \sigma\big(\Phi(Y)\big).$$

However, this is impossible because $S_2$ is weakly $m$-efficient. Therefore, $S$ is strongly $(n+3)$-efficient as well. This completes the inductive step and the proof.

## 1.8   India

### Problem 1

Let $ABC$ be a nonequilateral triangle. Suppose there is an interior point $P$ such that the three cevians through $P$ all have the same length $\lambda$ where $\lambda < \min\{AB, BC, CA\}$. Show that there is another interior point $P' \neq P$ such that the three cevians through $P'$ also are of equal length.

**Solution.**   Let the three given cevians be $\overline{AD}$, $\overline{BE}$, and $\overline{CF}$, and let the altitudes of the triangle be $\overline{AH_a}$, $\overline{BH_b}$, and $\overline{CH_c}$. Reflect each cevian across the corresponding altitude to obtain the segments $\overline{AD'}$, $\overline{BE'}$, and $\overline{CF'}$. If $\overline{AD'}$ were not inside the triangle, then either $\overline{AB}$ or $\overline{AC}$ would be contained in the triangle $ADD'$. However, this is impossible because $AD = AD' = \lambda < \min\{AB, AC\}$. Therefore, $\overline{AD'}$ is a cevian of triangle $ABC$, as are $\overline{BE'}$ and $\overline{CF'}$.

We now use directed lengths. Observe that

$$BD \cdot BD' = (BH_a + H_aD)(BH_a - H_aD) = BH_a^2 - H_aD^2$$
$$= (AB^2 - AH_a^2) - (AD^2 - AH_a^2) = AB^2 - AD^2$$
$$= AB^2 - \lambda^2.$$

Similarly, $EA \cdot E'A = AB^2 - \lambda^2$.

Thus, $EA \cdot E'A = BD \cdot BD'$. Likewise, $FB \cdot F'B = CE \cdot CE'$ and $DC \cdot D'C = AF \cdot AF'$. Now, applying Ceva's Theorem to the three concurrent cevians $\overline{AD}$, $\overline{BE}$, $\overline{CF}$ yields

$$\frac{BD}{DC} \cdot \frac{CE}{EA} \cdot \frac{AF}{FB} = 1.$$

Hence,

$$\frac{BD'}{D'C} \cdot \frac{CE'}{E'A} \cdot \frac{AF'}{F'B} = \left(\frac{BD' \cdot BD}{D'C \cdot DC}\right)\left(\frac{CE' \cdot CE}{E'A \cdot EA}\right)\left(\frac{AF' \cdot AF}{F'B \cdot FB}\right)$$
$$= \left(\frac{BD' \cdot BD}{E'A \cdot EA}\right)\left(\frac{CE' \cdot CE}{F'B \cdot FB}\right)\left(\frac{AF' \cdot AF}{D'C \cdot DC}\right)$$
$$= 1.$$

By Ceva's Theorem, $\overline{AD'}$, $\overline{BE'}$, and $\overline{CF'}$ concur at some point $P'$ inside triangle $ABC$. If $P$ were the same point as $P'$, then $P$ would also be the same as the orthocenter, and the altitudes of $ABC$ would have the same length $\lambda$. However, this is impossible because triangle $ABC$ is not equilateral. Therefore, $P' \neq P$, and the cevians through $P'$ are of equal length (namely, $\lambda$), as desired.

## Problem 2

Let $m, n$ be positive integers such that $m \leq n^2/4$ and every prime divisor of $m$ is less than or equal to $n$. Show that $m$ divides $n!$.

**Solution.**   It suffices to show that $p^k \mid n!$ for all primes $p$ and integers $k \geq 1$ such that $p^k \mid m$. If $k = 1$, we are done because we are given that $p \leq n$ and hence $p \mid n!$. Otherwise, because $m \leq n^2/4$, we have $p^k \leq n^2/4$. Thus, $n \geq 2\sqrt{p^k}$. If $n \geq kp$, then at least $k$ of the numbers $1, 2, \ldots, n$ are multiples of $p$, implying that $p^k \mid n!$. Hence, it suffices to show that $2\sqrt{p^k}/p \geq k$ or equivalently that

$$p^{(k-2)/2} \geq k/2, \qquad\qquad (*)$$

because this implies $n \geq 2\sqrt{p^k} \geq kp$.

Indeed, we can do so for most values $p$ and $k$. If $k = 2$, then $(*)$ reads $1 \geq 1$. If $k \geq 4$, then Bernoulli's inequality implies that

$$p^{(k-2)/2} = (1 + (p-1))^{(k-2)/2} \geq 1 + \frac{k-2}{2}(p-1) \geq \frac{k}{2},$$

as desired. Finally, if $k = 3$, then $(*)$ is clearly true unless $p = 2$.

If indeed $k = 3$ and $p = 2$, then $m \geq 8$, $n \geq 6$, and $n!$ is certainly divisible by 8. Thus, we have shown that $p^k \mid n!$ in all cases, completing the proof.

## Problem 3

Let $G$ be a graph with $n \geq 4$ vertices and $m$ edges. If $m > n(\sqrt{4n-3} + 1)/4$, show that $G$ has a 4-cycle.

**Solution.**   We count the number of triples of distinct vertices $(v, a, b)$ such that $v$ is adjacent to both $a$ and $b$. The number of such triples with $v$ fixed is $\deg(v)(\deg(v) - 1)$. Because the sum over vertices $v$ of $\deg(v)$ is $2m$, and $x(x-1)$ is a convex function of $x$, Jensen's Inequality implies that the sum over vertices $v$ of $\deg(v)(\deg(v) - 1)$ is at least $n \cdot (2m/n)((2m/n) - 1)) = 2m(2m/n - 1)$.

If there is no 4-cycle, then for any fixed $a$ and $b$ there can be at most one vertex adjacent to both of them, implying that there are at most $n(n - 1)$ triples of the above sort. Hence, there *is* a 4-cycle if

$2m(2m/n - 1) > n(n - 1)$, or equivalently if

$$4m^2 - (2n)m - n^2(n - 1) > 0.$$

This inequality holds if $m$ is greater than the larger of the two roots of the quadratic $4x^2 - (2n)x - n^2(n - 1)$, because this quadratic has positive leading coefficient. Applying the quadratic formula to calculate the larger root, we find that there is a 4-cycle if $m$ is greater than

$$\frac{2n + \sqrt{4n^2 + 16n^3 - 16n^2}}{8} = n \cdot \frac{\sqrt{4n - 3} + 1}{4},$$

as desired.

## Problem 4

Suppose $f : \mathbb{Q} \to \{0, 1\}$ is a function with the property that for $x, y \in \mathbb{Q}$, if $f(x) = f(y)$ then $f(x) = f((x + y)/2) = f(y)$. If $f(0) = 0$ and $f(1) = 1$ show that $f(q) = 1$ for all rational numbers $q$ greater than or equal to 1.

### Solution.

**Lemma.** *Suppose that $a$ and $b$ are rational numbers. If $f(a) \neq f(b)$, then $f(n(b - a) + a) = f(b)$ for all positive integers $n$.*

*Proof:* We prove the claim by strong induction on $n$. For $n = 1$, the claim is clear. Now assume that the claim is true for $n \leq k$. Let $(x_1, y_1, x_2, y_2) = (b, k(b - a) + a, a, (k + 1)(b - a) + a)$. By the induction hypothesis, $f(x_1) = f(y_1)$. We claim that $f(x_2) \neq f(y_2)$. Otherwise, setting $(x, y) = (x_1, y_1)$ and $(x, y) = (x_2, y_2)$ in the given condition, we would have $f(b) = f((x_1 + y_1)/2)$ and $f(a) = f((x_2 + y_2)/2)$. However, this is impossible because $x_1 + y_1 = x_2 + y_2$. Therefore, $f(y_2)$ must equal the value in $\{0, 1\} - \{f(a)\}$, namely $f(b)$. This completes the induction. ∎

Applying the lemma with $a = 0$ and $b = 1$, we see that $f(n) = 1$ for all positive integers $n$. Thus, $f(1 + r/s) \neq 0$ for all natural numbers $r$ and $s$, because otherwise applying the lemma with $a = 1$, $b = 1 + r/s$, and $n = s$ yields $f(1 + r) = 0$, a contradiction. Therefore, $f(q) = 1$ for all rational numbers $q \geq 1$.

# 1.9 Iran

## Problem 1

Call two circles in three-dimensional space *pairwise tangent* at a point $P$ if they both pass through $P$ and the lines tangent to each circle at $P$ coincide. Three circles not all lying in a plane are pairwise tangent at three distinct points. Prove that there exists a sphere which passes through the three circles.

**Solution.** Let the given circles be $\omega_1$, $\omega_2$, and $\omega_3$, and let $\mathcal{P}_k$ be the plane containing $\omega_k$ for $k = 1, 2, 3$.

The locus of points in three-dimensional space equidistant from the points on a circle (resp. from three points) is a line perpendicular to the plane containing those points, passing through the center of the circle that contains those points. Consider the two such lines corresponding to $\omega_1$ and the set of the three given tangent points. Because the plane containing $\omega_1$ cannot also contain all three tangent points, these two lines coincide in at most one point.

Suppose that $\omega_1$ and $\omega_2$ have common tangent $\ell$ and common tangent point $P$. Let $A$ and $B$ be the points diametrically opposite $P$ on $\omega_1$ and $\omega_2$, respectively. We claim that for $k = 1, 2$, plane $(PAB)$ is perpendicular to $\mathcal{P}_k$, i.e., to some line in $\mathcal{P}_k$. Indeed, $\ell$ lies in $\mathcal{P}_k$, and $(PAB)$ is perpendicular to $\ell$ because it contains two non-parallel lines perpendicular to $\ell$: lines $PA$ and $PB$.

Let $O_2$ be the circumcenter of triangle $PAB$. Because $(PAB) \perp \mathcal{P}_1$, the perpendicular from $O_2$ to $\mathcal{P}_1$ is the perpendicular from $O_2$ to line $PA$, intersecting $\mathcal{P}_1$ at the midpoint of $\overline{PA}$ — the center of $\omega_1$. It follows that $O_2$ is equidistant from every point on $\omega_1$. Because $P$ lies on $\omega_1$, the common distance between $O_2$ and any point on $\omega_1$ is $OP$. Similarly, $O_2$ is a distance $OP$ from every point on $\omega_2$.

Likewise, there exists $O_3$ which is equidistant from every point on both $\omega_1$ and $\omega_3$. Thus, each of $O_2$ and $O_3$ is equidistant from every point on $\omega_1$ and the three given tangent points. From our analysis at the beginning of this proof, there is at most one such point. Hence, $O_2$ and $O_3$ equal the same point, equidistant from every point on $\omega_1$, $\omega_2$, and $\omega_3$. Therefore, some sphere centered at this point passes through the three circles, as desired.

## Problem 2

We are given a sequence $c_1, c_2, \ldots$ of natural numbers. For any natural numbers $m, n$ with $1 \leq m \leq \sum_{i=1}^{n} c_i$, we can choose natural numbers $a_1, a_2, \ldots, a_n$ such that

$$m = \sum_{i=1}^{n} \frac{c_i}{a_i}.$$

For each $i$, find the maximum value of $c_i$.

**Solution.** Let $C_1 = 2$ and $C_i = 4 \cdot 3^{i-2}$ for $i \geq 2$. We claim that for each $i$, $C_i$ is the maximum possible value of $c_i$.

We first prove by induction on $i \geq 1$ that $c_i \leq C_i$. For $i = 1$, if $c_1 > 1$, then setting $(m, n) = (c_1 - 1, 1)$ shows that $a_1 = c_1/(c_1 - 1)$ is an integer. This happens exactly when $c_1 = 2$. Thus, $c_1 \leq C_1$.

Now suppose that $c_i \leq C_i$ for $i = 1, 2, \ldots, k - 1$, where $k \geq 2$. We find the $a_i$ corresponding to $(m, n) = (c_k, k)$. Clearly $a_n \geq 2$. Then,

$$c_n = \sum_{i=1}^{n} \frac{c_i}{a_i} \leq \frac{c_n}{2} + \sum_{i=1}^{n-1} C_i,$$

or $c_n \leq 2 \sum_{i=1}^{n-1} C_i = C_n$. This completes the inductive step and the proof of the claim.

It remains to be proven that for each $i$, it is possible to have $c_i = C_i$. Indeed, we prove by induction on $n$ that we can have $c_i = C_i$ for all $i$ simultaneously. For $n = 1$, if $m = 1$ we can set $a_1 = 2$; if $m = 2$ we can set $a_1 = 1$.

Assuming that the claim is true for $n = 1, 2, \ldots, k - 1$, where $k \geq 2$, we prove it for $n = k$. If $m = 1$, we may set $a_i = nC_i$ for $i = 1, 2, \ldots, n$.

If $2 \leq m \leq C_n/2 + 1$, set $a_n = C_n$; if $C_n/2 + 1 \leq m \leq C_n$, set $a_n = 2$; if

$$C_n + 1 \leq m \leq \frac{3C_n}{2} = \sum_{i=1}^{n} C_i,$$

set $a_n = 1$. In each case, $1 \leq m - C_n/a_n \leq C_n/2$. Thus, by the induction hypothesis, we may choose positive integers $a_1, a_2, \ldots, a_{n-1}$ such that

$$m - \frac{C_n}{a_n} = \sum_{i=1}^{n-1} \frac{C_n}{a_n},$$

as desired. This completes the proof.

## Problem 3

Circles $C_1$ and $C_2$ with centers $O_1$ and $O_2$, respectively, meet at points $A$ and $B$. Lines $O_1B$ and $O_2B$ intersect $C_2$ and $C_1$ at $F$ and $E$, respectively. The line parallel to $\overline{EF}$ through $B$ meets $C_1$ and $C_2$ at $M$ and $N$. Given that $B$ lies between $M$ and $N$, prove that $MN = AE + AF$.

**Solution.** All angles are directed modulo $\pi$. Let $X$ be the point on $C_1$ diametrically opposite $B$, and let $Y$ be the point on $C_2$ diametrically opposite $B$. Because

$$\angle XAB + \angle BAY = \pi/2 + \pi/2 = \pi,$$

the points $A$, $X$, and $Y$ are collinear.

$X$ and $F$ lie on line $BO_1$, and $Y$ and $E$ lie on line $BO_2$. Thus,

$$\angle XFY = \angle BFY = \pi/2 = \angle XEB = \angle XEY,$$

implying that points $E$, $F$, $X$, and $Y$ are concyclic. Hence,

$$\angle BEF = \angle YEF = \angle YXF = \angle AXB = \angle AMB.$$

Because $\overline{EF} \parallel \overline{MB}$, we have $\overline{BE} \parallel \overline{AM}$, and the perpendicular bisectors of $\overline{BE}$ and $\overline{AM}$ coincide. Thus, the reflection across this common perpendicular bisector sends $\overline{AE}$ to $\overline{MB}$, implying that $AE = MB$. Similarly, $AF = BN$. Because $B$ lies between $M$ and $N$, it follows that $MN = AE + AF$.

## Problem 4

Two triangles $ABC$ and $A'B'C'$ lie in three-dimensional space. The sides of triangle $ABC$ have lengths greater than or equal to $a$, and the sides of triangle $A'B'C'$ have lengths greater than or equal to $a'$. Prove that one can select one vertex from triangle $ABC$ and one vertex from triangle $A'B'C'$ such that the distance between them is at least

$$\sqrt{\frac{a^2 + a'^2}{3}}.$$

**Solution.** Let $O$ be an arbitrary point in space, and for any point $X$ let $\mathbf{x}$ denote the vector from $O$ to $X$. Let $\mathcal{S} = \{A, B, C\}$, $\sigma_1 = \mathbf{a} + \mathbf{b} + \mathbf{c}$, and $\sigma_2 = \mathbf{a} \cdot \mathbf{b} + \mathbf{b} \cdot \mathbf{c} + \mathbf{c} \cdot \mathbf{a}$. Define $\mathcal{S}'$, $\sigma_1'$, and $\sigma_2'$ analogously in terms of $A'$, $B'$, and $C'$.

Given $(P, P') \in \mathcal{S} \times \mathcal{S}'$, note that $PP'^2 = |\mathbf{p}|^2 - 2\mathbf{p} \cdot \mathbf{p}' + |\mathbf{p}'|^2$. Summing over all 9 possible pairs yields the total $t$ with

$$
\begin{aligned}
t &= \sum_{P \in \mathcal{S}} 3|\mathbf{p}|^2 + \sum_{P' \in \mathcal{S}'} 3|\mathbf{p}'|^2 - 2\sigma_1 \cdot \sigma_1' \\
&= \sum_{P \in \mathcal{S}} 3|\mathbf{p}|^2 + \sum_{P' \in \mathcal{S}'} 3|\mathbf{p}'|^2 + \left(|\sigma_1 - \sigma_1'|^2 - |\sigma_1|^2 - |\sigma_1'|^2\right) \\
&\geq \sum_{P \in \mathcal{S}} 3|\mathbf{p}|^2 + \sum_{P' \in \mathcal{S}'} 3|\mathbf{p}'|^2 - |\sigma_1|^2 - |\sigma_1'|^2 \\
&= \left(\sum_{P \in \mathcal{S}} 2|\mathbf{p}|^2 - 2\sigma_2\right) + \left(\sum_{P' \in \mathcal{S}'} 2|\mathbf{p}'|^2 - 2\sigma_2'\right) \\
&= |\mathbf{a} - \mathbf{b}|^2 + |\mathbf{b} - \mathbf{c}|^2 + |\mathbf{c} - \mathbf{a}|^2 \\
&\quad + |\mathbf{a}' - \mathbf{b}'|^2 + |\mathbf{b}' - \mathbf{c}'|^2 + |\mathbf{c}' - \mathbf{a}'|^2 \\
&= AB^2 + BC^2 + CA^2 + A'B'^2 + B'C'^2 + C'A'^2 \\
&\geq 3(a^2 + a'^2).
\end{aligned}
$$

Thus one of the nine distances is greater than or equal to

$$
\sqrt{\frac{t}{9}} \geq \sqrt{\frac{a^2 + a'^2}{3}},
$$

as desired. The equality case is never reached. Because $A$, $B$, and $C$ do not all lie on a line, the nine values $PP'^2$ we sum to find $t$ cannot all be equal to each other.

## Problem 5

The function $f : \mathbb{N} \to \mathbb{N}$ is defined recursively by $f(1) = 1$ and

$$
f(n+1) = \begin{cases} f(n) + 2 & \text{if } n = f(f(n) - n + 1) \\ f(n) + 1 & \text{otherwise.} \end{cases} \qquad (*)
$$

for all $n \geq 1$.

(a) Prove that $f(f(n) - n + 1) \in \{n, n + 1\}$.

(b) Find an explicit formula for $f$.

**Solution.** Because $f(1) = 1$, it follows easily from $(*)$ that $f(n) \leq 2n - 1$, or equivalently that $f(n) - n + 1 \leq n$, for all integers $n$. Thus, if $f(1), f(2), \ldots, f(n)$ have been determined, so has $f(f(n) - n + 1)$. Therefore, $f$ is well-defined.

Define the function $g : \mathbb{N} \to \mathbb{N}$ by $g(n) = \lfloor \varphi n \rfloor$, where $\varphi = (1+\sqrt{5})/2$. We claim that $g$ satisfies the recursion $(*)$.

To prove this, let $n$ be a positive integer, and define $\epsilon = \{\varphi n\} = \varphi n - \lfloor \varphi n \rfloor$. Observe that $\varphi(\varphi - 1) = 1$ and that $\lfloor \varphi \rfloor = 1$. Thus, writing $\alpha_n = g(n+1) - g(n)$ and $\beta_n = g(g(n) - n + 1)$, we have

$$g(n+1) - g(n) = \lfloor (\lfloor \varphi n \rfloor + \epsilon) + \varphi \rfloor - \lfloor \varphi n \rfloor = \lfloor \varphi + \epsilon \rfloor$$

and

$$g(g(n) - n + 1) = \lfloor \varphi((\varphi n - \epsilon) - n + 1) \rfloor$$
$$= \lfloor \varphi(\varphi - 1)n + \varphi(1 - \epsilon) \rfloor = n + \lfloor \varphi(1 - \epsilon) \rfloor.$$

We cannot have $\epsilon \neq 2 - \varphi$, for otherwise $n$ would equal

$$\frac{\lfloor \varphi n \rfloor + 2 - \varphi}{\varphi} = \frac{\lfloor \varphi n \rfloor + 2}{\varphi} - 1,$$

which is not an integer. If $0 \le \epsilon < 2 - \varphi$, then $(\alpha_n, \beta_n) = (1, n+1)$. Otherwise, $2 - \varphi < \epsilon < 1$ and $(\alpha_n, \beta_n) = (2, n)$.

Thus, $g$ satisfies the given recursion and hence the function $f$ defined by $f(n) = g(n) = \lfloor \varphi n \rfloor$ is the unique solution to the recursion, solving part (b). Furthermore, $f(f(n) - n + 1) = \beta_n \in \{n, n+1\}$ for all $n$, proving the claim in part (a).

## Problem 6

Find all functions $f : \mathbb{N} \to \mathbb{N}$ such that

(i) $f(m) = 1$ if and only if $m = 1$;

(ii) if $d = \gcd(m, n)$, then $f(mn) = \frac{f(m)f(n)}{f(d)}$; and

(iii) for every $m \in \mathbb{N}$, we have $f^{2000}(m) = m$.

**Solution.** If such a function existed, then $f(4) = f(2)f(2)/f(2) = f(2)$. Thus, $2 = f^{2000}(2) = f^{1999}(f(2)) = f^{1999}(f(4)) = f^{2000}(4) = 4$, a contradiction. Therefore, no function $f$ satisfies the given conditions.

## Problem 7

The $n$ tennis players $A_1, A_2, \ldots, A_n$ participate in a tournament. Before the start of the tournament, $k \le \frac{n(n-1)}{2}$ distinct pairs of players are chosen. During the tournament, any two players in a chosen pair compete against each other exactly once; no draws occur, and in each match the winner adds

1 point to his tournament score while the loser adds 0. Let $d_1, d_2, \ldots, d_n$ be nonnegative integers. Prove that after the $k$ preassigned matches, it is possible for $A_1, A_2, \ldots, A_n$ to obtain the tournament scores $d_1, d_2, \ldots, d_n$, respectively, if and only if the following conditions are satisfied:

(i) $\sum_{i=1}^{n} d_i = k$.

(ii) For every subset $X \subseteq \{A_1, \ldots, A_n\}$, the number of matches taking place among the players in $X$ is at most $\sum_{A_j \in X} d_j$.

**Solution.**   Let $\mathcal{A} = \{A_1, \ldots, A_n\}$.

Suppose that such a tournament exists. Consider any set $X \subseteq \mathcal{A}$, and let $G$ be the set of matches among players in $X$. A total of $|G|$ points are scored during the matches in $G$. We can also calculate this total by summing over players: during these matches, each player not in $X$ scores 0 points; and each player $A_j$ in $X$ scores at most $d_j$ points during these matches. Hence, $|G| \geq \sum_{A_j \in X} d_j$, proving (ii). Furthermore, each player $A_j$ in $X$ scores *exactly* $d_j$ points if $X = \mathcal{A}$, because in this case $G$ is the set of all matches. Thus, if $X = \mathcal{A}$ then $k = |G| = \sum_{A_j \in \mathcal{A}} d_j$, proving (i).

We now prove the "if" direction. For any $X, Y \subseteq \mathcal{A}$, we define two quantities $\sigma(X)$ and $\Gamma(X, Y)$ which vary throughout the tournament. Let $\sigma(X)$ equal $\sum_{A_j \in X} d_j$ (the total number of points we wish the players in $X$ to score in total), minus the number of points players in $X$ have already scored. In other words, $\sigma(X)$ is the total number of points we wish the players in $X$ to score in the future. Also, let $\Gamma(X, Y)$ denote the number of matches remaining which take place between some player in $X$ and some player in $Y$.

At any point in the tournament, say that tournament is *feasible* if $\sigma(X) \geq \Gamma(X, X)$ for all $X \subseteq \mathcal{A}$. We are given that the tournament is initially feasible; we show that as long as games remain unplayed, one game can occur so that the tournament remains feasible.

Without loss of generality, suppose that $A_1$ and $A_2$ must still play against each other. We claim that for some $j \in \{1, 2\}$, the following statement holds:

$$\sigma(X_j) > \Gamma(X_j, X_j) \text{ for all } X_j \subseteq \mathcal{A}$$
$$\text{containing } A_j \text{ but not } A_{3-j}. \tag{$*$}$$

Suppose, for the sake of contradiction, that the claim fails for some sets $X_1$ and $X_2$. We prove that $\sigma(X_1 \cup X_2) < \Gamma(X_1 \cup X_2, X_1 \cup X_2)$, a contradiction. Indeed, $\Gamma(X_1 \cup X_2, X_1 \cup X_2)$ equals

$$\Gamma(X_1, X_1) + \Gamma(X_2, X_2) + \Gamma(X_1 \setminus X_2, X_2 \setminus X_1) - \Gamma(X_1 \cap X_2, X_1 \cap X_2)$$
$$\geq \sigma(X_1) + \sigma(X_2) + \Gamma(\{A_s\}, \{A_t\}) - \sigma(X_1 \cap X_2)$$
$$= [\sigma(X_1) + \sigma(X_2) - \sigma(X_1 \cap X_2)] + \Gamma(\{A_s\}, \{A_t\})$$
$$= \sigma(X_1 \cup X_2) + 1,$$

as desired.

Hence, $(*)$ holds for either $j = 1$ or $j = 2$; without loss of generality, assume that it holds for $j = 1$. Let $A_1$ beat $A_2$ in the next match M. We prove that after M, we have $\sigma(X) \geq \Gamma(X, X)$ for each $X \subseteq \mathcal{A}$.

If $X$ does not contain $A_1$, then $\sigma(X)$ and $\Gamma(X, X)$ do not change after M is played. If $X$ contains both $A_1$ and $A_2$, then $\sigma(X)$ and $\Gamma(X, X)$ both decrease by 1 after M is played. In both cases, we still have $\sigma(X) \geq \Gamma(X, X)$.

Otherwise, $X$ contains $A_1$ but not $A_2$. In this case, because we assumed that $(*)$ holds for $j = 1$, we know that $\sigma(X) > \Gamma(X, X)$ before M is played. After the match is played, $\sigma(X)$ decreases by 1 while $\Gamma(X, X)$ remains constant — implying that $\sigma(X) \geq \Gamma(X, X)$.

Therefore, another match may indeed occur in such a way that $\sigma(X)$ continues to be at least $\Gamma(X, X)$ for each subset $X \subseteq \mathcal{A}$. Let all the matches occur in this manner. Observe first that $\sigma(\mathcal{A}) = 0$ — because $\sigma(\mathcal{A})$ initially equalled $k$, and $k$ points are scored during the tournament. Next observe that $\sigma(X) \geq \Gamma(X, X) = 0$ for all $X \subseteq \mathcal{A}$. Thus,

$$0 = \sigma(\mathcal{A}) = \sigma(\{A_j\}) + \sigma(\mathcal{A} - \{A_j\}) \geq 0 + 0$$

for all $j$, implying that $\sigma(\{A_j\}) = 0$ for all $j$. Hence, $A_j$ has scored $d_j$ points for all $j$. This completes the proof.

## Problem 8

Isosceles triangles $A_3 A_1 O_2$ and $A_1 A_2 O_3$ are constructed externally along the sides of a triangle $A_1 A_2 A_3$ with $O_2 A_3 = O_2 A_1$ and $O_3 A_1 = O_3 A_2$. Let $O_1$ be a point on the opposite side of line $A_2 A_3$ as $A_1$, with $\angle O_1 A_3 A_2 = \frac{1}{2} \angle A_1 O_3 A_2$ and $\angle O_1 A_2 A_3 = \frac{1}{2} \angle A_1 O_2 A_3$, and let $T$ be the foot of the perpendicular from $O_1$ to $\overline{A_2 A_3}$. Prove that $\overline{A_1 O_1} \perp \overline{O_2 O_3}$ and that

$$\frac{A_1 O_1}{O_2 O_3} = 2 \frac{O_1 T}{A_2 A_3}.$$

**Solution.** Without loss of generality, assume that triangle $A_1A_2A_3$ is oriented counterclockwise (i.e., angle $A_1A_2A_3$ is oriented clockwise). Let $P$ be the reflection of $O_1$ across $T$.

We use complex numbers with origin $O_1$, where each point denoted by an uppercase letter is represented by the complex number with the corresponding lowercase letter. Let $\zeta_k = a_k/p$ for $k = 1, 2$, so that $z \mapsto z_0 + \zeta_k(z - z_0)$ is a spiral similarity through angle $\angle PO_1A_k$ with ratio $O_1A_3/O_1P$ about the point corresponding to $z_0$.

Because $O_1$ and $A_1$ lie on opposite sides of line $A_2A_3$, angles $A_2A_3O_1$ and $A_2A_3A_1$ have opposite orientations — i.e., the former is oriented counterclockwise. Thus, angles $PA_3O1$ and $A_2O_3A_1$ are both oriented counterclockwise. Because $\angle PA_3O_1 = 2\angle A_2A_3O_1 = \angle A_2O_3A_1$, it follows that isosceles triangles $PA_3O_1$ and $A_2O_3A_1$ are similar and have the same orientation. Hence, $o_3 = a_1 + \zeta_3(a_2 - a_1)$.

Similarly, $o_2 = a_1 + \zeta_2(a_3 - a_1)$. Hence,

$$o_3 - o_2 = (\zeta_2 - \zeta_3)a_1 + \zeta_3a_2 - \zeta_2a_3$$

$$= \zeta_2(a_2 - a_3) + \zeta_3(\zeta_2p) - \zeta_2(\zeta_3p) = \zeta_2(a_2 - a_3),$$

or (recalling that $o_1 = 0$ and $t = 2p$)

$$\frac{o_3 - o_2}{a_1 - o_1} = \zeta_2 = \frac{a_2 - a_3}{p - o_1} = \frac{1}{2}\frac{a_2 - a_3}{t - o_1}.$$

Thus, the angle between $\overline{O_1A_1}$ and $\overline{O_2O_3}$ equals the angle between $\overline{O_1T}$ and $\overline{A_3A_2}$, which is $\pi/2$. Furthermore, $O_2O_3/O_1A_1 = \frac{1}{2}A_3A_2/O_1T$, or $O_1A_1/O_2O_3 = 2O_1T/A_2A_3$. This completes the proof.

## Problem 9

Given a circle $\Gamma$, a line $d$ is drawn not intersecting $\Gamma$. $M, N$ are two points varying on line $d$ such that the circle with diameter $\overline{MN}$ is externally tangent to $\Gamma$. Prove that there exists a point $P$ in the plane such that for any such segment $\overline{MN}$, $\angle MPN$ is constant.

**Solution.** Let $\Gamma$ have center $O_1$ and radius $r$, and let $A$ be the foot of the perpendicular from $O_1$ to $d$. Let $P$ be the point on $\overline{AO_1}$ such that $AO_1^2 - AP^2 = r^2$.

Fix $M$ and $N$ such that the circle $\omega$ with diameter $\overline{MN}$ is externally tangent to $\Gamma$, and let $O_2$ be its center. For any point $X$, let $p(X) = O_2X^2 - (MN/2)^2$ denote the power of point $X$ with respect to $\omega$.

Let line $PN$ intersect $\omega$ again at $Q$. Then $\angle PQM = \angle NQM = \pi/2$, and

$$p(P) = PQ \cdot PN = (PM \cos \angle MPN) \cdot PN$$
$$= 2[MPN] \cot \angle MPN.$$

Also,

$$p(P) = (O_2 A^2 + AP^2) - (MN/2)^2$$
$$= \left((O_2 A^2 + AO_1^2) - (MN/2)^2\right) + AP^2 - AO_1^2$$
$$= p(O_1) + (AP^2 - AO_1^2) = r(r + MN) - r^2 = r \cdot MN.$$

Equating these two expressions for $p(P)$, we find that

$$\tan \angle MPN = \frac{2[MPN]}{r \cdot MN} = \frac{AP \cdot MN}{r \cdot MN} = \frac{AP}{r},$$

implying that $\angle MPN = \tan^{-1}\left(\frac{AP}{r}\right)$ is constant as $M$ and $N$ vary.

## Problem 10

Suppose that $a, b, c$ are real numbers such that for any positive real numbers $x_1, x_2, \ldots, x_n$, we have

$$\left(\frac{\sum_{i=1}^{n} x_i}{n}\right)^a \cdot \left(\frac{\sum_{i=1}^{n} x_i^2}{n}\right)^b \cdot \left(\frac{\sum_{i=1}^{n} x_i^3}{n}\right)^c \geq 1.$$

Prove that the vector $(a, b, c)$ has the form $p(-2, 1, 0) + q(1, -2, 1)$ for some nonnegative real numbers $p$ and $q$.

**Solution.** First, set $n = 1$. Then $x_1^{a+2b+3c} = (x_1)^a (x_1^2)^b (x_1^3)^c \geq 1$ for all $x_1 > 0$. In particular, because this holds for both $x_1 < 1$ and $x_1 > 1$, we must have $a + 2b + 3c = 0$. Setting $p = b + 2c$ and $q = c$, we find that

$$p(-2, 1, 0) + q(1, -2, 1) = (-2p + q, p - 2q, q)$$
$$= (-2b - 3c, b, c) = (a, b, c).$$

It remains to show $p = b + 2c \geq 0$ and $q = c \geq 0$.

To show that $p \geq 0$, set $n = 2$, $x_1 = 1$, and $x_2 = \epsilon > 0$. By the given inequality,

$$f(\epsilon) := \left(\frac{1 + \epsilon}{2}\right)^a \left(\frac{1 + \epsilon^2}{2}\right)^b \left(\frac{1 + \epsilon^3}{2}\right)^c \geq 1.$$

On the other hand, as $\epsilon \to 0$, $f(\epsilon) \to 1/2^{a+b+c}$, implying that $1/2^{a+b+c} \geq 1$. Therefore, $a + b + c \leq 0$, so

$$p = b + 2c = (a + 2b + 3c) - (a + b + c) \geq 0.$$

To show that $q \geq 0$, set $n = k + 1$, $x_1 = x_2 = \cdots = x_k = 1 - \epsilon$, and $x_{k+1} = 1 + k\epsilon$, where $k$ is an arbitrary positive integer and $\epsilon$ is an arbitrary real number in $(0, 1)$. Then

$$\sum_{i=1}^{n} x_i = k(1 - \epsilon) + (1 + k\epsilon) = k + 1,$$

$$\sum_{i=1}^{n} x_i^2 = k(1 - \epsilon)^2 + (1 + k\epsilon)^2 = (k + 1)(1 + k\epsilon^2),$$

$$\sum_{i=1}^{n} x_i^3 = k(1 - \epsilon)^3 + (1 + k\epsilon)^3 = (k + 1)(1 + 3k\epsilon^2 + (k^2 - k)\epsilon^3).$$

Hence, we may apply the given inequality to find that

$$g(k, \epsilon) := 1^a \left(1 + k\epsilon^2\right)^b \left(1 + 3k\epsilon^2 + (k^2 - k)\epsilon^3\right)^c \geq 1$$

for all $t$, $k$, $\epsilon$. Now take $\epsilon = k^{-1/2}$, so that

$$g(k, k^{-1/2}) = 2^b \left(4 + \frac{k - 1}{\sqrt{k}}\right)^c \geq 1$$

for all positive integers $k$. Because $4 + (k - 1)/\sqrt{k}$ can be made arbitrarily large for sufficiently large values of $k$, $q = c$ must be non-negative.

Thus, $p$ and $q$ are non-negative, as desired.

## 1.10 Israel

### Problem 1

Define $f(n) = n!$. Let

$$a = 0.f(1)f(2)f(3)\ldots.$$

In other words, to obtain the decimal representation of $a$ write the decimal representations of $f(1), f(2), f(3), \ldots$ in a row. Is $a$ rational?

**Solution.** If $a$ were rational, then the digits in the decimal must eventually appear cyclicly. Because $f(n)$ always contains a nonzero digit, the cyclic portion of the decimal could not consist solely of zeroes. However, when $n$ is large, the number of zeros contained in $f(n)$ tends to infinity, so the cyclic part of the decimal *must* contain all zeroes — a contradiction. Therefore, $a$ is irrational.

### Problem 2

The vertices of triangle $ABC$ are lattice points. Two of its sides have lengths which belong to the set $\{\sqrt{17}, \sqrt{1999}, \sqrt{2000}\}$. What is the maximum possible area of triangle $ABC$?

**Solution.** Without loss of generality, assume that the lengths $AB$ and $BC$ are in $\{\sqrt{17}, \sqrt{1999}, \sqrt{2000}\}$. Then

$$[ABC] = \frac{1}{2}AB \cdot BC \sin \angle BCA \leq \frac{1}{2}\sqrt{2000} \cdot \sqrt{2000} \sin(\pi/2) = 1000.$$

Equality can hold, for instance in the triangle whose vertices are $(0,0)$, $(44,8)$ and $(-8,44)$ — exactly two sides have length $\sqrt{2000}$ because $44^2 + 8^2 = 2000$, and the angle between these sides is $\pi/2$. Thus, the maximum possible area is 1000.

### Problem 3

The points $A, B, C, D, E, F$ lie on a circle, and the lines $AD$, $BE$, $CF$ concur. Let $P, Q, R$ be the midpoints of $\overline{AD}, \overline{BE}, \overline{CF}$, respectively. Two chords $\overline{AG}, \overline{AH}$ are drawn such that $\overline{AG} \parallel \overline{BE}$ and $\overline{AH} \parallel \overline{CF}$. Prove that triangles $PQR$ and $DGH$ are similar.

**Solution.** All angles are directed modulo $\pi$. Let $\overline{AD}, \overline{BE}, \overline{CF}$ intersect at $X$ and let $O$ be the center of the given circle. Angles $OPX$, $OQX$, and $ORX$ measure $\pi/2$, implying that $O, P, Q, R$, and $X$ are concyclic. Therefore,

$$\angle DGH = \angle DAH = \angle DXC$$

$$= \pi - \angle CXP = \pi - \angle RXP = \angle PQR.$$

Similarly, $\angle DHG = \angle PRQ$, implying that $\triangle PQR \sim \triangle DGH$.

## Problem 4

A square $ABCD$ is given. A *triangulation* of the square is a partition of the square into triangles such that any two triangles are either disjoint, share only a common vertex, or share only a common side. (In particular, no vertex of a triangle can lie on the interior of the side of another triangle.) A *good triangulation* of the square is a triangulation in which all the triangles are acute.

(a) Give an example of a good triangulation of the square.

(b) What is the minimal number of triangles required for a good triangulation?

**Solution.**  (a) We provide an example of a good triangulation with 8 triangles. Orient the square so that $\overline{AB}$ is horizontal and $A$ is in the upper-left corner. Let $M$ and $N$ be the midpoints of sides $\overline{AB}$ and $\overline{CD}$, respectively, and let $P$ be a point on the interior of $\overline{MN}$ distinct from its midpoint. Angles $MPA$, $APD$, and $DPN$ — and their reflections across $\overline{MN}$ — are all acute. Now choose $Q$ and $R$ on the horizontal line through $P$, so that $Q$, $P$, and $R$ lie in that order from left to right and so that $\overline{QP}$ and $\overline{PR}$ are of negligible length. Partition the square into the triangles by drawing the segments $\overline{QA}, \overline{QM}, \overline{QN}, \overline{QD}, \overline{RB}, \overline{RM}, \overline{RN}, \overline{RC}$, and $\overline{QR}$. If we choose $Q$ so that $PQ$ is sufficiently small, then the measures of angles $MQA$, $AQD$, and $DQN$ will remain sufficiently close to those of $MPA$, $APD$, $DPN$, so that these angles will be acute. Similarly, if we choose $R$ so that $PR$ is sufficiently small, then angles $MRB$, $BRC$, and $CRN$ will be acute as well. It is easy to verify that the remaining angles in the partition are acute, as needed.

(b) We will prove that the minimal number is 8. We have already shown above that 8 is achievable, so it suffices to show that no good triangulation exists with fewer than 8 triangles. Observe that in a good triangulation, each corner of $ABCD$ must be a vertex of at least two triangles because the right angle there must be divided into acute angles. Likewise, any vertex on a side of $ABCD$ must be part of at least 3 triangles, and any vertex in the interior must be part of at least 5 triangles.

In fact, we can prove a stronger statement about each corner of square $ABCD$: there must be a triangle edge emanating from that corner whose other endpoint lies strictly inside square $ABCD$. Without loss of generality, assume the corner in question is $A$. Some edge $\overline{AX}$ of a triangle splits the right angle at $A$; assume, for the sake of contradiction, that $X$ lies does not lie strictly inside square $ABCD$. Without loss of generality, assume that $X$ lies on $\overline{BC} - \{B\}$. By the given definition of "triangulation," no other vertex of a triangle in the triangulation lies on $\overline{AX}$. Hence, there exists a point $Y$ in triangle $ABX$ such that triangle $AXY$ is a member of the good triangulation. But then $\angle AYX \geq \angle ABX = \pi/2$, a contradiction.

Now, consider an arbitrary good triangulation of $ABCD$. Let $i$ be the number of *interior vertices* — vertices in the triangulation which lie in the interior of square $ABCD$. From above, $i \geq 1$. First suppose that there is one interior vertex, $P$. The result in the previous paragraph implies that $\overline{PA}$, $\overline{PB}$, $\overline{PC}$, and $\overline{PD}$ must be edges of triangles in the triangulation. One of $\angle APB$, $\angle BPC$, $\angle CPD$, $\angle DPA$ must be at least $\pi/2$ — say, $\angle APB$. This angle must be divided in this triangulation by some edge $\overline{PQ}$, where $Q$ is on the interior of $\overline{AB}$. But then either angle $AQP$ or angle $BQP$ measures at least $\pi/2$, so $Q$ must lie on some triangle edge that does not lie on $\overline{QA}$, $\overline{QB}$, or $\overline{QP}$. However, it is impossible to construct such an edge that does not intersect $\overline{AP}$ or $\overline{BP}$ and that does not end in a second interior vertex.

Next suppose that $i \geq 2$. On each of the $n$ triangles, we may count 3 sides for a total of $3n$; each side which lies on the square's boundary is counted once, and the other sides are each counted twice. If $i = 2$, then for each of the two interior vertices, at least 5 triangle sides have that vertex as an endpoint; at most 1 triangle side contains both interior vertices, so there are at least 9 triangle sides which do not lie on the square's boundary. If $i \geq 3$, then take any three of the interior vertices. Each lies on at least 5 triangle sides, and at most 3 triangle sides contain some two of these three vertices. Hence, at least $3 \cdot 5 - 3 = 12$ triangle sides do not lie on the square's boundary. In both cases, then, at least 9 triangle sides do not lie on the square's boundary, and furthermore at least 4 triangle sides *do* lie on the square's boundary. Therefore, $3n \geq 9 \cdot 2 + 4 = 22$, or $n \geq 8$.

Thus, in all cases there must be at least 8 triangles, as desired.

# 1.11   Italy

## Problem 1

Let $ABCD$ be a convex quadrilateral, and write $\alpha = \angle DAB$; $\beta = \angle ADB$; $\gamma = \angle ACB$; $\delta = \angle DBC$; and $\epsilon = \angle DBA$. Assuming that $\alpha < \pi/2$, $\beta + \gamma = \pi/2$, and $\delta + 2\epsilon = \pi$, prove that

$$(DB + BC)^2 = AD^2 + AC^2.$$

**Solution.** Let $D'$ be the reflection of $D$ across line $AB$. We have $\angle D'BA = \angle DBA = \epsilon$, so

$$\angle D'BC = \angle D'BA + \angle ABD + \angle DBC = 2\epsilon + \delta = \pi.$$

Thus, $D'$, $B$, and $C$ are collinear. Also,

$$\angle AD'C + \angle ACD' = \angle ADB + \angle ACB = \beta + \gamma = \pi/2,$$

so $\angle D'AC = \pi/2$ and triangle $D'AC$ is right. By the Pythagorean Theorem, $D'C^2 = AD'^2 + AC^2$, implying that

$$(DB + BC)^2 = (D'B + BC)^2$$
$$= D'C^2 = AD'^2 + AC^2 = AD^2 + AC^2,$$

as desired.

## Problem 2

Given a fixed integer $n > 1$, Alberto and Barbara play the following game, starting with the first step and then alternating between the second and third:

- Alberto chooses a positive integer.
- Barbara picks an integer greater than 1 which is a multiple or divisor of Alberto's number, possibly choosing Alberto's number itself.
- Alberto adds or subtracts 1 from Barbara's number.

Barbara wins if she succeeds in picking $n$ by her fiftieth move. For which values of $n$ does she have a winning strategy?

**Solution.** We claim that Barbara has a winning strategy if and only if at least one of these conditions is met:

- $n = 2$;

- $4 \mid n$;

- for some integer $m > 1$, $(m^2 - 1) \mid n$.

First we show that if one of these three conditions holds, then Barbara has a winning strategy. If Alberto's first choice $a$ is even, then Barbara can choose 2 on her first turn. If instead $a$ is odd, then Barbara can choose $a$ as well. If $a = n$, she wins; otherwise, Alberto's second choice must be even, and Barbara can choose 2 on her second turn. Let $a_1, b_1, a_2, b_2, \ldots,$ be the numbers chosen after Barbara chooses 2 for the first time.

*Case 1: $n = 2$.* In this case, Barbara has already won.

*Case 2: $4 \mid n$.* If $a_1 = 1$, then Barbara can choose $b_1 = n$ and win. Otherwise, $a_1 = 3$, Barbara can let $b_1 = 3$, $a_2$ equals 2 or 4, and Barbara can let $b_2 = n$.

*Case 3: for some integer $m > 1$, $(m^2 - 1) \mid n$.* As in case 2, Alberto must choose $a_1 = 3$ in order to prevent Barbara from winning. Now, exactly one of the integers $m - 1$, $m$, and $m + 1$ is divisible by 3, implying that either 3 divides $m$, or else 3 divides $m^2 - 1$ and hence $n$. In the first case, Barbara can let $b_1 = m$, forcing $a_2 = m \pm 1$ and allowing Barbara to choose $b_2 = n$. In the latter case, Barbara can let $b_1 = n$.

We now know that Barbara has a winning strategy if at least one of the conditions holds. Now we assume that none of the conditions is true for some $n > 1$ and prove that Alberto can always keep Barbara from winning. Because the first and second conditions fail, and because the third condition fails for $m = 2$, we have $n \neq 2, 3, 4$. Hence, $n > 4$.

Call a positive integer $a$ *hopeful* if $a \nmid n$ and $n \nmid a$. We prove below that for any integer $b > 1$, there exists $a \in \{b - 1, b + 1\}$ such that $a$ is hopeful. It follows that Alberto can initially choose some hopeful number and also choose a hopeful number on every subsequent turn, preventing Barbara from winning for at least 50 turns.

Suppose for the sake of contradiction that the above claims fails for some $b > 1$. If $b > n$, then $b - 1$ and $b + 1$ must be multiples of $n$. Then $n$ divides their difference, 2, which is impossible.

If $b \leq n$, then because $n$ does not divide $n + 1$ or $n + 2$ for $n > 2$, we must have $(b - 1) \mid n$ and $(b + 1) \mid n$. If $b - 1$ and $b + 1$ were even, then one is divisible by 4 — but then $4 \mid n$, a contradiction. Thus, $b - 1$ and $b + 1$ are odd. It follows that they are relatively prime and that their product $b^2 - 1$ divides $n$, contradicting the assumption that the third condition fails. This completes the proof.

## Problem 3

Let $p(x)$ be a polynomial with integer coefficients such that $p(0) = 0$ and $0 \leq p(1) \leq 10^7$, and such that there exist integers $a, b$ satisfying $p(a) = 1999$ and $p(b) = 2001$. Determine the possible values of $p(1)$.

**Solution.**  If $p(x) = 2000x^2 - x$, then $p(0) = 0$, $p(1) = 1999$, and $p(-1) = 2001$. If $p(x) = 2000x^2 + x$, then $p(0) = 0$, $p(1) = 2001$, and $p(-1) = 1999$. Therefore, it is possible for $p(1) = 1999$ or $2001$.

Now assume that $p(1) \neq 1999, 2001$. Then $a, b \neq 1$. Because $p(0) = 0$, we may write $p(x) = xq(x)$ for some polynomial $q(x)$ with integer coefficients. Because $q$ has integer coefficients, $q(a)$ is an integer, and we may write $q(x) - q(a) = (x - a)r(x)$ for some polynomial $r$ with integer coefficients. And because $r$ has integer coefficients, $r(b)$ is an integer, and we may write $r(x) - r(b) = (x - b)s(x)$ for some polynomial $s$ with integer coefficients. Therefore,

$$
\begin{aligned}
p(x) &= xq(x) = xq(a) + x(x-a)r(x) \\
&= xq(a) + x(x-a)r(b) + x(x-a)(x-b)s(x).
\end{aligned} \tag{$*$}
$$

Specifically, plugging in $x = a$ and $x = b$, we find that

$$1999 = aq(a),$$

$$2001 = bq(a) + b(b-a)r(b).$$

Because $p(0)$, $p(a)$, and $p(b)$ are pairwise distinct, so are $0$, $a$, and $b$. Therefore, we can solve the above two equations to find

$$
\begin{aligned}
q(a) &= \tfrac{1999}{a}, \\
r(b) &= \tfrac{2001 - bq(a)}{b(b-a)}.
\end{aligned} \tag{$\dagger$}
$$

Because $a \neq b$, we have that $|a - b|$ divides $p(a) - p(b)$. Hence, $|a - b|$ equals 1 or 2. Also, for all $x \in \mathbb{Z}$, we have $p(x) = xq(x)$ and hence $x \mid p(x)$. In particular, $a \mid 1999$, so that

$$|a| \in \{1, 1999\}.$$

This restriction, combined with the conditions $|a - b| \in \{1, 2\}$, $b \mid 2001$, $a \neq 1$, and $b \neq 1$, imply that $(a, b)$ equals one of the following pairs:

$$(-1999, -2001), (-1, -3), (1999, 2001).$$

Fix $(a, b)$ from among these three pairs. From ($\dagger$) we know that $q(a)$ must equal $\tilde{q} = 1999/a$ and that $r(b)$ must equal $\tilde{r} = (2001 - b\tilde{q})/b(b-a)$. We then set $x = 1$ into $(*)$ to find $p(1)$:

| $(a, b)$ | $q(a)$ | $r(a)$ | $p(1)$ |
|---|---|---|---|
| $(-1999, -2001)$ | $-1$ | $0$ | $-1 + (2000 \cdot 2002)s(1)$ |
| $(-1, -3)$ | $-1999$ | $-666$ | $-3331 + 8s(1)$ |
| $(1999, 2001)$ | $1$ | $0$ | $1 + (1998 \cdot 2000)s(1)$. |

Hence, $p(1)$ is of the form $m + ns(1)$ for some fixed integers $m, n$. Indeed, suppose that we have any number of this form $m + n\tilde{s}$ between 0 and $10^7$, where $s$ is an integer. Then writing

$$p(x) = \tilde{q}x + \tilde{r}x(x - a) + \tilde{s}x(x - a)(x - b),$$

we have $p(0) = 0$, $p(a) = 1999$, $p(b) = 2001$, and $p(1) = m + n\tilde{s}$.

Therefore, the possible values of $p(1)$ are 1999 and 2001, and the numbers between 0 and $10^7$ congruent to $-1 \pmod{2000 \cdot 2002}$, $-3331 \equiv 5 \pmod 8$, or $1 \pmod{1998 \cdot 2000}$.

# 1.12   Japan

## Problem 1

We *shuffle* a line of cards labelled $a_1, a_2, \ldots, a_{3n}$ from left to right by rearranging the cards into the new order

$$a_3, a_6, \ldots, a_{3n}, a_2, a_5, \ldots, a_{3n-1}, a_1, a_4, \ldots, a_{3n-2}.$$

For example, if six cards are labelled $1, 2, \ldots, 6$ from left to right, then shuffling them twice changes their order as follows:

$$1, 2, 3, 4, 5, 6 \longrightarrow 3, 6, 2, 5, 1, 4 \longrightarrow 2, 4, 6, 1, 3, 5.$$

Starting with 192 cards labelled $1, 2, \ldots, 192$ from left to right, is it possible to obtain the order $192, 191, \ldots, 1$ after a finite number of shuffles?

**Solution.** For each $n$, let $f(n)$ be the position in the line whose card goes to the $n$th position during each shuffle. Observe that after $k$ shuffles, $f^k(n)$ is in the $n$th position. We are given that $f(1), \ldots, f(192)$ equals $3, 6, \ldots, 192, 2, 5, \ldots, 191, 1, 4, \ldots, 190$. In this sequence, the difference between any term and the preceding term is congruent to 3 modulo 193. Because $f(1) \equiv 3 \pmod{193}$, we have $f(n) \equiv 3n \pmod{193}$ for each $n$.

In the sequence $(3^3)^{2^0}, (3^3)^{2^1}, (3^3)^{2^2}, \ldots, (3^3)^{2^6}$, each term is the square of the last. At least one term (the first, 27) is not congruent to 1 modulo 193; suppose that $N = 3^d$ (where $d$ is a positive integer) is the largest term with this property. Because 193 is prime, Fermat's Little Theorem implies that $(3^3)^{2^6} \equiv 3^{192} \equiv 1 \pmod{193}$, so $3^d$ is not the last term in the sequence. Hence, $N^2$ — the term following $N$ in the sequence — is congruent to 1 modulo 193. Because 193 divides $N^2 - 1$ but not $N - 1$, it must divide $(N^2 - 1)/(N - 1) = N + 1 = 3^d + 1$, implying that $3^d \equiv -1 \pmod{193}$.

For $n = 1, 2, \ldots, 193$, we have $f^d(n) \equiv 3^d n \equiv -n \pmod{193}$. Thus, $f^d(n) = 193 - n$, implying that the order $192, 191, \ldots, 1$ appears after $d$ shuffles.

**Note.** The value $d$ found above actually equals 24. The smallest positive integer $k$ such that $3^k \equiv -1 \pmod{193}$ is 8, implying that the order $192, 191, \ldots, 1$ first appears after 8 shuffles.

## Problem 2

In the plane are given distinct points $A, B, C, P, Q$, no three of which are collinear. Prove that

$$AB + BC + CA + PQ < AP + AQ + BP + BQ + CP + CQ.$$

**Solution.**   In this solution, we call a polygon $V_1 \ldots V_n$ convex if $V_1, \ldots, V_n$ form a convex polygon in that order. (For instance, if we say that a square $ABCD$ is convex, then we do not say that the quadrilateral $ACBD$ is convex.)

We say that *condition (a) holds* if quadrilateral $XYPQ$ is convex for some $X, Y \in \{A, B, C\}$. We prove that in this case, the desired inequality holds. Without loss of generality, we may assume that quadrilateral $ABPQ$ is convex. If $\overline{AP}$ and $\overline{BQ}$ intersect at $O$, then the triangle inequality gives $AB \leq AO + OB$ and $PQ \leq PO + OQ$. Adding these two inequalities yields

$$AB + PQ \leq AO + OP + BO + OQ = AP + BQ.$$

Because no three of the five given points are collinear, the triangle inequality also implies that $BC < BP + PC$ and $CA < CQ + QA$. Summing the last three inequalities yields the desired result.

Next, we say that *condition (b) holds* if $X$ lies inside triangle $YZM$ for some permutation $(X, Y, Z)$ of $(A, B, C)$ and some $M \in \{P, Q\}$. We prove that the desired inequality holds in this case as well. Without loss of generality, assume that $A$ lies inside triangle $BCQ$. The maps which send an arbitrary point $P$ to each of the lengths $PB$ and $PC$ are strictly convex functions, implying that $P \mapsto PB + PC$ is a strictly convex function as well. Hence, over all points $P$ on or inside triangle $BCQ$, this function can only attain its maximum when $P$ equals $B$, $C$, or $Q$. Thus,

$$AB + AC < \max\{BB + BC, CB + CC, QB + QC\} = QB + QC.$$

Adding this inequality to the inequalities $BC < BP + PC$ and $PQ < PA + AQ$ — as given by the triangle inequality — yields the desired result.

Up to the relabelling of points, the convex hull of the five given points must either be triangle $ABC$, triangle $ABP$, triangle $APQ$, convex quadrilateral $ABCP$, convex quadrilateral $ABPQ$, convex quadrilateral $APBQ$, convex pentagon $ABCPQ$, or convex pentagon $ABPCQ$.

If triangle $ABC$ is the convex hull, then $Q$ must lie in the interior of one of the triangles $APB$, $BPC$, $CPA$. Without loss of generality, suppose that $Q$ lies inside triangle $APB$. Because $C$ is not inside triangle $APB$ but lies on the same side of $\overline{AB}$ as $Q$, $\overline{QC}$ must intersect one of the two segments $\overline{AP}$ and $\overline{PB}$. If $\overline{QC}$ intersects $\overline{AP}$, then quadrilateral $ACPQ$ is convex and condition (a) holds; similarly, condition (a) holds if $\overline{QC}$ intersects $\overline{PB}$.

If triangle $ABP$ is the convex hull, then $C$ must lie inside triangle $ABP$, and condition (b) holds.

If triangle $APQ$ is the convex hull, then we may assume without loss of generality that $C$ is no closer to line $PQ$ than $B$ is. Then $C$ must lie inside one of the triangles $ABP$, $ABQ$, $BPQ$. If it lies in either of the first two triangles, condition (b) holds; and $C$ cannot lie in the third triangle $BPQ$ because $C$ is not closer to line $PQ$ than $B$. Hence, condition (b) holds.

If convex quadrilateral $ABCP$ is the convex hull, then $Q$ lies inside either triangle $APB$ or triangle $CPB$; in the former case, quadrilateral $BCPQ$ is convex, and in the latter case, quadrilateral $BAPQ$ is convex. Hence, condition (a) holds.

If convex quadrilateral $ABPQ$, convex pentagon $ABCPQ$, or convex pentagon $ABPCQ$ is the convex hull, then quadrilateral $ABPQ$ is convex and condition (a) holds.

Finally, if convex quadrilateral $APBQ$ is the convex hull, then $C$ lies inside either triangle $ABP$ or triangle $ABQ$; in both cases, condition (b) holds.

Hence, in all cases, either condition (a) or (b) holds; it follows that the desired inequality is true.

## Problem 3

Given a natural number $n \geq 3$, prove that there exists a set $A_n$ with the following two properties:

(i) $A_n$ consists of $n$ distinct natural numbers.

(ii) For any $a \in A_n$, the product of all the other elements in $A_n$ has remainder 1 when divided by $a$.

**Solution.** Suppose that $a_1, a_2, \ldots, a_k$ (with $k \geq 2$) are distinct integers greater than 1 such that $a_1 a_2 \cdots a_{i-1} a_{i+1} a_{i+2} \cdots a_k \equiv -1 \pmod{a_i}$ whenever $1 \leq i \leq k$. Suppose that $\epsilon \in \{-1, 1\}$ and define $a_{k+1} = a_1 a_2 \cdots a_k - \epsilon$. Because $a_{k+1} \geq 2a_i - 1 > a_i$ for all $i$, the integers $a_1, a_2, \ldots, a_{k+1}$ are still distinct integers greater than 1. Consider the equation $a_1 a_2 \cdots a_{i-1} a_{i+1} a_{i+2} \cdots a_{k+1} \equiv \epsilon \pmod{a_i}$. It clearly holds for $i = k + 1$. For $i < k$, it holds because

$$(a_1 a_2 \cdots a_{i-1} a_{i+1} a_{i+2} \cdots a_k) a_{k+1} \equiv (-1)(-\epsilon) \equiv \epsilon \pmod{a_i}.$$

Beginning with the numbers $a_1 = 2$, $a_2 = 3$, we apply this construction $n - 3$ times setting $\epsilon = -1$ and then one additional time setting $\epsilon = 1$. The set $A_n$ consisting of the resulting numbers $a_1, a_2, \ldots, a_n$ then satisfies the given conditions.

## Problem 4

We are given finitely many lines in the plane. Let an *intersection point* be a point where at least two of these lines meet, and let a *good intersection point* be a point where exactly two of these lines meet. Given that there are at least two intersection points, find the minimum number of good intersection points.

**First Solution.** Assume, for the sake of contradiction, that we are given lines satisfying the stated conditions such that no good intersection points exist. We prove that this is impossible using a technique similar to that of the classic solution to Sylvester's Line Problem. (Sylvester's Line Problem asks for a proof that given finitely many points in the plane which do not lie on a single line, there exists a line which passes through exactly two of the given points.)

There exist finitely many pairs consisting of a intersection point and a given line; for each, find the distance between the point and the line. Given any intersection point $P$, there is at least one other intersection point $Q$ by assumption and hence some given line passing through $Q$ but not $P$. Thus, at least one of the finitely many distances found is positive; let $d > 0$ be the minimum such distance.

Given any intersection point $A$ and any given line $\ell$ such that the distance between them is $d$, at least two given lines pass through $A$ that intersect $\ell$, say at $B$ and $C$. Assume, for the sake of contradiction, that $B$ and $C$ are not separated by (i.e., are not on opposite sides of) the perpendicular from $A$ to $\ell_1$. Let $A'$ be the foot of the perpendicular from $A$ to $\ell_1$ and assume, without loss of generality, that $BA' > CA'$. Letting $C'$ be the foot of the perpendicular from $C$ to line $AB$, right triangles $AA'B$ and $CC'B$ are similar. Hence, $CC' = AA' \cdot \frac{BC}{AB} < AA'$, a contradiction.

Now fix a specific intersection point $A_0$ and a given line $\ell_0$ such that the distance between them is $d$. Because $A_0$ is not good, at least three given lines pass through $A_0$. If all these lines passed through $A_0$ and intersected

$\ell_0$, two of the resulting intersection points would not be separated by the perpendicular from $A_0$ to $\ell_0$, contradicting our above analysis. Hence, one of the given lines passing through $A_0$ — say, $\ell_1$ — is parallel to $\ell_0$. Let $A_1$ be the leftmost intersection point on either line, and suppose that $A_1$ is on $\ell_k$. Because $A_1$ is not good, at least three given lines pass through $A_1$, two of which must intersect $\ell_{1-k}$ at two intersection points $B_1$ and $C_1$. However, the perpendicular from $A_1$ to $\ell_{1-k}$ does not separate $B_1$ and $C_1$, contradicting our above analysis. Therefore our original assumption was false, and at least one good intersection point exists.

To finish the proof, we show that it is possible to have exactly one good intersection point. Take four vertices of a parallelogram and draw the six lines which pass through pairs of the vertices. There are five intersection points, the four vertices themselves and the intersection of the parallelogram's diagonals. Of these, only the last point is a good intersection point.

**Second Solution.** We present a sketch of an alternative proof. We again assume, for the sake of contradiction, that we are given lines satisfying the stated conditions such that no good intersection points exist. Choose a point in the plane to be the origin. Because there are at least two intersection points, there exists a triangle formed by the given lines. Of all such triangles, consider those with minimal area. Of all these triangles, consider one whose centroid is farthest from the origin. Suppose that this triangle has vertices $A$, $B$, and $C$. Because these points are not good intersection points, they must lie on three sides $\overline{A_0B_0}$, $\overline{B_0C_0}$, $\overline{C_0A_0}$ of a larger triangle formed by the given lines. This larger triangle is partitioned into four triangles by $\overline{AB}$, $\overline{BC}$, and $\overline{CA}$ — namely, triangle $ABC$ is surrounded by three outer triangles. It is possible to show that the area of triangle $ABC$ is greater than or equal to the minimum of the areas of the other three triangles, with equality if and only if triangle $ABC$ is the medial triangle of triangle $A_0B_0C_0$. Indeed, equality *must* hold because of the minimality requirement on triangle $ABC$. Hence, each of the outer triangles is formed by the given lines and has the same area as triangle $ABC$. However, one of these triangles has a centroid farther from the origin than the centroid of triangle $ABC$, giving a contradiction.

## 1.13  Korea

### Problem 1

Show that given any prime $p$, there exist integers $x, y, z, w$ satisfying $x^2 + y^2 + z^2 - wp = 0$ and $0 < w < p$.

**Solution.** For the case $p = 2$, we may take $x = 0$ and $y = z = w = 1$. Now assume $p > 2$. We first consider the case where $-1$ is a quadratic residue modulo $p$. Then there exists an integer $a$ between $0$ and $p-1$ such that $a^2 \equiv -1 \pmod{p}$. Set $(x, y, z) = (0, 1, a)$. Because $x^2 + y^2 + z^2 = a^2 + 1$ is divisible by $p$ but is at most $1 + (p-1)^2 < p^2$, there exists $w \in \{1, 2, \ldots, p-1\}$ such that $x^2 + y^2 + z^2 - wp = 0$.

Next suppose that $-1$ is not a quadratic residue modulo $p$. We claim that $k$ and $p - 1 - k$ are both quadratic residues for some $k$. If $\frac{p-1}{2}$ is a quadratic residue, then we may set $k = \frac{p-1}{2}$. Otherwise, each of the $\frac{p-1}{2}$ nonzero quadratic residues modulo $p$ is in one of the pairs $\{1, p-2\}, \{2, p-3\}, \ldots, \{\frac{p-3}{2}, \frac{p+1}{2}\}$. By the Pigeonhole Principle, two of the numbers in some pair $\{k, p-k\}$ are quadratic residues, as desired.

Thus, we may choose $x, y \in \{0, 1, \ldots, \frac{p-1}{2}\}$ such that $x^2 \equiv k \pmod{p}$ and $y^2 \equiv p - k \pmod{p}$. Letting $z = 1$, we see that $x^2 + y^2 + z^2$ is divisible by $p$ and in the interval $(0, p^2)$. The value of $w$ then follows as before.

### Problem 2

Find all functions $f : \mathbb{R} \to \mathbb{R}$ satisfying

$$f(x^2 - y^2) = (x - y)(f(x) + f(y))$$

for all $x, y \in \mathbb{R}$.

**Solution.** Setting $x = y$ gives $f(0) = 0$. Setting $x = -1, y = 0$ yields $f(1) = -f(-1)$. Setting $x = a, y = 1$, then $x = a, y = -1$, we find that

$$f(a^2 - 1) = (a - 1)(f(a) + f(1)),$$
$$f(a^2 - 1) = (a + 1)(f(a) - f(1)).$$

Setting the right-hand sides of these equations equal and solving for $f(a)$ yields $f(a) = f(1)a$ for all $a$.

Therefore, any function satisfying the given relation is of the form $f(x) = kx$ for some constant $k$. Conversely, any such function clearly satisfies the given functional equation.

## Problem 3

We are given a convex cyclic quadrilateral $ABCD$. Let $P$, $Q$, $R$, $S$ be
the intersections of the exterior angle bisectors of angles $ABD$ and $ADB$,
$DAB$ and $DBA$, $ACD$ and $ADC$, $DAC$ and $DCA$, respectively. Show
that the four points $P, Q, R, S$ are concyclic.

**Solution.** All angles are directed modulo $\pi$ except where otherwise
stated.

Suppose that we have an arbitrary triangle $XYZ$ with incenter $I$ and
excenter $I_x$ opposite $X$. Points $X, I, I_x$ are collinear. Also, $\angle IYI_x =
\pi/2 = \angle IZI_x$, so quadrilateral $IYI_xZ$ is cyclic and $\angle XI_xY = \angle II_xY =
\angle IZY$, or equivalently $\angle YI_xX = \angle YZI$.

Let $I_1$ be the incenter of triangle $ABD$ and $I_2$ be the incenter of triangle
$ACD$. The given conditions imply that $P$ and $Q$ are the excenters in
triangle $ABD$ opposite $A$ and $D$, respectively, and that $R$ and $S$ are the
excenters in triangle $ACD$ opposite $A$ and $D$, respectively. Applying the
result in the previous paragraph with $(X, Y, Z, I_x)$ equal to $(A, D, B, P)$,
$(D, A, B, Q)$, $(A, D, C, R)$, and $(D, A, C, S)$, we find that $\angle APD =
\angle I_1BD$, $\angle AQD = \angle ABI_1$, $\angle ARD = \angle I_2CD$, and $\angle ASD = \angle ACI_2$.

Using undirected angles, we know that $\angle I_1BD$, $\angle ABI_1$, $\angle I_2CD$, and
$\angle ACI_2$ all equal $\angle ABD/2 = \angle ACD/2$. Furthermore, they all have the
same orientation, implying that they are equal as directed angles. Therefore
(again using directed angles), $\angle APD = \angle AQD = \angle ARD = \angle ASD$,
and $P, Q, R, S$ lie on a single circle passing through $A$ and $D$.

## Problem 4

Let $p$ be a prime number such that $p \equiv 1 \pmod 4$. Evaluate

$$\sum_{k=1}^{p-1} \left( \left\lfloor \frac{2k^2}{p} \right\rfloor - 2 \left\lfloor \frac{k^2}{p} \right\rfloor \right).$$

**Solution.** For all real $x$, let $\{x\} = x - \lfloor x \rfloor \in [0, 1)$. Writing $\lfloor 2k^2/p \rfloor =
2k^2/p - \{2k^2/p\}$ and $\lfloor k^2/p \rfloor = k^2/p - \{k^2/p\}$, we find that

$$\left\lfloor \frac{2k^2}{p} \right\rfloor - 2 \left\lfloor \frac{k^2}{p} \right\rfloor = 2 \left\{ \frac{k^2}{p} \right\} - \left\{ \frac{2k^2}{p} \right\}.$$

When $\{x\} < \frac{1}{2}$, $2\{x\} - \{2x\} = 2\{x\} - 2\{x\} = 0$. When $\{x\} \geq \frac{1}{2}$,
$2\{x\} - \{2x\} = 2\{x\} - (2\{x\} - 1) = 1$. Therefore, the desired sum equals
the number $\alpha$ of $k$ in $[1, p-1]$ such that $\{k^2/p\} \geq \frac{1}{2}$, or equivalently, the

number of nonzero residues $k$ modulo $p$ such that $k^2$ is congruent to some number in $[(p+1)/2, p-1]$ modulo $p$.

Because $p$ is a prime congruent to 1 modulo 4, it is well known that $-1 \equiv d^2 \pmod{p}$ for some integer $d$. Partition the nonzero residues modulo $p$ into $\frac{p-1}{2}$ pairs of the form $\{a, da\}$, so that $a^2 \equiv -(da)^2 \pmod{p}$ in each pair. Thus, exactly one residue in each pair has a square congruent to some number in $\left[\frac{p+1}{2}, p-1\right]$, for a total of $\frac{p-1}{2}$ such residues. It follows that the given sum equals $\frac{p-1}{2}$.

## Problem 5

Consider the following L-shaped figures, each made of four unit squares:

Let $m$ and $n$ be integers greater than 1. Prove that an $m \times n$ rectangular region can be tiled with such figures if and only if $mn$ is a multiple of 8.

**Solution.** First we prove that if $8 \mid mn$, then an $m \times n$ rectangular region can be tiled by the given figures.

*Case 1:* Both $m$ and $n$ are even. Without loss of generality, assume that $4 \mid m$ and $2 \mid n$. Two of the given figures can be joined into a $4 \times 2$ rectangle, and $mn/8$ such rectangles can be joined into an $m \times n$ rectangular region (with $n/2$ rows and $m/4$ columns of such rectangles).

*Case 2:* Either $m$ or $n$ is odd. Without loss of generality, assume that $m$ is odd. Then $8 \mid n$. Because $m > 1$, we must have $m \geq 3$. We can tile a $3 \times 8$ region as in the following diagram:

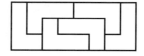

Such $3 \times 8$ regions can further be combined into a $3 \times n$ region. If $m = 3$, this suffices; otherwise, the remaining $(m-3) \times n$ region can be tiled as shown in Case 1 because $2 \mid (m-3)$.

Now we prove that if an $m \times n$ rectangular region can be tiled, then $8 \mid mn$. Because each of the given L-shaped figures has area 4, $4 \mid mn$. Without loss of generality, assume that $2 \mid n$, and color the $m$ rows of the $m \times n$ grid alternatingly black and white. Any L-shaped figure in a tiling of the rectangle would cover an odd number of black squares. Because there are an even number ($n \times \lceil m/2 \rceil$) of black squares, any tiling must contain an even number of L-shaped figures — say, $2k$. Then $mn = 8k$, so that $8 \mid mn$.

## Problem 6

The real numbers $a, b, c, x, y, z$ satisfy $a \geq b \geq c > 0$ and $x \geq y \geq z > 0$. Prove that

$$\frac{a^2x^2}{(by+cz)(bz+cy)} + \frac{b^2y^2}{(cz+ax)(cx+az)} + \frac{c^2z^2}{(ax+by)(ay+bx)}$$

is at least $\frac{3}{4}$.

**Solution.**  Denote the left-hand side of the given inequality by $S$. Because $a \geq b \geq c$ and $x \geq y \geq z$, by the rearrangement inequality we have $bz + cy \leq by + cz$ so

$$(by+cz)(bz+cy) \leq (by+cz)^2 \leq 2\big((by)^2 + (cz)^2\big).$$

Setting $\alpha = (ax)^2$, $\beta = (by)^2$, $\gamma = (cz)^2$, we obtain

$$\frac{a^2x^2}{(by+cz)(bz+cy)} \geq \frac{a^2x^2}{2\,((by)^2 + (cz)^2)} = \frac{\alpha}{2(\beta+\gamma)}.$$

Adding to this the two analogous inequalities for the other summands, we find that

$$S \geq \frac{1}{2}\left(\frac{\alpha}{\beta+\gamma} + \frac{\beta}{\gamma+\alpha} + \frac{\gamma}{\alpha+\beta}\right).$$

By the Cauchy-Schwarz inequality,

$$\left(\frac{\alpha}{\beta+\gamma} + \frac{\beta}{\gamma+\alpha} + \frac{\gamma}{\alpha+\beta}\right)(\alpha(\beta+\gamma) + \beta(\gamma+\alpha) + \gamma(\alpha+\beta))$$

is at least $(\alpha+\beta+\gamma)^2$, which in turn equals

$$\frac{1}{2}\big((\alpha-\beta)^2 + (\beta-\gamma)^2 + (\gamma-\alpha)^2\big) + 3(\alpha\beta + \beta\gamma + \gamma\alpha)$$

$$\geq \frac{3}{2}(2\alpha\beta + 2\beta\gamma + 2\gamma\alpha).$$

Therefore,

$$S \geq \frac{1}{2}\left(\frac{\alpha}{\beta+\gamma} + \frac{\beta}{\gamma+\alpha} + \frac{\gamma}{\alpha+\beta}\right) \geq \frac{1}{2}\frac{(\alpha+\beta+\gamma)^2}{(2\alpha\beta + 2\beta\gamma + 2\gamma\alpha)} \geq \frac{3}{4},$$

as desired.

# 1.14 Mongolia

## Problem 1

Let $\text{rad}(1) = 1$, and for $k > 1$ let $\text{rad}(k)$ equal the product of the prime divisors of $k$. A sequence of natural numbers $a_1, a_2, \ldots$ with arbitrary first term $a_1$ is defined recursively by the relation $a_{n+1} = a_n + \text{rad}(a_n)$. Show that for any positive integer $N$, the sequence $a_1, a_2, \ldots$ contains some $N$ consecutive terms in arithmetic progression.

**Solution.**

**Lemma 1.** *In the sequence* $\text{rad}(a_1), \text{rad}(a_2), \ldots,$ *each term divides the next.*

*Proof:* Because $\text{rad}(a_n)$ divides both $a_n$ and $\text{rad}(a_n)$, it also divides $a_n + \text{rad}(a_n) = a_{n+1}$, so all prime factors of $\text{rad}(a_n)$ divide $a_{n+1}$. Because $\text{rad}(a_n)$ and $\text{rad}(a_{n+1})$ are square-free, this implies that $\text{rad}(a_n)$ divides $\text{rad}(a_{n+1})$. ∎

For all positive integers $n$, define $b_n = a_n/\text{rad}(a_n)$ and $z_n = \text{rad}(a_{n+1})/\text{rad}(a_n)$. Because $\text{rad}(a_n) \mid a_n$, $b_n$ is an integer for all $n$, and because of the above lemma, the same holds for $z_n$. Note that $z_n$ is relatively prime to $\text{rad}(a_n)$ because $\text{rad}(a_{n+1})$ is square-free. Also observe that

$$b_{n+1} = a_{n+1}/\text{rad}(a_{n+1}) = \frac{[a_n + \text{rad}(a_n)]/\text{rad}(a_n)}{\text{rad}(a_{n+1})/\text{rad}(a_n)}$$

$$= \frac{b_n + 1}{\text{rad}(a_{n+1})/\text{rad}(a_n)} = \frac{b_n + 1}{z_n}.$$

**Lemma 2.** *For any $N$, there exists an integer $M$ such that*

$$z_M = z_{M+1} = z_{M+2} = \cdots = z_{M+N-2} = 1.$$

*Proof:* There are some primes $p$ less than $2N$ for which there exists an $n$ such that $p \mid a_n$. By our first lemma, there exists an $m$ sufficiently large so that $a_m$ is divisible by all such primes. Let $M$ be a number greater than $m$ so that $b_M$ is minimal. We claim that this $M$ satisfies the condition of this lemma.

Suppose for the purpose of contradiction that this is not true. Then we can pick the smallest positive $k$ for which $z_{M+k-1} \neq 1$. Note that $k \leq N - 1$ and that $z_M = z_{M+1} = \cdots = z_{M+k-2} = 1$, so that $b_{M+k-1} = b_M + k - 1$.

We claim that no primes less than $2N$ can divide $z_{M+k-1}$. This is true because $z_{M+k-1}$ is the product of the primes dividing $a_{M+k}$ but not $a_{M+k-1}$, and because $a_{M+k-1}$ is divisible by $\mathrm{rad}(a_M)$, which is divisible by all the primes less than $2N$ that divide any $a_n$. Thus, $z_{M+k-1} \geq 2N$.

Therefore,

$$b_{M+k} = \frac{b_{M+k-1}+1}{z_{M+k-1}} = \frac{b_M+k}{z_{M+k-1}}$$
$$\leq \frac{b_M+k}{2N} \leq \frac{b_M+N-1}{2N} < b_M.$$

This contradicts our assumption that $M$ is a number greater than $m$ for which $b_M$ is minimal. Thus, the lemma is proved.  ∎

By the second lemma, for any $N$, there exists an integer $M$ such that

$$\mathrm{rad}(a_M) = \mathrm{rad}(a_{M+1}) = \mathrm{rad}(a_{M+2}) = \cdots = \mathrm{rad}(a_{M+N-1}).$$

Then $a_M, a_{M+1}, \ldots, a_{M+N-1}$ is an arithmetic progression (with common difference $\mathrm{rad}(a_M)$), as desired.

## Problem 2

The circles $\omega_1, \omega_2, \omega_3$ in the plane are pairwise externally tangent to each other. Let $P_1$ be the point of tangency between circles $\omega_1$ and $\omega_3$, and let $P_2$ be the point of tangency between circles $\omega_2$ and $\omega_3$. $A$ and $B$, both different from $P_1$ and $P_2$, are points on $\omega_3$ such that $\overline{AB}$ is a diameter of $\omega_3$. Line $AP_1$ intersects $\omega_1$ again at $X$, line $BP_2$ intersects $\omega_2$ again at $Y$, and lines $AP_2$ and $BP_1$ intersect at $Z$. Prove that $X$, $Y$, and $Z$ are collinear.

**Solution.** All angles are directed modulo $\pi$.

Let $P_3$ be the point of tangency of $\omega_1$ and $\omega_2$, and let $O_1$, $O_2$, and $O_3$ be the centers of $\omega_1$, $\omega_2$, and $\omega_3$, respectively.

Let $\omega_4$ be the circumcircle of triangle $P_1P_2P_3$. Let $O_4$ be the radical center of $\omega_1$, $\omega_2$, and $\omega_3$. Note that $O_4P_1 = O_4P_2 = O_4P_3$, so $O_4$ is the center of $\omega_4$. Because $\overline{O_4P_1} \perp \overline{O_1O_3}$, $\omega_4$ is tangent to line $O_1O_3$. Likewise, $\omega_4$ is tangent to lines $O_1O_2$ and $O_2O_3$.

Because $O_3$ lies on $\overline{AB}$, we have

$$\angle P_2P_1Z = \angle P_2AO_3 = \angle O_3P_2A.$$

If we let $Z'$ be the second intersection of line $AP_2$ with $\omega_4$, then, because line $O_3P_2$ is tangent to $\omega_4$,

$$\angle O_3 P_2 A = \angle O_3 P_2 Z' = \angle P_2 P_1 Z'.$$

Hence, $\angle P_2 P_1 Z = \angle P_2 P_1 Z'$, and $Z'$ lies on the line $BZ$. Because $Z$ and $Z'$ both lie on line $AP_2$, which is different from line $BZ$, we must have $Z = Z'$. Thus, $Z$ lies on $\omega_4$.

Because angles $O_4 P_1 O_3$ and $X P_1 Z$ are both right,

$$\angle Z P_1 O_3 = \angle Z P_1 O_4 + \angle O_4 P_1 O_3 = \angle X P_1 Z + \angle Z P_1 O_4 = \angle X P_1 O_4.$$

Because line $P_1 O_4$ is tangent to $\omega_1$, we have $\angle X P_1 O_4 = \angle X P_3 P_1$. Therefore, $\angle Z P_1 O_3 = \angle X P_3 P_1$.

Let $\ell$ be the line $Z P_3$ if $Z$ does not coincide with $P_3$, or the line tangent to $\omega_4$ at $P_3$ otherwise. Then $\angle(\ell, P_3 P_1) = \angle Z P_1 O_3$ because $O_3 P_1$ is tangent to $\omega_4$. Combining this with the above result yields $\angle(\ell, P_3 P_1) = \angle X P_3 P_1$. Thus, $X$ lies on $\ell$. Similarly, $Y$ lies on $\ell$. Because $Z$ also lies on $\ell$, the points $X$, $Y$, and $Z$ are collinear, as desired.

## Problem 3

A function $f : \mathbb{R} \to \mathbb{R}$ satisfies the following conditions:

(i) $|f(a) - f(b)| \leq |a - b|$ for any real numbers $a, b \in \mathbb{R}$.

(ii) $f(f(f(0))) = 0$.

Prove that $f(0) = 0$.

**Solution.** We shall use the notation

$$f^k(x) = \underbrace{f(f(\cdots f(x) \cdots))}_{k \ f\text{'s}}.$$

From

$$|f(0)| = |f(0) - 0| \geq |f^2(0) - f(0)| \geq |f^3(0) - f^2(0)| = |f^2(0)|$$

and

$$|f^2(0)| = |f^2(0) - 0| \geq |f^3(0) - f(0)| = |f(0)|,$$

we have

$$|f(0)| = |f^2(0)|.$$

There are two cases to consider. If $f(0) = f^2(0)$, then

$$f(0) = f^2(0) = f^3(0) = 0,$$

as desired. If $f(0) = -f^2(0)$,

$$|f(0)| = |f(0) - 0| \geq |f^2(0) - f(0)| = 2|f(0)|,$$

from which we again conclude that $f(0) = 0$.

## Problem 4

The bisectors of angles $A, B, C$ of a triangle $ABC$ intersect its sides at points $A_1, B_1, C_1$. Prove that if the quadrilateral $BA_1B_1C_1$ is cyclic, then

$$\frac{BC}{AC + AB} = \frac{AC}{AB + BC} - \frac{AB}{BC + AC}.$$

**Solution.**  Let the circumcircle $\omega$ of quadrilateral $BA_1B_1C_1$ intersect line $AC$ again at $X$. We claim that $X$ must lie on the *segment* $\overline{AC}$. First, because $A$ lies on the line $BC_1$ but not the segment $\overline{BC_1}$, it must lie outside $\omega$. Similarly, $C$ lies outside $\omega$. Any point on $\overline{B_1X}$ lies in $\omega$. Therefore, $\overline{B_1X}$ contains neither $A$ nor $C$. Because $B_1$ lies on $\overline{AC}$, so must $X$.

Let $a = BC$, $b = AC$, $c = AB$. By the power of a point theorem applied to $A$ with respect to $\omega$, we have $AC_1 \cdot AB = AX \cdot AB_1$. By the angle bisector theorem, $AC_1 = bc/(a + b)$ and $AB_1 = bc/(a + c)$. Therefore,

$$AX = \frac{AC_1 \cdot AB}{AB_1} = \frac{bc}{a+b} \cdot c \cdot \frac{a+c}{bc} = (a+c) \cdot \frac{c}{a+b}.$$

Similarly,

$$CX = (a+c) \cdot \frac{a}{b+c}.$$

Therefore, because $X$ lies on $\overline{AC}$,

$$b = AC = AX + XC = (a+c)\left(\frac{c}{a+b} + \frac{a}{b+c}\right).$$

The desired result follows immediately.

## Problem 5

Which integers can be represented in the form

$$\frac{(x+y+z)^2}{xyz},$$

where $x$, $y$, and $z$ are positive integers?

**Solution.**   Note that

$$1 = (9+9+9)^2/(9 \cdot 9 \cdot 9), \qquad 2 = (4+4+8)^2/(4 \cdot 4 \cdot 8),$$
$$3 = (3+3+3)^2/(3 \cdot 3 \cdot 3), \qquad 4 = (2+2+4)^2/(2 \cdot 2 \cdot 4),$$
$$5 = (1+4+5)^2/(1 \cdot 4 \cdot 5), \qquad 6 = (1+2+3)^2/(1 \cdot 2 \cdot 3),$$
$$8 = (1+1+2)^2/(1 \cdot 1 \cdot 2), \qquad 9 = (1+1+1)^2/(1 \cdot 1 \cdot 1).$$

We prove that no other solutions are possible by using the following lemma.

**Lemma.** *If $n$ can be expressed as $(x+y+z)^2/(xyz)$, then $n$ can be written as $(x'+y'+z')^2/(x'y'z')$, where $x' \le y'+z'$, $y' \le x'+z'$, and $z' \le x'+y'$.*

*Proof:*   Let $x$, $y$, $z$ be the positive integers such that $n = (x+y+z)^2/(xyz)$ and $x+y+z$ is minimal. Because $n$ is an integer, $x$ divides $(x+y+z)^2$. Therefore, $x$ divides $(y+z)^2$. Let $x' = (y+z)^2/x$.

$$\frac{(x'+y+z)^2}{x'yz} = \frac{(y+z)^2 \left(\frac{y+z}{x}+1\right)^2}{\frac{(y+z)^2}{x}yz}$$

$$= \frac{x \left(\frac{y+z}{x}+1\right)^2}{yz} = \frac{(x+y+z)^2}{xyz} = n.$$

Because $x+y+z$ is minimal, $x+y+z \le x'+y+z$. Therefore, $x \le x' = (y+z)^2/x$, and it follows that $x \le y+z$. Similarly, $y \le x+z$ and $z \le x+y$.                                                  ∎

Now, suppose that $n = (x+y+z)^2/xyz$. By the lemma, we may assume without loss of generality that $y+z \ge x \ge y \ge z$. We consider the following cases.

*Case 1: $x = y \ge z = 1$.*   In this case, $n = (2x+1)^2/(x^2)$. Thus, $x$ divides $2x+1$, so $x = 1$ and $n = 9$.

*Case 2: $x = y+1 > z = 1$.*   Here, $n = (2x)^2/(x(x-1)) = 4x/(x-1)$. Thus, $x-1$ divides $4x$, implying that $x-1$ divides 4. Therefore, $x \in \{2, 3, 5\}$, and $n \in \{8, 6, 5\}$.

*Case 3: $y+z \ge x \ge y \ge z > 1$.*   Here,

$$y \cdot z - (y+z) = (y-1)(z-1) - 1 \ge 0,$$

implying that $yz \ge y+z \ge x$. Because $x \ge y \ge z$, we also have $xy \ge z$

and $xz \geq y$. Thus,

$$n = \frac{(x+y+z)^2}{xyz} = 2\left(\frac{1}{x} + \frac{1}{y} + \frac{1}{z}\right) + \frac{x}{yz} + \frac{y}{xz} + \frac{z}{xy}$$

$$\leq 2 \cdot \frac{3}{2} + 1 + 1 + 1 = 6.$$

Therefore, 1, 2, 3, 4, 5, 6, 8, and 9 are the only solutions.

## Problem 6

In a country with $n$ towns the cost of travel from the $i$th town to the $j$th town is $x_{ij}$. Suppose that the total cost of any route passing through each town exactly once and ending at its starting point does not depend on which route is chosen. Prove that there exist numbers $a_1, \ldots, a_n$ and $b_1, \ldots, b_n$ such that $x_{ij} = a_i + b_j$ for all integers $i, j$ with $1 \leq i < j \leq n$.

**Solution.**  Let $f(a, b) = x_{a1} + x_{1b} - x_{ab}$ for $a$, $b$, and 1 all distinct.

**Lemma.** $f(a, b)$ *is independent of* $a$ *and* $b$.

*Proof:* For $n \leq 2$ this is trivial because $f$ is defined for no $a$ and $b$. For $n = 3$, we need to show that $f(2, 3) = f(3, 2)$, or that $x_{21} + x_{13} + x_{32} = x_{31} + x_{12} + x_{23}$. Because these are the total costs of two routes which each pass through every town exactly once, they are equal.

For $n \geq 4$, the route

$$a, 1, b, c, 2, 3, \ldots, a-1, a+1, \ldots, b-1, b+1, \ldots, c-1, c+1, \ldots, n$$

and the route

$$a, b, 1, c, 2, 3, \ldots, a-1, a+1, \ldots, b-1, b+1, \ldots, c-1, c+1, \ldots, n.$$

must have equal total costs. The routes are nearly identical, and it is easy to see that the difference of their total costs is

$$(x_{a1} + x_{1b} + x_{bc}) - (x_{ab} + x_{b1} + x_{1c}).$$

Therefore, $f(a, b) = f(b, c)$ for any $a$, $b$, $c$ distinct from each other and from 1.

Furthermore, the sum of the total costs of the three routes

$$1, a, b, 2, \ldots, n; \quad b, 1, a, 2, \ldots, n; \quad a, b, 1, 2, \ldots, n$$

must equal the sum of the total costs of the three routes

$$1, b, a, 2, \ldots, n; \quad a, 1, b, 2, \ldots, n; \quad b, a, 1, 2, \ldots, n.$$

Hence,

$$2 \cdot (x_{1a} + x_{ab} + x_{b1}) = 2 \cdot (x_{1b} + x_{ba} + x_{a1}),$$

implying that $f(a, b) = f(b, a)$.

For $c$, $d$ not equal to $a$ and $b$, we find that $f(a, b) = f(b, c) = f(c, d)$, $f(a, b) = f(b, c) = f(c, b)$, and $f(a, b) = f(b, a) = f(a, c) = f(c, a)$. This establishes the lemma. ∎

For all $a, b$ distinct from each other and from 1, we have $f(a, b) = F$ for some constant $F$.

Let $a_1 = 0$ and $b_1 = F$, and let $b_k = x_{1k}$ and $a_k = x_{k1} - F$. For $i$ and $j$ both not equal to 1,

$$x_{ij} = x_{i1} - x_{i1} - x_{1j} + x_{ij} + x_{1j}$$
$$= x_{i1} - F + x_{1j} = a_i + b_j,$$

as desired.

## 1.15  Poland

### Problem 1

Let $n \geq 2$ be a given integer. How many solutions does the system of equations

$$
\begin{cases}
x_1 + x_n^2 &= 4x_n \\
x_2 + x_1^2 &= 4x_1 \\
&\vdots \\
x_n + x_{n-1}^2 &= 4x_{n-1}
\end{cases}
$$

have in nonnegative real numbers $x_1, \ldots, x_n$?

**Solution.**  We take the indices of the $x_i$ modulo $n$. Let $f(x) = 4x - x^2$, so that $x_i = f(x_{i-1})$ for each $i$, We have $4 - f(x_{i-1}) = (x_{i-1} - 2)^2 \geq 0$, implying that $x_i \leq 4$ for each $i$. Also, we are given that $x_i \geq 0$. Thus, in particular, we can write $x_1 = 2 - 2\cos\theta$ for a unique $\theta \in [0, \pi]$. Then,

$$
x_2 = f(x_1) = 4(2 - 2\cos\theta) - (2 - 2\cos\theta)^2
$$
$$
= 4 - 4\cos^2\theta = 2 - 2\cos 2\theta.
$$

The same argument proves (inductively) that $x_i = 2 - 2\cos 2^{i-1}\theta$ for each $i \geq 1$. In particular, $x_1 = x_{n+1} = 2 - 2\cos 2^n\theta$. Thus, $\cos\theta = \cos 2^n\theta$; conversely, every such value of $\theta \in [0, \pi]$ gives a different solution to the system of equations.

Note that $\cos\theta = \cos 2^n\theta$ holds if and only if $2^n\theta = 2k\pi \pm \theta$ for some integer $k$, or equivalently if $\theta = 2k\pi/(2^n \pm 1)$. Thus, the desired $\theta \in [0, \pi]$ are $2k_1\pi/(2^n - 1)$ for $k_1 = 0, 1, \ldots, 2^{n-1} - 1$ and $2k_2\pi/(2^n + 1)$ for $k_2 = 1, 2, \ldots, 2^{n-1}$. We claim that these $2^n$ values for $\theta$ are distinct. Indeed, suppose that $2k_1\pi/(2^n - 1) = 2k_2\pi/(2^n + 1)$ for some $k_1$ and $k_2$, so that $k_1(2^n + 1) = k_2(2^n - 1)$. Because $2^n + 1$ is relatively prime to $2^n - 1$, we must have $(2^n + 1) \mid k_2$, which is impossible. Therefore, there are $2^n$ possible values for $\theta$ and hence $2^n$ solutions to the given system of equations.

### Problem 2

The sides $\overline{AC}$ and $\overline{BC}$ of a triangle $ABC$ have equal length. Let $P$ be a point inside triangle $ABC$ such that $\angle PAB = \angle PBC$ and let $M$ be the midpoint of $\overline{AB}$. Prove that $\angle APM + \angle BPC = \pi$.

**Solution.** Extend $\overline{AP}$ through $P$ to $D$ such that $CD \parallel AB$. Let $\angle CAP = x$ and $\angle PAB = y$. Then $\angle ABP = x$ and $\angle PBC = \angle PDC = \angle ADC = y$. Hence $CPBD$ is cyclic. Consequently,

$$\angle BPC + \angle CDB = \pi \tag{1}$$

and

$$\angle DCB = \angle DPB = \angle BAP + \angle ABP = x + y. \tag{2}$$

By (1), it suffices to prove that $\angle CDB = \angle APM$.

Extend $\overline{PM}$ through $M$ to $Q$ such that $PM = MQ$. Then $APBQ$ is a parallelogram. Hence

$$\angle BAQ = \angle ABP = x$$

and

$$\angle PAQ = \angle BAP + \angle BAQ = x + y = \angle DCB$$

by (2). Hence it suffices to prove that triangles $APQ$ and $CDB$ are similar, or, $AP/AQ = CD/CB$ (by SAS), which is equivalent to proving $AP/PB = CD/CB$ as $APBQ$ is a parallelogram.

Recall that $\angle PAB = y = \angle ADC$ and $\angle ABP = x = \angle CAP = \angle CAD$, implying that triangles $ABP$ and $DAC$ are similar. Hence $AP/BP = DC/AC = DC/BC$, as desired.

## Problem 3

A sequence $p_1, p_2, \ldots$ of prime numbers satisfies the following condition: for $n \geq 3$, $p_n$ is the greatest prime divisor of $p_{n-1} + p_{n-2} + 2000$. Prove that the sequence is bounded.

**Solution.** Let $b_n = \max\{p_n, p_{n+1}\}$ for $n \geq 1$. We first prove that $b_{n+1} \leq b_n + 2002$ for all such $n$. Certainly $p_{n+1} \leq b_n$, so it suffices to show that $p_{n+2} \leq b_n + 2002$. If either $p_n$ or $p_{n+1}$ equals 2, then we have $p_{n+2} \leq p_n + p_{n+1} + 2000 = b_n + 2002$. Otherwise, $p_n$ and $p_{n+1}$ are both odd, so $p_n + p_{n+1} + 2000$ is even. Because $p_{n+2} \neq 2$ divides this number, we have

$$p_{n+2} \leq \frac{p_n + p_{n+1} + 2000}{2} = \frac{p_n + p_{n+1}}{2} + 1000 \leq b_n + 1000.$$

This proves the claim.

Choose $k$ large enough so that $b_1 \leq k \cdot 2003! + 1$. We prove by induction that $b_n \leq k \cdot 2003! + 1$ for all $n$. If this statement holds for some $n$, then $b_{n+1} \leq b_n + 2002 \leq k \cdot 2003! + 2003$. If $b_{n+1} > k \cdot 2003! + 1$, then let $m = b_{n+1} - k \cdot 2003!$. We have $1 < m \leq 2003$, implying that $m \mid 2003!$.

Hence, $m$ is a proper divisor of $k \cdot 2003! + m = b_{n+1}$, which is impossible because $b_{n+1}$ is prime.

Thus, $p_n \leq b_n \leq k \cdot 2003! + 1$ for all $n$.

## Problem 4

For an integer $n \geq 3$, consider a pyramid with vertex $S$ and the regular $n$-gon $A_1 A_2 \ldots A_n$ as a base, such that all the angles between lateral edges and the base equal $\pi/3$. Points $B_2, B_3, \ldots$ lie on $\overline{A_2 S}, \overline{A_3 S}, \ldots, \overline{A_n S}$, respectively, such that $A_1 B_2 + B_2 B_3 + \cdots + B_{n-1} B_n + B_n A_1 < 2 A_1 S$. For which $n$ is this possible?

**Solution.**   We claim that this is possible for any $n \geq 3$. The shortest path between any two points in a plane is the length of the straight line segment connecting them. Although we cannot simply draw a straight line segment around the pyramid connecting $A_1$ with itself, we can translate this three-dimensional problem into a two-dimensional problem where we *can* draw such a segment. We develop (i.e. flatten) the lateral surface of the pyramid to form triangles $S' A_i' A_{i+1}'$ congruent to (and with the same orientation as) triangle $SA_i A_{i+1}$ for $i = 1, 2, \ldots, n$, where we write $A_{n+1} = A_1$. Any broken-line path from $A_1'$ to $A_{n+1}'$, consisting of segments connecting $\overline{S' A_i'}$ and $\overline{S' A_{i+1}'}$ for $i = 1, 2, \ldots, n$, can be transformed into a broken-line path of the same length from $A_1$ to $A_{n+1} = A_1$, consisting of segments connecting $\overline{SA_i}$ and $\overline{SA_{i+1}}$ for $i = 1, 2, \ldots, n$. Thus, it suffices to prove that one such broken-line path from $A_1'$ to $A_{n+1}'$ has perimeter less than $2A_1 S$.

Indeed, we claim that the straight line path connecting $A_1'$ to $A_{n+1}'$, with length $A_1' A_{n+1}'$, is such a path. By the triangle inequality, $A_1' A_{n+1}' < A_1' S' + S' A_{n+1}' = 2A_1 S$. We need only verify, then, that this path consists of segments connecting $\overline{S' A_i'}$ with $\overline{S' A_{i+1}'}$ for $i = 1, 2, \ldots, n$.

Let $O$ be the center of the base of the given pyramid. Suppose that $1 \leq i \leq n$. Then $\angle A_i O S = \pi/2$ and $\angle S A_i O = \pi/3$, implying that $SA_i = 2OA_i$. Let $M$ be the midpoint of $\overline{A_i A_{i+1}}$. Because triangle $OA_i A_{i+1}$ is isosceles, we have $\overline{OM} \perp \overline{A_i A_{i+1}}$, so that $A_i M = OA_i \sin \angle A_i OM = OA_i \sin \pi/n$. Similarly, we have $A_i M = SA_i \sin \angle A_i SM = 2OA_i \sin \angle A_i SM$. Thus,

$$\angle A_i S A_{i+1} = 2\angle A_i SM = 2\sin^{-1}\left(\frac{1}{2}\sin\frac{\pi}{n}\right)$$

$$= 2\sin^{-1}\left(\frac{1}{2} \cdot 2\sin\frac{\pi}{2n}\cos\frac{\pi}{2n}\right) < 2\sin^{-1}\left(\sin\frac{\pi}{2n}\right) = \frac{\pi}{n}.$$

Hence,

$$\sum_{i=1}^{n} \angle A_i' S' A_{i+1}' = \sum_{i=1}^{n} \angle A_i S A_{i+1} < \pi,$$

implying that $\overline{A_1' A_{n+1}'}$ intersects each of the segments $\overline{S'A_2'}$, $\overline{S'A_3'}$, ..., $\overline{S'A_n'}$. Therefore, the straight line path from $A_1'$ to $A_{n+1}'$ indeed consists of segments connecting $\overline{S'A_i'}$ with $\overline{S'A_{i+1}'}$ for $i = 1, 2, \ldots, n$, as desired. This completes the proof.

## Problem 5

Given a natural number $n \geq 2$, find the smallest integer $k$ with the following property: Every set consisting of $k$ cells of an $n \times n$ table contains a nonempty subset $S$ such that in every row and in every column of the table, there is an even number of cells belonging to $S$.

**Solution.** The answer is $2n$. To see that $2n - 1$ cells do not suffice, consider the "staircase" of cells consisting of the main diagonal and the diagonal immediately below it. Number these cells from upper-left to lower-right. For any subset $S$ of the staircase, consider its lowest-numbered cell; this cell is either the only cell of $S$ in its row or the only cell of $S$ in its column, so $S$ cannot have the desired property.

To see that $2n$ suffices, we use the following lemma:

**Lemma.** *If a graph is drawn whose vertices are the cells of an $m \times n$ grid, where two vertices are connected by an edge if and only if they lie in the same row or the same column, then any set $T$ of at least $m + n$ vertices includes the vertices of some cycle whose edges alternate between horizontal and vertical.*

*Proof:* We induct on $m + n$. If $m = 1$ or $n = 1$, the statement is vacuously true; this gives us the base case. Otherwise, we construct a trail as follows. We arbitrarily pick a starting vertex in $T$ and, if possible, proceed horizontally to another vertex of $T$. We then continue vertically to another vertex of $T$. We continue this process, alternating between horizontal and vertical travel. Eventually, we must either (a) be unable to proceed further, or (b) return to a previously visited vertex. In case (a), we must have arrived at a vertex which is the only element of $T$ in its row or in its column; then remove this row or column from the grid, remove the appropriate vertex from $T$, and apply the induction hypothesis. In case (b), we have formed a cycle. If there are two consecutive horizontal edges (resp. vertical edges), as is the case if our cycle contains an odd number of vertices, then we replace these two edges by a single horizontal (resp.) vertical edge. We thus obtain a cycle which alternates between horizontal and vertical edges. By construction, our cycle does not visit any vertex twice.                                                                        ∎

   To see that this solves the problem, suppose we have a set of $2n$ cells from our $n \times n$ grid. It then contains a cycle; let $S$ be the set of vertices of this cycle. Consider any row of the grid. Every square of $S$ in this row belongs to exactly one horizontal edge, so if the row contains $m$ horizontal edges, then it contains $2m$ cells of $S$. Thus, every row (and similarly, every column) contains an even number of cells of $S$.

## Problem 6

Let $P$ be a polynomial of odd degree satisfying the identity

$$P(x^2 - 1) = P(x)^2 - 1.$$

Prove that $P(x) = x$ for all real $x$.

**Solution.**   Setting $x = y$ and $x = -y$ in the given equation, we find that $P(y)^2 = P(-y)^2$ for all $y$. Thus, one of the polynomials $P(x) - P(-x)$ or $P(x) + P(-x)$ vanishes for infinitely many, and hence all, $x$. Because $P$ has odd degree, the latter must be the case, and $P$ is an odd polynomial. In particular, $P(0) = 0$, which in turn implies that $P(-1) = P(0^2 - 1) = P(0)^2 - 1 = -1$ and that $P(1) = -P(-1) = 1$.
   Set $a_0 = 1$, and let $a_n = \sqrt{a_{n-1} + 1}$ for $n \geq 1$; note that $a_n > 1$ when $n \geq 1$. We claim that $P(a_n) = a_n$ for all $n$. This is true for $n = 0$. If it holds for $a_n$, then

$$P(a_{n+1})^2 = P(a_{n+1}^2 - 1) + 1 = P(a_n) + 1 = a_n + 1,$$

implying that $P(a_{n+1}) = \pm a_{n+1}$. If $P(a_{n+1}) = -a_{n+1}$, then

$$P(a_{n+2})^2 = P(a_{n+1}) + 1 = 1 - a_{n+1} < 0,$$

a contradiction. Thus, $P(a_{n+1}) = a_{n+1}$, and the claim holds by induction.
   If $-1 \leq x \leq (1 + \sqrt{5})/2$, then it is easy to check that $x \leq \sqrt{1 + x} \leq (1 + \sqrt{5})/2$. Thus, the $a_i$ form an increasing sequence. Thus, there are infinitely many values of $x$ — namely, the $a_i$ — such that $P(x) = x$. Because $P$ is a polynomial, we in fact have $P(x) = x$ for all $x$.

## 1.16  Romania

### Problem 1

A function $f : \mathbb{R}^2 \to \mathbb{R}$ is called *olympic* if it has the following property: given $n \geq 3$ distinct points $A_1, A_2, \ldots, A_n \in \mathbb{R}^2$, if $f(A_1) = f(A_2) = \cdots = f(A_n)$ then the points $A_1, A_2, \ldots, A_n$ are the vertices of a convex polygon. Let $P \in \mathbb{C}[X]$ be a non-constant polynomial. Prove that the function $f : \mathbb{R}^2 \to \mathbb{R}$, defined by $f(x,y) = |P(x+iy)|$, is olympic if and only if all the roots of $P$ are equal.

**Solution.** First suppose that all the roots of $P$ are equal, and write $P(x) = a(z - z_0)^n$ for some $a, z_0 \in \mathbb{C}$ and $n \in \mathbb{N}$. If $A_1, A_2, \ldots, A_n$ are distinct points in $\mathbb{R}^2$ such that $f(A_1) = f(A_2) = \cdots = f(A_n)$, then $A_1, \ldots, A_n$ are situated on a circle with center $(\mathrm{Re}(z_0), \mathrm{Im}(z_0))$ and radius $\sqrt[n]{|f(A_1)/a|}$, implying that the points are the vertices of a convex polygon.

Conversely, suppose that not all the roots of $P$ are equal, and write $P(x) = (z - z_1)(z - z_2)Q(z)$ where $z_1$ and $z_2$ are distinct roots of $P(x)$ such that $|z_1 - z_2|$ is minimal. Let $\ell$ be the line containing $Z_1 = (\mathrm{Re}(z_1), \mathrm{Im}(z_1))$ and $Z_2 = (\mathrm{Re}(z_2), \mathrm{Im}(z_2))$, and let $z_3 = \frac{1}{2}(z_1 + z_2)$ so that $Z_3 = (\mathrm{Re}(z_3), \mathrm{Im}(z_3))$ is the midpoint of $\overline{Z_1 Z_2}$. Also, let $s_1, s_2$ denote the rays $Z_3 Z_1$ and $Z_3 Z_2$, and let $r = f(Z_3) \geq 0$. We must have $r > 0$, because otherwise $z_3$ would be a root of $P$ such that $|z_1 - z_3| < |z_1 - z_2|$, which is impossible. Because $f(Z_3) = 0$,

$$\lim_{\substack{ZZ_3 \to \infty, \\ Z \in s_1}} f(Z) = +\infty,$$

and $f$ is continuous, there exists $Z_4 \in s_1$ — on the side of $Z_1$ away from $Z_3$ — such that $f(Z_4) = r$. Similarly, there exists $Z_5 \in s_2$ — on the side of $Z_2$ away from $Z_3$ — such that $f(Z_5) = r$. Thus, $f(Z_3) = f(Z_4) = f(Z_5)$ and $Z_3, Z_4, Z_5$ are not vertices of a convex polygon. Hence, $f$ is not olympic.

### Problem 2

Let $n \geq 2$ be a positive integer. Find the number of functions $f : \{1, 2, \ldots, n\} \to \{1, 2, 3, 4, 5\}$ which have the following property:

$$|f(k+1) - f(k)| \geq 3 \quad \text{for } k = 1, 2, \ldots, n-1.$$

**Solution.** We let $n \geq 2$ vary and find the number of functions in terms of $n$. If a function $f : \{1, 2, \ldots, n\} \to \{1, 2, 3, 4, 5\}$ satisfies the required property, then $f(n) \neq 3$ because otherwise either $f(n-1) \leq 0$ or $f(n-1) \geq 6$, a contradiction. Denote by $a_n, b_n, d_n, e_n$ the number of functions $f : \{1, 2, \ldots, n\} \to \{1, 2, 3, 4, 5\}$ satisfying the required property such that $f(n)$ equals 1, 2, 4, 5, respectively. Then $a_2 = e_2 = 2$ and $b_2 = d_2 = 1$, and the following recursive relations hold for $n \geq 2$:

$$a_{n+1} = e_n + d_n, \quad b_{n+1} = e_n,$$
$$e_{n+1} = a_n + b_n, \quad d_{n+1} = a_n.$$

We wish to find $a_n + b_n + d_n + e_n$ for all $n \geq 2$. We know that $a_2 = e_2$ and $b_2 = d_2$; it follows by induction that $a_n = e_n$ and $b_n = d_n$ for all $n \geq 2$. Hence, for all such $n$, we have

$$a_{n+2} = e_{n+1} + d_{n+1} = a_{n+1} + b_{n+1} = a_{n+1} + e_n = a_{n+1} + a_n.$$

Thus, $\{a_n\}_{n \geq 2}$ satisfies the same recursive relation as the Fibonacci sequence $\{F_n\}_{n \geq 0}$, where indices of the Fibonaccis are chosen such that $F_1 = 0$ and $F_1 = 1$. Because $a_2 = 2 = F_2$ and $a_3 = e_2 + d_2 = 3 = F_3$, it follows that $a_n = F_n$ for all $n$. Therefore, $a_n + b_n + d_n + e_n = 2(a_n + b_n) = 2e_{n+1} = 2a_{n+1} = 2F_{n+1}$ for all $n \geq 2$, and there are $2F_{n+1}$ functions with the required property.

## Problem 3

Let $n \geq 1$ be an odd positive integer and $x_1, x_2, \ldots, x_n$ be real numbers such that $|x_{k+1} - x_k| \leq 1$ for $k = 1, 2, \ldots, n - 1$. Show that

$$\sum_{k=1}^{n} |x_k| - \left| \sum_{k=1}^{n} x_k \right| \leq \frac{n^2 - 1}{4}.$$

**Solution.** Without loss of generality, assume that there are more $k$ such that $x_k$ is negative than there are $k$ such that $x_k$ is positive. Let $(a_1, \ldots, a_n)$ be a permutation of $(x_1, \ldots, x_n)$ such that $a_1, \ldots, a_n$ is a nondecreasing sequence. By construction, $|P| \leq (n-1)/2$.

Suppose that $1 \leq i \leq n - 1$. In the sequence $x_1, \ldots, x_n$, there must be two adjacent terms $x_k$ and $x_{k+1}$ which are separated by the interval $(a_i, a_{i+1})$ — that is, such that either $x_k \leq a_i, a_{i+1} \leq x_{k+1}$ or $x_{k+1} \leq a_i$, $a_{i+1} \leq x_k$. Thus, $a_{i+1} - a_i \leq |x_k - x_{k+1}| \leq 1$. That is, $a_1, \ldots, a_n$ is a nondecreasing sequence of terms, such that any two adjacent terms differ by at most 1.

Let $\sigma_P$ denote the sum of the numbers in $P$. We claim that $\sigma_P \leq (n^2 - 1)/8$. This is certainly true if $P$ is empty.

If $P$ is nonempty, then the elements of $P$ are $a_\ell \leq a_{\ell+1} \leq \cdots \leq a_n$ for some $2 \leq \ell \leq n$. Because $a_{\ell-1} \leq 0$ by assumption and $a_\ell \leq a_{\ell-1}+1$ from the previous paragraph, we have $a_\ell \leq 1$. Similarly, $a_{\ell+1} \leq a_\ell + 1 \leq 2$, and so on up to $a_n \leq |P|$. Therefore, $\sigma_P \leq 1+2+\cdots+|P|$. Recalling that $|P| \leq (n-1)/2$, we have $\sigma_P \leq (n^2 - 1)/8$, as claimed.

Let $\sigma_N$ denote the sums of the numbers in $N$. The left-hand side of the required inequality then equals

$$|\sigma_P - \sigma_N| - |-\sigma_P - \sigma_N| \leq |2\sigma_P| \leq 2 \cdot \frac{n^2 - 1}{8} = \frac{n^2 - 1}{4},$$

as needed.

## Problem 4

Let $n, k$ be arbitrary positive integers. Show that there exist positive integers $a_1 > a_2 > a_3 > a_4 > a_5 > k$ such that

$$n = \pm \binom{a_1}{3} \pm \binom{a_2}{3} \pm \binom{a_3}{3} \pm \binom{a_4}{3} \pm \binom{a_5}{3},$$

where $\binom{a}{3} = \big(a(a-1)(a-2)\big)/6$.

**Solution.** Observe that $n + \binom{m}{3} > 2m + 1$ for all $m$ larger than some value $N$, because the left-hand side is a cubic in $m$ with positive leading coefficient while the right-hand side is linear in $m$.

If $m \equiv 0 \pmod 4$, then $\binom{m}{3} = \big(m(m-1)(m-2)\big)/6$ is even because the numerator is divisible by 4 while the denominator is not. If $m \equiv 3 \pmod 4$, then $\binom{m}{3} = \big(m(m-1)(m-2)\big)/6$ is odd because both the numerator and denominator are divisible by 2 but not 4. Hence, we may choose $m > \max\{k, N\}$ such that $n + \binom{m}{3}$ is odd.

Write $2a + 1 = n + \binom{m}{3} > 2m + 1$. Observe that

$$\left(\binom{a+3}{3} - \binom{a+2}{3}\right) - \left(\binom{a+1}{3} - \binom{a}{3}\right) = \binom{a+2}{2} - \binom{a}{2}$$

$$= 2a + 1.$$

Hence,

$$n = (2a+1) - \binom{m}{3} = \binom{a+3}{3} - \binom{a+2}{3} - \binom{a+1}{3} + \binom{a}{3} - \binom{m}{3},$$

which is of the desired form because $a+3 > a+2 > a+1 > a > m > k$.

## Problem 5

Let $P_1 P_2 \cdots P_n$ be a convex polygon in the plane. Assume that for any pair of vertices $P_i, P_j$, there exists a vertex $V$ of the polygon such that $\angle P_i V P_j = \pi/3$. Show that $n = 3$.

**Solution.** Throughout this solution, we use the following facts: Given a triangle $XYZ$ such that $\angle XYZ \leq \pi/3$, either the triangle is equilateral or else $\max\{YX, YZ\} > XZ$. Similarly, if $\angle XYZ \geq \pi/3$, then either the triangle is equilateral or else $\min\{YX, YZ\} < XZ$.

We claim that there exist vertices $A, B, C$ and $A_1, B_1, C_1$ such that (i) triangles $ABC$ and $A_1 B_1 C_1$ are equilateral, and (ii) $AB$ (resp., $A_1 B_1$) is the minimal (resp., maximal) nonzero distance between two vertices. Indeed, let $A, B$ be distinct vertices such that $\overline{AB}$ has minimal length, and let $C$ be a vertex which satisfies the condition $\angle ACB = \pi/3$. Then $\max\{AC, CB\} \leq AC$, so that triangle $ABC$ must be equilateral. Similarly, if we choose $A_1, B_1$ such that $\overline{A_1 B_1}$ has maximal length, and a vertex $C_1$ for which $\angle A_1 C_1 B_1 = \pi/3$, then triangle $A_1 B_1 C_1$ is equilateral.

We claim that $\triangle ABC \cong \triangle A_1 B_1 C_1$. The lines $AB, BC, CA$ divide the plane into seven open regions. Let $D_A$ consist of the region distinct from triangle $ABC$ and bounded by $\overline{BC}$, plus the boundaries of this region except for the points $B$ and $C$. Define $D_B$ and $D_C$ analogously. Because the given polygon is convex, each of $A_1, B_1, C_1$ is either in one of these regions or coincides with one of $A, B, C$.

If any two of $A_1, B_1, C_1$ — say, $A_1$ and $B_1$ — are in the same region $D_X$, then $\angle A_1 X B_1 < \pi/3$. Hence, $\max\{A_1 X, X B_1\} > A_1 B_1$, contradicting the maximal definition of $A_1 B_1$.

Therefore, no two of $A_1, B_1, C_1$ are in the same region. Suppose now that even one of $A_1, B_1, C_1$ (say, $A_1$) lies in one of the regions (say, $D_A$). Because $\min\{A_1 B, A_1 C\} \geq BC$, we have that $\angle B A_1 C \leq \pi/3$. We have that $B_1$ does not lie in $D_A$. Because the given polygon is convex, $B$ does not lie in the interior of triangle $A A_1 B_1$, and similarly $C$ does not lie in the interior of triangle $A A_1 B_1$. It follows that $B_1$ lies on the closed region bounded by rays $A_1 B$ and $A_1 C$. Similarly, so does $C_1$. Therefore, $\pi/3 = \angle B_1 A_1 C_1 \leq \angle B A_1 C = \pi/3$, with equality if $B_1$ and $C_1$ lie on rays $A_1 B$ and $A_1 C$ in some order. Because the given polygon is convex, this is possible only if $B_1$ and $C_1$ equal $B$ and $C$ in some order — in which case $BC = B_1 C_1$, implying that triangles $ABC$ and $A_1 B_1 C_1$ are congruent.

Otherwise, none of $A_1, B_1, C_1$ lies in $D_A \cup D_B \cup D_C$, implying that they coincide with $A, B, C$ in some order. In this case, triangles $ABC$ and $A_1B_1C_1$ are congruent as well.

Hence, any two distinct vertices of the polygon are separated by the same distance, namely $AB = A_1B_1$. It is impossible for more than three points in the plane to have this property, implying that $n = 3$.

## Problem 6

Show that there exist infinitely many 4-tuples of positive integers $(x, y, z, t)$ such that the four numbers' greatest common divisor is 1 and such that

$$x^3 + y^3 + z^2 = t^4.$$

**Solution.**   Setting $a = k^3$ for any even $k > 0$ into the identity

$$(a+1)^4 - (a-1)^4 = 8a^3 + 8a$$

yields

$$(2k^3)^3 + (2k)^3 + [(k^3 - 1)^2]^2 = (k^3 + 1)^4.$$

Because $k^3 + 1$ is odd, $\gcd(2k^3, k^3 + 1) = \gcd(k^3, k^3 + 1) = 1$. Hence, there are infinitely many quadruples of the form $(x, y, z, t) = (2k^3, 2k, (k^3 - 1)^2, k^3 + 1)$, for $k > 0$ even, satisfying the required conditions.

## Problem 7

Given the binary representation of an odd positive integer $a$, find a simple algorithm to determine the least positive integer $n$ such that $2^{2000}$ is a divisor of $a^n - 1$.

**Solution.**   Because $a$ is odd, $\gcd(a, 2^k) = 1$ for all $k \geq 0$. Hence, by Euler's Theorem, $a^{2^{k-1}} \equiv a^{\phi(2^k)} \equiv 1 \pmod{2^k}$ for all such $k$. Thus, the order $n$ of $a$ modulo $2^{2000}$ divides $2^{2000-1} = 2^{1999}$.

If $a \equiv 1 \pmod{2^{2000}}$, it follows that $n = 1$. We now assume that $a \not\equiv 1 \pmod{2^{2000}}$. For any $m \geq 1$, we write

$$a^{2^m} - 1 = (a-1)(a+1)\underbrace{(a^2 + 1)(a^{2^2} + 1)\dots(a^{2^{m-1}} + 1)}_{(\dagger)}. \qquad (*)$$

The binary representation of $a$ either ends in the two digits 01 or the two digits 11. In either case, we have $a \equiv \pm 1 \pmod 4$ and hence $a^{2^k} \equiv 1 \pmod 4$ for all $k \geq 1$. Thus, in the above decomposition $(*)$

for some fixed $m \geq 1$, $2^1$ is the greatest power of 2 dividing each of the $m - 1$ expressions in parentheses above the label (†).

If $a \equiv 1 \,(\mathrm{mod}\ 4)$, then because $a \neq 1$, the binary representation of $a$ ends in the digits

$$1 \underbrace{00 \ldots 01}_{s \text{ digits}}$$

for some $s$ — namely, the largest integer such that $2^s \mid (a - 1)$. In this case, the greatest power of 2 dividing $a - 1$ is $2^s$ while the greatest power of 2 dividing $a + 1$ is 2.

If instead $a \equiv -1 \,(\mathrm{mod}\ 4)$, then because $a \neq 1$, the binary representation of $a$ ends in the digits

$$0 \underbrace{11 \ldots 1}_{s \text{ digits}}$$

for some $s$ — namely, the largest integer such that $2^s \mid (a + 1)$. In this case, the greatest power of 2 dividing $a + 1$ is $2^s$ while the greatest power of 2 dividing $a - 1$ is 2.

In either case, we find using $(*)$ and these results that the greatest power of 2 dividing $a^{2^m} - 1$ is $2^{s+m}$. It follows that the smallest $m \geq 1$ such that $a^{2^m} - 1$ is divisible by $2^{2000}$ is either $2000 - s$ (if $s < 2000$) or 1 (if $s \geq 2000$). In these cases, we have $n = 2^{1999-s}$ or $n = 2$, respectively. Because we can easily use the binary representation of $a$ to deduce which of the two cases holds and what the value of $s$ is, we can use the binary representation of $a$ to find $n$.

## Problem 8

Let $ABC$ be an acute triangle and let $M$ be the midpoint of $\overline{BC}$. There exists a unique interior point $N$ such that $\angle ABN = \angle BAM$ and $\angle ACN = \angle CAM$. Prove that $\angle BAN = \angle CAM$.

**Solution.**   Let $B'$ be the point on ray $AC$ such that $\angle ABB' = \angle BAM$, and let $C'$ be the point on ray $AB$ such that $\angle ACC' = \angle CAM$. Then $N$ is the unique intersection of lines $BB'$ and $CC'$.

Reflect line $AM$ across the angle bisector of angle $BAC$, and let this reflection intersect line $BB'$ at $P$. Also, let $D$ be the reflection of $A$ across $M$, so that quadrilateral $ABDC$ is a parallelogram. Because $\angle PAB = \angle CAM = \angle CAD$ and $\angle ABP = \angle MAB = \angle DAB = \angle ADC$, triangles $ABP$ and $ADC$ are similar. Hence, $\frac{AB}{AD} = \frac{AP}{AC}$. Because $\angle BAD = \angle PAC$, it follows that triangles $BAD$ and $PAC$

are similar. Therefore, $\angle ACP = \angle ADB = \angle CAM$. It follows that $P$ lies on line $CC'$ as well as line $BB'$, and hence that $N = P$. Therefore, $\angle BAN = \angle BAP = \angle CAM$, as desired.

## Problem 9

Determine, with proof, whether there exists a sphere with interior $\mathcal{S}$, a circle with interior $\mathcal{C}$, and a function $f : \mathcal{S} \to \mathcal{C}$ such that the distance between $f(A)$ and $f(B)$ is greater than or equal to $AB$ for all points $A$ and $B$ in $\mathcal{S}$.

**Solution.** We shall prove that such a function does not exist. Assume the contrary, that such a function $f$ exists, and let $r$ be the radius of $\mathcal{C}$. Construct a cube in the interior of the sphere with side length $s$. Partition the cube into $(n-1)^3$ small congruent cubes for some arbitrary integer $n \geq 2$. The set of vertices of these small cubes consists of $n^3$ points $A_1, A_2, \ldots, A_{n^3}$.

Write $A'_i = f(A_i)$ for each $i$. For any $i \neq j$ we have $A'_i A'_j \geq A_i A_j \geq \frac{s}{n}$. It follows that the disks $D_i$ of centers $A'_i$ and radius $\frac{s}{2n}$ are all disjoint, and they are contained in a disk $\mathcal{C}'$ with the same center as $\mathcal{C}$ but with radius $r + \frac{s}{2n}$. Therefore,

$$n^3 \cdot \frac{\pi s^2}{4n^2} = \sum_{i=1}^{n^3} \text{Area}(D_i) \leq \text{Area}(\mathcal{C}') = \pi \left( r + \frac{s}{2n} \right)^2.$$

This inequality cannot hold for arbitrary integers $n \geq 2$, a contradiction.

## Problem 10

Let $n \geq 3$ be an odd integer and $m \geq n^2 - n + 1$ be an integer. The sequence of polygons $P_1, P_2, \ldots, P_m$ is defined as follows:

(i) $P_1$ is a regular polygon with $n$ vertices.

(ii) For $k > 1$, $P_k$ is the regular polygon whose vertices are the midpoints of the sides of $P_{k-1}$.

Find, with proof, the maximum number of colors which can be used such that for every coloring of the vertices of these polygons, one can find four vertices $A$, $B$, $C$, $D$ which have the same color, form an isosceles trapezoid (perhaps a degenerate one), and do not lie on a single line passing through the center of $P_1$.

**Solution.** We claim that the maximum number of colors is $n - 1$. Let $V_1, V_2, \ldots, V_n$ be the vertices of $P_1$ in counterclockwise order, with indices of the $V_i$ taken modulo $n$.

We first show that the desired maximum is less than $n$. Let $O$ be the common center of the polygons. Because $n$ is odd, every vertex lies on one of the lines $VA_i$. Given $n' \geq n$ colors, let $c_1, c_2, \ldots, c_n$ be $n$ of them and color every vertex on line $OV_i$ with color $c_i$ for $i = 1, 2, \ldots, n$. Any four vertices of the same color lie on a line passing through the center of $P_1$, implying that no set of four vertices has the required property.

We now claim that in any coloring of the vertices in $n - 1$ colors, we can find four vertices with the desired properties. Draw $n$ lines through $V_1$: the line $\ell_1$ tangent to the circumcircle of $P_1$ at $V_1$, and the lines $\ell_i$ passing through $V_1$ and $V_i$ for $2 \leq i \leq n$. Any line $m$ passing along an edge or a diagonal $\overline{V_i V_j}$ of $P_1$ is parallel to one of these lines, namely $\ell_{i+j-1}$.

Suppose that we have a diagonal or edge of $P_2$, connecting the midpoint $W_1$ of $\overline{V_i V_{i+1}}$ with the midpoint $W_2$ of $\overline{V_j V_{j+1}}$. Then lines $V_i V_{j+1}$ and $V_{i+1} V_j$ are parallel. Because $\overline{W_1 W_2}$ lies halfway between these two lines, it is parallel to each of them. Thus, $\overline{W_1 W_2}$ is parallel to one of the $n$ lines $\ell_i$. Similarly, we find that *any* diagonal or edge in one of the polygons $P_j$ is parallel to one of the $\ell_i$.

In each polygon $P_j$, because there are $n$ vertices in $n - 1$ colors, the Pigeonhole Principle implies that some edge or diagonal has two endpoints of the same color — we call such a segment *monochromatic*. As $j$ varies from 1 to $m$, we find $m \geq n^2 - n + 1 > n(n - 1)$ such segments. By the Pigeonhole Principle, one of the $n$ lines $\ell_i$ is parallel to more than $n - 1$ monochromatic segments, each from a different $P_j$. Applying the Pigeonhole Principle one final time, some two of these $n$ segments — say, $\overline{AB}$ and $\overline{CD}$ — correspond to the same color. It follows that either quadrilateral $ABCD$ or quadrilateral $ABDC$ is an isosceles trapezoid (perhaps a degenerate one) formed by four vertices of the same color. Because the $m$ polygons are inscribed in concentric circles and $\overline{AB}, \overline{CD}$ cannot be diameters, we have that $A, B, C, D$ do not lie on a single line passing through the center of $P_1$.

## 1.17 Russia

### Problem 1

Sasha tries to determine some positive integer $X \leq 100$. He can choose any two positive integers $M$ and $N$ that are less than 100 and ask the question, "What is the greatest common divisor of the numbers $X + M$ and $N$?" Prove that Sasha can determine the value of $X$ after 7 questions.

**Solution.** For $n = 0, 1, \ldots, 6$, let $a_n$ be the unique integer in $[0, 2^n)$ such that $2^n \mid (X - a_n)$. Clearly $a_0 = 0$. For $n \leq 5$, $a_{n+1}$ equals either $a_n$ or $a_n + 2^n$, where the former holds if and only if $\gcd(X + 2^n - a_n, 2^{n+1}) = 2^n$. Because $2^n - a_n < 2^{n+1} < 100$, it follows that if Sasha knows the value of $a_n$, he can determine $a_{n+1}$ with one additional question by setting $(M, N) = (2^n - a_n, 2^{n+1})$. Hence, after six questions, Sasha can determine $a_1, a_2, \ldots, a_6$ and conclude that $X$ equals $a_6$ or $a_6 + 64$. Because $a_6 \not\equiv a_6 + 64 \pmod{3}$, Sasha can determine $X$ if he can discover whether or not $X \equiv a_6 \pmod{3}$ with his final question. Indeed, he can: if he sets $N = 3$ and $M \in \{1, 2, 3\}$ such that $3 \mid (a_6 + M)$, he will obtain the answer "3" if and only if $X \equiv a_6 \pmod{3}$.

### Problem 2

Let $O$ be the center of the circumcircle $\omega$ of an acute-angled triangle $ABC$. The circle $\omega_1$ with center $K$ passes through the points $A, O, C$ and intersects sides $\overline{AB}$ and $\overline{BC}$ at points $M$ and $N$. Let $L$ be the reflection of $K$ across line $MN$. Prove that $\overline{BL} \perp \overline{AC}$.

**Solution.** We use directed angles modulo $\pi$. Let $\alpha, \beta, \gamma$ equal $\angle CAB$, $\angle ABC$, $\angle BCA$ respectively. Because $A, M, C$ lie on a circle with center $K$, we have $\angle CKM = 2\angle CAM = 2\alpha$. Similarly, $\angle NKA = 2\gamma$.

Because $A, B, C$ lie on a circle with center $O$, we have $\angle AOC = 2\angle ABC = 2\beta$. Because $A, O, C$ lie on a circle with center $K$, we have $\angle AKC = 2\angle AOC = 4\beta$.

Therefore,

$$\angle NKM = \angle CKM + \angle NKA - \angle CKA = 2\alpha + 2\gamma + 4\beta = 4\pi + 2\beta,$$

which equals $2\beta$ (modulo $\pi$, of course).

By the definition of $L$, we have

$$\angle MLN = \angle NKM = 2\beta \quad \text{and} \quad LM = LN$$

(because $LM = KM = KN = LN$). Also,

$$\angle AOC = 2\angle ABC = 2\beta \quad \text{and} \quad OC = OA.$$

Thus,

$$\angle MLN = \angle AOC \quad \text{and} \quad LM/LN = 1 = OC/OA.$$

Hence, $\triangle LMN \sim \triangle OCA$ (with opposite orientations). On the other hand, we also have $\triangle MBN \sim \triangle CBA$ (with opposite orientations) because quadrilateral $ACNM$ is cyclic.

These pairs of similar triangles show that the configurations of points $C, B, A, O$ and $M, B, N, L$ are similar (with opposite orientations). Thus, $\angle MBL = \angle CBO$, which (because triangle $ABC$ is acute) equals $\pi/2 - \alpha$. Because $\angle BAC = \alpha$, it follows that $\overline{BL} \perp \overline{AC}$, as desired.

## Problem 3

There are several cities in a state and a set of roads, where each road connects two cities and no two roads connect the same pair of cities. It is known that at least 3 roads go out of every city. Prove that there exists a cyclic path (that is, a path where the last road ends where the first road begins) such that the number of roads in the path is not divisible by 3.

**Solution.** We use the natural translation into graph theory: in a graph where every vertex has degree at least 3, we wish to prove that there exists a cycle whose length is not divisible by 3. Perform the following algorithm: Pick an initial vertex $v_1$. Then, given $v_1, v_2, \ldots, v_i$, if there exists a vertex distinct from these $i$ vertices and adjacent to $v_i$, then let $v_{i+1}$ be that vertex. Because the graph is finite, and all the vertices obtained by this algorithm are distinct, the process must eventually terminate at some vertex $v_n$. We know that every vertex has degree at least 3, and by assumption, every vertex adjacent to $v_n$ occurs somewhere earlier in the sequence; thus, $v_n$ is adjacent to $v_a, v_b, v_{n-1}$ for some $a < b < n - 1$. We then have three cycles:

$$v_a \to v_{a+1} \to v_{a+2} \to \cdots \to v_{n-1} \to v_n \to v_a;$$

$$v_b \to v_{b+1} \to v_{b+2} \to \cdots \to v_{n-1} \to v_n \to v_b;$$

$$v_a \to v_{a+1} \to \cdots \to v_{b-1} \to v_b \to v_n \to v_a.$$

These cycles have lengths $n - a + 1$, $n - b + 1$, and $b - a + 2$, respectively. Because $(n - a + 1) - (n - b + 1) - (b - a + 2) = -2$ is not divisible by 3, we conclude that one of these cycles' lengths is not divisible by 3.

## Problem 4

Let $x_1, x_2, \ldots, x_n$ be real numbers ($n \geq 2$), satisfying the conditions $-1 < x_1 < x_2 < \cdots < x_n < 1$ and

$$x_1^{13} + x_2^{13} + \cdots + x_n^{13} = x_1 + x_2 + \cdots + x_n.$$

Prove that

$$x_1^{13} y_1 + x_2^{13} y_2 + \cdots + x_n^{13} y_n < x_1 y_1 + x_2 y_2 + \cdots + x_n y_n$$

for any real numbers $y_1 < y_2 < \cdots < y_n$.

**Solution.** For $-1 < x < 1$, let $f(x) = x - x^{13}$. We must have $x_1 < 0$; otherwise, $f(x_1) \geq 0$ and $f(x_2), f(x_3), \ldots, f(x_n) > 0$, which is impossible because $f(x_1) + f(x_2) + \cdots + f(x_n) = 0$ by the given equation. Similarly, $x_n > 0$.

Suppose that $2 \leq i \leq n$. If $x_i \leq 0$, then $x_1 < x_2 < \cdots < x_{i-1} < x_i \leq 0$ and

$$\sum_{j=i}^{n} f(x_j) = -\sum_{j=1}^{i-1} f(x_j) > 0.$$

If instead $x_i > 0$, then $0 < x_i < x_{i+1} < \cdots < x_n$, and we again find that $\sum_{j=i}^{n} f(x_i) > 0$.

Using the Abel summation formula (although the below manipulations can also be verified by inspection) and the above result, we have

$$\sum_{i=1}^{n} x_i y_i - \sum_{i=1}^{n} x_i^{13} y_i$$

$$= \sum_{i=1}^{n} y_i f(x_i)$$

$$= y_1 \sum_{i=1}^{n} f(x_i) + \sum_{i=2}^{n} (y_i - y_{i-1})\big(f(x_i) + f(x_{i+1}) + \cdots f(x_n)\big)$$

$$= \sum_{i=2}^{n} (y_i - y_{i-1})\big(f(x_i) + f(x_{i+1}) + \cdots f(x_n)\big) > 0,$$

as desired.

## Problem 5

Let $\overline{AA_1}$ and $\overline{CC_1}$ be altitudes of an acute-angled nonisosceles triangle $ABC$. The bisector of the acute angles between lines $AA_1$ and $CC_1$

intersects sides $\overline{AB}$ and $\overline{BC}$ at $P$ and $Q$, respectively. Let $H$ be the orthocenter of triangle $ABC$ and let $M$ be the midpoint of $\overline{AC}$; and let the bisector of angle $ABC$ intersect $\overline{HM}$ at $R$. Prove that quadrilateral $PBQR$ is cyclic.

**Solution.**  Let the perpendiculars to sides $\overline{AB}$ and $\overline{BC}$ at points $P$ and $Q$, respectively, intersect at $R'$. Let line $R'P$ intersect line $HA$ at $S$, and let line $R'Q$ intersect line $HC$ at $T$. Let the perpendicular from $M$ to line $AB$ intersect line $HA$ at $U$, and let the perpendicular from $M$ to line $BC$ intersect line $HC$ at $V$. Because the sides of triangles $PSH$ and $HTQ$ are parallel to each other, $\angle PSH = \angle HTQ$. Also, because line $PQ$ bisects the acute angles between lines $AA_1$ and $CC_1$, $\angle PHS = \angle QHT$. Hence, $\triangle PHS \sim \triangle QHT$. Also, $\angle HAP = \pi/2 - \angle ABC = \angle QCH$ and $\angle PHA = \angle QHC$, so $\triangle PHA \sim \triangle QHC$. Therefore,

$$\frac{HT}{HS} = \frac{HQ}{HP} = \frac{HC}{HA} = \frac{2MU}{2MV} = \frac{UM}{VM} = \frac{HV}{HU}.$$

Thus, the homothety about $H$ taking line $PS$ to line $MU$ also takes line $QT$ to line $MV$ and hence $R' = PS \cap QT$ to $M = MU \cap MV$. Therefore, $H$, $R'$, and $M$ are collinear.

Again using the fact that $\triangle PHA \sim \triangle QHC$, we have that angles $HPB$ and $HQB$ are congruent because they are supplements to the congruent angles $HPA$ and $HQC$. Thus, $BP = BQ$, right triangles $BR'P$ and $BR'Q$ are congruent, and $\angle PBR' = \angle QBR'$. Hence, $R'$ lies on both line $HM$ and the angle bisector of angle $ABC$, implying that $R' = R$. It follows immediately that quadrilateral $PBQR$ is cyclic, because $\angle BPR = \pi/2 = \angle BQR$.

## Problem 6

Five stones which appear identical all have different weights; Oleg knows the weight of each stone. Given any stone $x$, let $m(x)$ denote its weight. Dmitrii tries to determine the order of the weights of the stones. He is allowed to choose any three stones $A, B, C$ and ask Oleg the question, "Is it true that $m(A) < m(B) < m(C)$?" Oleg then responds "yes" or "no." Can Dmitrii determine the order of the weights with at most nine questions?

**Solution.**  We will show that it is impossible for Dmitrii to determine the order with at most nine questions. Assume, for the sake of contradiction, that Dmitrii has a method which guarantees that he can find the order in nine or fewer questions. Suppose that after Oleg answers Dmitrii's $i$th question, there are exactly $x_i$ orderings of the weights of the stones which fit the responses to the first $i$ questions. We show that it is possible that $x_{i+1} \geq \max\{x_i - 20, \frac{1}{2}x_i\}$ for $i = 1, 2, \ldots, 8$.

Observe that there are $5! = 120$ ways that the weights of the stones can be ordered. Then for any three stones $A, B, C$, exactly $\frac{1}{6}$ of the possible orderings will have $m(A) < m(B) < m(C)$. Thus, if Dmitrii asks whether $m(A) < m(B) < m(C)$ and Oleg answers "no," then Dmitrii can eliminate at most 20 of the 120 possibilities. In this case, $x_{i+1} \geq x_i - 20$ for each $i$.

Of the $x_i$ orderings which fit the responses to the first $i$ questions, some subset $S_1$ of these possibilities will be eliminated if Oleg answers "yes" to the $(i+1)$th question, while the complement of that subset $S_2$ will be eliminated if Oleg answers "no." If $|S_1| \leq x_i/2$ and Oleg answers "yes," then we have $x_{i+1} = x_i - |S_1| \geq x_i/2$. Otherwise, $|S_2| \leq x_i/2$; if Oleg answers "no," we again have $x_{i+1} \geq x_i/2$.

We have that $x_0 = 10$; from above, it is possible that $x_1 \geq 100$, $x_2 \geq 80$, $x_3 \geq 60$, $x_4 \geq 40$, $x_5 \geq 20$, $x_6 \geq 10$, $x_7 \geq 5$, $x_8 \geq 3$, and $x_9 \geq 2$. Therefore, it is impossible for Dmitrii to ensure that he finds out the ordering in nine questions.

## Problem 7

Find all functions $f : \mathbb{R} \to \mathbb{R}$ that satisfy the inequality

$$f(x + y) + f(y + z) + f(z + x) \geq 3f(x + 2y + 3z)$$

for all $x, y, z \in \mathbb{R}$.

**Solution.**  Let $t \in \mathbb{R}$. Plugging in $x = t$, $y = 0$, $z = 0$ gives

$$f(t) + f(0) + f(t) \geq 3f(t), \quad \text{or} \quad f(0) \geq f(t).$$

Plugging in $x = t/2$, $y = t/2$, $z = -t/2$ gives

$$f(t) + f(0) + f(0) \geq 3f(0), \quad \text{or} \quad f(t) \geq f(0).$$

Hence, $f(t) = f(0)$ for all $t$, so $f$ must be constant. Conversely, any constant function $f$ clearly satisfies the given condition.

## Problem 8

Prove that the set of all positive integers can be partitioned into 100 nonempty subsets such that if three positive integers $a, b, c$ satisfy $a + 99b = c$, then at least two of them belong to the same subset.

**Solution.** Let $f(n)$ denote the largest nonnegative integer $k$ such that $2^k \mid n$. We claim that if $a, b, c$ satisfy $a + 99b = c$, then at least two of $f(a), f(b), f(c)$ are equal. If $f(a) = f(b)$, we are done. If $f(a) < f(b)$, then $2^{f(a)}$ divides both $a$ and $99b$, while $2^{(f(a)+1)}$ divides $99b$ but not $a$, so $f(c) = f(a)$. Similarly, if $f(a) > f(b)$, then $f(c) = f(b)$.

Thus, the partition $\mathbb{N}$ into the sets $S_i = \{n \mid f(n) \equiv i \,(\mathrm{mod}\ 100)\}$ for $i = 1, 2, \ldots, 100$ (with indices taken modulo 100) suffices: given three positive integers $a, b, c$ such that $a + 99b = c$, two of $f(a)$, $f(b)$, $f(c)$ equal the same number $k$, implying that two of $a, b, c$ are in $S_k$.

## Problem 9

Let $ABCDE$ be a convex pentagon on the coordinate plane. Each of its vertices are lattice points. The five diagonals of $ABCDE$ form a convex pentagon $A_1 B_1 C_1 D_1 E_1$ inside of $ABCDE$. Prove that this smaller pentagon contains a lattice point on its boundary or within its interior.

**Solution.** Suppose the statement is false. By Pick's Theorem, the area of any lattice polygon is either an integer or a half-integer, implying that there is a counterexample to the statement with minimal area. Label the intersections of the diagonals so that $A_1$ is the intersection of $\overline{BD}$ with $\overline{CE}$ and so forth.

We claim that the triangle $AC_1 D_1$ has no lattice points on its edges or in its interior, except for $A$. By assumption, no lattice point lies on $\overline{C_1 D_1}$. Suppose, for the sake of contradiction, that there is a lattice point $A'$ either on $\overline{AC_1}$ or $\overline{AD_1}$, or inside triangle $AC_1 D_1$. Then pentagon $A'BCDE$ is a convex pentagon with smaller area than pentagon $ABCDE$. Furthermore, the corresponding inner pentagon is contained within pentagon $A_1 B_1 C_1 D_1 E_1$, implying that it contains no lattice points. But then pentagon $A'BCDE$ has smaller area than pentagon $ABCDE$ and satisfies the same requirements, contradicting the minimal definition of pentagon $ABCDE$.

Hence, no lattice points besides $A$ lie on the closed region bounded by triangle $AC_1 D_1$. Similarly, none of the lattice points inside pentagon

$ABCDE$ lie on or inside triangles $BD_1E_1, CE_1A_1, \ldots, EB_1C_1$, implying that any such lattice points lie on one of the triangles $A_1CD, B_1DE$,
$\ldots, E_1BC$.

Let $e_1$, $e_2$, and $e_3$ denote the numbers of lattice points on the interiors
of edges $\overline{BC}$, $\overline{CD}$, and $\overline{DE}$, and let $i_1$, $i_2$, and $i_3$ denote the numbers of
lattice points inside triangles $E_1BC$, $A_1CD$, and $B_1DE$. Then, by Pick's
Theorem,

$$[ACD] = i_2 + \frac{1}{2}(e_2 + 3) - 1,$$

$$[BCD] = (i_1 + i_2) + \frac{1}{2}(e_1 + e_2 + 3) - 1,$$

$$[ECD] = (i_2 + i_3) + \frac{1}{2}(e_2 + e_3 + 3) - 1.$$

Therefore, $[BCD] \geq [ACD]$ and $[ECD] \geq [ACD]$. Now, consider the
distances from $A$, $B$, and $E$ to line $CD$. Because pentagon $ABCDE$ is
convex, the distance from $A$ must exceed the distance from at least one of
$B$ and $E$, say $B$. But then $[ACD] > [BCD]$, a contradiction.

## Problem 10

Let $a_1, a_2, \ldots, a_n$ be a sequence of nonnegative real numbers, not all zero.
For $1 \leq k \leq n$, let

$$m_k = \max_{1 \leq i \leq k} \frac{a_{k-i+1} + a_{k-i+2} + \cdots + a_k}{i}.$$

Prove that for any $\alpha > 0$, the number of integers $k$ which satisfy $m_k > \alpha$
is less than $(a_1 + a_2 + \cdots + a_n)/\alpha$.

**Solution.** We prove this statement by induction on $n$. For $n = 1$, we
have $m_1 = a_1$. If $\alpha > a_1$, then there are no $k$ with $m_k > \alpha$, so the claim
holds trivially. If $\alpha < a_1$, then there is exactly one such $k$, and $1 < a_1/\alpha$.
Thus, the claim holds for $n = 1$.

Now suppose $n > 1$ and assume the claim holds for all smaller $n$. Let
$r$ be the number of integers $k$ for which $m_k > \alpha$.

If $m_n \leq \alpha$, then the sequence $a_1, a_2, \ldots, a_{n-1}$ also contains $r$ values
of $k$ for which $m_k > \alpha$. By the inductive hypothesis,

$$r < \frac{a_1 + a_2 + \cdots + a_{n-1}}{\alpha} \leq \frac{a_1 + a_2 + \cdots + a_n}{\alpha},$$

as desired.

If instead $m_n > \alpha$, then there is some $1 \le i \le n$ such that

$$\frac{a_{n-i+1} + a_{n-i+2} + \cdots + a_n}{i} > \alpha.$$

Fix such an $i$. The sequence $a_1, a_2, \ldots, a_{n-i}$ contains at least $r - i$ values of $k$ for which $m_k > \alpha$, so by the inductive hypothesis

$$r - i < \frac{a_1 + a_2 + \cdots + a_{n-i}}{\alpha}.$$

Then

$$(a_1 + a_2 + \cdots + a_{n-i}) + (a_{n-i+1} + \cdots + a_n) > (r - i)\alpha + i\alpha = r\alpha.$$

Dividing by $\alpha$ yields the desired inequality.

Hence, the statement holds for all integers $n$.

## Problem 11

Let $a_1, a_2, a_3, \ldots$ be a sequence with $a_1 = 1$ satisfying the recursion

$$a_{n+1} = \begin{cases} a_n - 2 & \text{if } a_n - 2 \notin \{a_1, a_2, \ldots, a_n\} \text{ and } a_n - 2 > 0 \\ a_n + 3 & \text{otherwise.} \end{cases}$$

Prove that for every positive integer $k > 1$, we have $a_n = k^2 = a_{n-1} + 3$ for some $n$.

**Solution.** We use induction to prove that for all nonnegative $n$, $a_{5n+1} = 5n + 1$, $a_{5n+2} = 5n + 4$, $a_{5n+3} = 5n + 2$, $a_{5n+4} = 5n + 5$, and $a_{5n+5} = 5n + 3$. The base case $n = 0$ can be verified easily from the recursion. Now assume that the claim is true for all $n < m$ for some positive $m$. We will prove that it is also true for $n = m$. Observe that by the induction hypothesis, $(a_1, a_2, \ldots, a_{5m})$ is a permutation of $1, 2, \ldots, 5m$. Thus, $a_{5m} - 2 = 5m - 4$ is included in this set, and hence $a_{5m+1} = a_{5m} + 3 = 5m + 1$. Similarly, $a_{5m+2} = a_{5m+1} + 3 = 5m + 4$. On the other hand, $a_{5m+2} - 2 = 5m + 2$ is not in $\{a_1, a_2, \ldots, a_{5m+2}\}$, so $a_{5m+3} = 5m + 2$. Continuing in this fashion, we find that $a_{5m+4} = a_{5m+3} + 3 = 5m + 5$ and $a_{5m+5} = a_{5m+4} - 2 = 5m + 3$. This completes the induction.

Each positive integer greater than 1 is included in the sequence $a_2, a_3, \ldots$ exactly once. Also, all squares are congruent to either 0, 1, or 4 (mod 5), which appear in the sequence only at indices congruent to 4, 1, and 2 (mod 5), respectively. From above, for any $n > 1$ in one of these residue classes, we have $a_n = a_{n-1} + 3$, and this completes the proof.

## Problem 12

There are black and white checkers on some squares of a $2n \times 2n$ board, with at most one checker on each square. First, we remove every black checker that is in the same column as any white checker. Next, we remove every white checker that is in the same row as any remaining black checker. Prove that for some color, at most $n^2$ checkers of this color remain.

**Solution.** After we remove the checkers, no column or row can contain both black and white checkers. Thus, we may call any row or column *black* if it contains only black checkers, and *white* otherwise. Let $r_b$ be the number of black rows, $r_w$ be the number of white rows, $c_b$ be the number of black columns, and $c_w$ be the number of white columns.

Because $r_w + r_b + c_w + c_b = 4n$, we have by the AM-GM inequality $r_w c_w r_b c_b \leq n^4$. Thus, either $r_w c_w \leq n^2$ or $r_b c_b \leq n^2$. In the first case, at most $n^2$ white checkers remain, and in the second, at most $n^2$ black checkers remain.

## Problem 13

Let $E$ be a point on the median $\overline{CD}$ of triangle $ABC$. Let $S_1$ be the circle passing through $E$ and tangent to line $AB$ at $A$, intersecting side $\overline{AC}$ again at $M$; let $S_2$ be the circle passing through $E$ and tangent to line $AB$ at $B$, intersecting side $\overline{BC}$ again at $N$. Prove that the circumcircle of triangle $CMN$ is tangent to circles $S_1$ and $S_2$.

**Solution.** Because tangents $\overline{DA}$ to $S_1$ and $\overline{DB}$ to $S_2$ have equal length, $D$ has equal powers with respect to circles $S_1$ and $S_2$ and hence lies on their radical axis. Also, $E$ certainly lies on the radical axis, so this radical axis is line $DE$. Because $C$ also lies on this line, $C$ also has equal powers with respect to the two circles. Therefore (using signed lengths), $CA \cdot CM = CB \cdot CN$, $CA \cdot CM = CB \cdot CN$, implying that $ABNM$ is cyclic. Now, if $\overline{MT}$ and $\overline{MT_1}$ are tangents to the circumcircle of $CMN$ and circle $S_1$ respectively, then (using directed angles modulo $\pi$)

$$\angle AMT = \angle CNM = \angle BAM = \angle AMT_1.$$

Hence, lines $MT$ and $MT_1$ coincide, implying that the circumcircle of triangle $CMN$ is tangent to $S_1$ at $M$. Similarly, this circle is tangent to $S_2$ at $N$.

## Problem 14

One hundred positive integers, with no common divisor greater than one, are arranged in a circle. To any number, we can add the greatest common divisor of its neighboring numbers. Prove that using this operation, we can transform these numbers into a new set of pairwise coprime numbers.

**Solution.** We begin by proving that if $a$ and $b$ are neighboring numbers such that $a$ is coprime to $\alpha$ and $b$ is coprime to $\beta$, then we may apply the operation to $b$ finitely many times so that $b$ is coprime to $\alpha\beta$. Let $\alpha_0$ be the product of the primes which divide $\alpha$ but not $\beta$, so that $\alpha_0$ is coprime to $\beta$. Also, letting $c$ be the other neighbor of $b$, observe that $\gcd(a, c)$ is coprime to $\alpha_0$. Hence, $\gcd(a, c)\beta$ is coprime to $\alpha_0$, and there exists a nonnegative integer $k$ such that

$$k \cdot \gcd(a, c)\beta \equiv -b + 1 \pmod{\alpha_0}.$$

If we apply the operation to $b$ exactly $k$ times, then $a$ and $c$ remain constant while $b$ changes to a number $b'$ congruent to

$$b + k \cdot \gcd(a, c)\beta \equiv 1 \pmod{\alpha_0}.$$

Thus, $b'$ is coprime to $\alpha_0$. Also, because $b' \equiv b \pmod{\beta}$ and $b$ is coprime to $\beta$, we know that $b'$ is coprime to $\beta$ as well. Therefore, $b'$ is coprime to $\alpha_0\beta$ and hence $b'$ is coprime to $\alpha\beta$, as desired.

For $i = 1, 2, \ldots, 99$, let $y_i$ equal the product of primes $p$ such that before applying any operations, we have $p \mid x_{100}$ and $p \nmid x_i$. Using the above result with $(a, b, \alpha, \beta) = (x_1, x_2, y_1, y_2)$, we find that we can apply the operation finitely many times to $x_2$ to make $x_2$ coprime to $y_1 y_2$. Next, using the above result with $(a, b, \alpha, \beta) = (x_2, x_3, y_1 y_2, y_3)$, we can apply the operation finitely many times to $x_3$ to make $x_3$ coprime to $y_1 y_2 y_3$. Continuing similarly, we find we can make $x_{99}$ coprime to $y_1 y_2 \cdots y_{99}$. Finally, using the above result with $(a, b) = (x_{99}, x_{98})$, we can make $x_{98}$ coprime to $y_1 y_2 \cdots y_{99}$. In doing so, we have applied the operation to $x_1, x_2, \ldots, x_{99}$ but not $x_{100}$, implying that $x_{100}$ has remained constant.

Because the numbers in the circle have no common divisor greater than 1, each prime $p$ that divides $x_{100}$ is coprime to one of $x_1, x_2, \ldots, x_{99}$; hence, one of $y_1, y_2, \ldots, y_{99}$ is divisible by $p$. Thus, $y_1 y_2 \cdots y_{99}$ is divisible by each prime $p$ dividing $x_{100}$. Because $x_{98}$ is coprime to $y_1 y_2 \cdots y_{99}$, it must then be coprime to each $p$ dividing $x_{100}$. Hence, $\gcd(x_{98}, x_{100}) = 1$.

Applying the operation to $x_{99}$ multiple times, we can transform $x_{99}$ into a prime which is larger than any other number on the circle. We now have that $\gcd(x_{97}, x_{99}) = 1$. Applying the operation to $x_{98}$ multiple times, we can transform $x_{98}$ into a prime which is larger than any other number on the circle. Continuing similarly, we transform $x_1, x_2, \ldots, x_{100}$ into 100 distinct primes, that is, into a new set of 100 pairwise coprime numbers.

## Problem 15

$M$ is a finite set of real numbers such that given three distinct elements from $M$, we can choose two of them whose sum also belongs to $M$. What is the largest number of elements that $M$ can have?

**Solution.** $M = \{-3, -2, -1, 0, 1, 2, 3\}$ contains 7 elements, and we claim that it has the required properties. Given three distinct elements, if any two of them are of opposite sign, then their sum is in $M$; if any of them is zero, then its sum with either of the other elements is in $M$. The only other triples of elements of $M$ are $\{-3, -2, -1\}$ and $\{1, 2, 3\}$; in the first case, we may choose $-2$ and $-1$, and in the second case, we may choose 1 and 2.

We claim that 7 is the largest number of elements that $M$ can have. Suppose that there are at least three positive elements, and let $b < c$ be the largest two. Given any other positive element $a$, one of the sums $a + b$, $a + c$, and $b + c$ must lie in $M$. However, the latter two cannot because they exceed $c$, the maximal element. Hence, $(a + b) \in M$, and because this sum exceeds $b$, it must equal $c$. Therefore, $a = c - b$, implying that there is at most one other positive element in $M$ besides $b$ and $c$.

Thus, $M$ cannot have more than three positive elements. Likewise, it contains at most three negative elements. $M$ might also contain 0, but this yields an upper bound of seven elements.

## Problem 16

A positive integer $n$ is called *perfect* if the sum of all its positive divisors, excluding $n$ itself, equals $n$. For example, 6 is perfect because $6 = 1+2+3$. Prove that

(a) if a perfect number larger than 6 is divisible by 3, then it is also divisible by 9.

(b) if a perfect number larger than 28 is divisible by 7, then it is also divisible by 49.

**Solution.** For a positive integer $n$, let $\sigma(n)$ denote the sum of all positive divisors of $n$. It is well-known that $\sigma$ is multiplicative, i.e., that if $a, b$ are relatively prime then $\sigma(ab) = \sigma(a) = \sigma(b)$. Note that $n$ is perfect if and only if $\sigma(n) = 2n$, and $\sigma(n) \geq n$ with equality if and only if $n = 1$.

Suppose that $p \in \{3, 7\}$, and that $n$ is a perfect number divisible by $p$ but not by $p^2$. Write $n = 2^a pm$ for integers $a, m$ such that $a \geq 0$ and $\gcd(m, 2p) = 1$. Then

$$2^{a+1} pm = 2n = \sigma(n) = \sigma(2^a)\sigma(p)\sigma(m) = (2^{a+1} - 1)(p+1)\sigma(m).$$

Because $p + 1$ is a power of 2, we have $2^{a+1} \geq p + 1$. Hence,

$$2^{a+1} pm = (2^{a+1} - 1)(p+1)\sigma(m)$$
$$\geq \left( 2^{a+1} \cdot \left( 1 - \frac{1}{2^{a+1}} \right) \right)(p+1)m$$
$$\geq 2^{a+1} \left( 1 - \frac{1}{p+1} \right)(p+1)m = 2^{a+1} pm,$$

where equality holds only if $\sigma(m) = m$ (i.e., $m = 1$) and $2^{a+1} = p + 1$. Equality must indeed hold. If $p = 3$, then we find that $(m, a) = (1, 1)$ and $n = 2^1 \cdot 3 \cdot 1 = 6$. If $p = 7$, then we find that $(m, a) = (1, 2)$ and $n = 2^2 \cdot 7 \cdot 1 = 28$. Therefore, if a perfect number greater than 6 (resp., 28) is divisible by 3 (resp., 7), then it is also divisible by 9 (resp., 49).

## Problem 17

Circles $\omega_1$ and $\omega_2$ are internally tangent at $N$, with $\omega_1$ larger than $\omega_2$. The chords $\overline{BA}$ and $\overline{BC}$ of $\omega_1$ are tangent to $\omega_2$ at $K$ and $M$, respectively. Let $Q$ and $P$ be the midpoints of the arcs $AB$ and $BC$ not containing the point $N$. Let the circumcircles of triangles $BQK$ and $BPM$ intersect at $B$ and $B_1$. Prove that $BPB_1Q$ is a parallelogram.

**Solution.** All angles are directed modulo $\pi$. The homothety about $N$ that sends $\omega_2$ to $\omega_1$ sends line $BC$ to a line $\ell$ tangent to the arc $BC$ not containing $N$. Because $\ell$ is parallel to line $BC$, it must be tangent at the midpoint $P$ of arc $BC$. Thus, the homothety sends $M$ to $P$, implying that $N$, $M$, and $P$ are collinear. Hence, $\angle MPB = \angle NPB$. Similarly, $\angle BQK = \angle BQN$.

Because $BB_1MP$ and $BB_1KQ$ are cyclic, we have

$$\angle BB_1M + \angle KB_1B = \angle BPM + \angle KQB = \angle BPN + \angle NQB = \pi.$$

Thus, $B_1$ is on line $MK$.

It follows that $\angle BQB_1 = \angle BKB_1 = \angle BKM$. Because $\overline{BK}$ is tangent to $\omega_1$, $\angle BKM$ in turn equals $\angle KNM = \angle QNP = \angle QBP$. Hence, $\angle BQB_1 = \angle QBP$, implying that $\overline{BP}$ and $\overline{QB_1}$ are parallel. Similarly, $\overline{BQ}$ and $\overline{PB_1}$ are parallel. This completes the proof.

## Problem 18

There is a finite set of congruent square cards, placed on a rectangular table with their sides parallel to the sides of the table. Each card is colored in one of $k$ colors. For any $k$ cards of different colors, it is possible to pierce some two of them with a single pin. Prove that all the cards of some color can be pierced by $2k - 2$ pins.

**Solution.** We prove the claim by induction on $k$. If $k = 1$, then we are told that given any set containing one card (of the single color), two cards in the set can be pierced with one pin. This is impossible unless there are no cards, in which case all the cards can be pierced by $0 = 2k - 2$ pins.

Assume that the claim is true for $k = n - 1$, and consider a set of cards colored in $n$ colors. Orient the table such that the sides of the cards are horizontal and vertical. Let $X$ be a card whose top edge has minimum distance to the top edge of the table. Because all of the cards are congruent and identically oriented, any card that overlaps with $X$ must overlap either $X$'s lower left corner or $X$'s lower right corner. Pierce pins $P_1$ and $P_2$ through these two corners.

Let $S$ be the set of cards which are not pierced by either of these two pins and which are colored differently than $X$. None of the cards in $S$ intersects $X$, and they are each colored in one of $k - 1$ colors. Given a set $T \subseteq S$ of $n - 1$ cards of different colors, it is possible to pierce some two of the cards in $T \cup \{X\}$ with a single pin. Because no card in $T$ overlaps with $X$, this single pin actually pierces two cards in $T$.

Therefore, we may apply our induction hypothesis to $S$ and pierce all the cards of some color $c$ in $S$ with $2n - 4$ pins. Combined with the pins $P_1$ and $P_2$, we find that all the cards of color $c$ can be pierced with $2n - 2$ pins. This completes the inductive step and the proof.

## Problem 19

Prove the inequality

$$\sin^n(2x) + (\sin^n x - \cos^n x)^2 \le 1.$$

**Solution.** Write $a = \sin x$ and $b = \cos x$; then $a^2 + b^2 = 1$. The left-hand side of the desired inequality equals

$$(2ab)^n + (a^n - b^n)^2 = a^{2n} + b^{2n} + (2^n - 2)a^n b^n,$$

while the right-hand side equals

$$1 = (a^2 + b^2)^n = a^{2n} + b^{2n} + \sum_{j=1}^{n-1} \binom{n}{j} a^{2(n-j)} b^{2j}.$$

It thus suffices to prove that $\sum_{j=1}^{n-1} \binom{n}{j} a^{2(n-j)} b^{2j} \geq (2^n - 2)a^n b^n$. We can do so by viewing $\sum_{j=1}^{n-1} \binom{n}{j} a^{2(n-j)} b^{2j}$ as a sum of $\sum_{j=1}^{n-1} \binom{n}{j} = 2^n - 2$ terms of the form $a^{2(n-j)} b^{2j}$, and then applying the arithmetic mean-geometric mean inequality to these terms.

## Problem 20

The circle $\omega$ is inscribed in the quadrilateral $ABCD$, where lines $AB$ and $CD$ are not parallel and intersect at a point $O$. The circle $\omega_1$ is tangent to side $\overline{BC}$ at $K$ and is tangent to lines $AB$ and $CD$ at points lying outside $ABCD$; the circle $\omega_2$ is tangent to side $\overline{AD}$ at $L$ and is also tangent to lines $AB$ and $CD$ at points lying outside $ABCD$. If $O, K, L$ are collinear, prove that the midpoint of side $\overline{BC}$, the midpoint of side $\overline{AD}$, and the center of $\omega$ are collinear.

**Solution.** Let $I, I_1$, and $I_2$ be the centers of circles $\omega, \omega_1$, and $\omega_2$, respectively. Let $P$ and $Q$ be the points of tangency of $\omega$ with sides $\overline{BC}$ and $\overline{AD}$, respectively, and let $S$ and $R$ be the points diametrically opposite $P$ and $Q$ on $\omega$. Because lines $IP$ and $I_1K$ are both perpendicular to line $BC$, $\overline{I_1K} \parallel \overline{IS}$. Furthermore, the homothety about $O$ that takes $\omega_1$ to $\omega$ takes $I_1$ to $I$. Hence, this homothety takes line $I_1K$ to line $IS$, and, in particular, it takes $K$ to $S$. Thus, $O$ is collinear with $K$ and $S$, implying that $S$ lies on the line through $O, K$, and $L$. Similarly, we find that $R$ lies on the same line.

Now, let $M$ and $N$ be the midpoints of $\overline{BC}$ and $\overline{AD}$, respectively. Because $K$ and $P$ are the points of tangency of the incircle and an excircle of triangle $OBC$ with side $\overline{BC}$, a standard computation using equal tangents shows that $CK = BP$. It follows that $M$ is also the midpoint of $\overline{KP}$, and similarly, $N$ is the midpoint of $\overline{LQ}$. Because $\overline{PS}$ and $\overline{QR}$ are diameters of $\omega$, quadrilateral $PQSR$ is a rectangle. Hence, line $PQ$ is parallel to line $RS$, which (from our previous work) is the same as the line through $K$ and $L$.

Because lines $PQ$ and $KL$ are parallel, $M$ is the midpoint of $\overline{PK}$, and $N$ is the midpoint of $\overline{QL}$, we find that $M$ and $N$ lie on the line parallel to, and halfway between, lines $PQ$ and $RS$. On the other hand, $I$ clearly also lies on this line. Thus, $M, I$, and $N$ are collinear, as desired.

## Problem 21

Every cell of a $100 \times 100$ board is colored in one of 4 colors so that there are exactly 25 cells of each color in every column and in every row. Prove that one can choose two columns and two rows so that the four cells where they intersect are colored in four different colors.

**Solution.** Let the colors used be $A, B, C, D$. We call an unordered pair of squares *sanguine* if the two squares lie in the same row and are of different colors. Every row gives rise to $6 \cdot 25^2$ sanguine pairs (given by $\binom{4}{2}$ possible pairs of colors and 25 squares of each color). Thus, summing over all the rows, there is a total of $100 \cdot 6 \cdot 25^2$ sanguine pairs. On the other hand, each such pair is simply the intersection of one row with a pair of distinct columns. Because there are $\binom{100}{2} = 100 \cdot 99/2$ pairs of columns, some pair of columns contains at least

$$\frac{100 \cdot 6 \cdot 25^2}{100 \cdot 99/2} = \frac{2 \cdot 6 \cdot 25^2}{99} > \frac{12 \cdot 25^2}{4 \cdot 25} = 75$$

sanguine pairs. Thus, some two fixed columns form sanguine pairs in at least 76 rows. We henceforth ignore all other rows and columns; we may as well assume that we have only a $76 \times 2$ board colored in four colors, in which each row contains two different colors and no color occurs more than 25 times in each column.

For each row, consider the pair of colors it contains. If the pairs $\{A, B\}$ and $\{C, D\}$ each occur in some row, we are done; likewise for $\{A, C\}, \{B, D\}$ and $\{A, D\}, \{B, C\}$. Thus, suppose that at most one pair of colors from each of these three sets occurs; we now seek a contradiction. We have only two possibilities, up to a possible relabelling of colors: either $\{A, B\}, \{A, C\}, \{A, D\}$ are the only pairs that occur, or $\{A, B\}, \{A, C\}, \{B, C\}$ are. In the first case, each of the 76 rows contains a square of color $A$, implying that one column has more than 25 squares of color $A$, a contradiction. In the second case, each column can contain only the letters $A, B, C$. There can only be 25 squares of each color $A, B, C$ in each column, for a total of at most 150 squares, but there are 152 squares in total, a contradiction. This completes the proof.

## Problem 22

The nonzero real numbers $a$, $b$ satisfy the equation

$$a^2 b^2 (a^2 b^2 + 4) = 2(a^6 + b^6).$$

Prove that $a$ and $b$ are not both rational.

**Solution.** We rewrite the given equation as

$$a^4 b^4 - 2a^6 - 2b^6 + 4a^2 b^2 = 0,$$

or

$$(a^4 - 2b^2)(b^4 - 2a^2) = 0.$$

It follows that either $a^4 = 2b^2$ or $b^4 = 2a^2$, that is, $\pm\sqrt{2} = a^2/b$ or $\pm\sqrt{2} = b^2/a$. Neither of these equations has solutions in nonzero rational numbers $a$ and $b$.

## Problem 23

Find the smallest odd integer $n$ such that some $n$-gon (not necessarily convex) can be partitioned into parallelograms whose interiors do not overlap.

**Solution.** Take a regular hexagon $ABCDEF$ with center $O$. Let $G$ be the reflection of $O$ across $A$, and let $H$ be the reflection of $C$ across $B$. Then the (concave) hexagon $AGHCDEF$ can be partitioned into the parallelograms $ABHG$, $ABCO$, $CDEO$, and $EFAO$.

We now show that $n = 7$ is minimal. Suppose, for the sake of contradiction, that a partition into parallelograms exists for some triangle or pentagon. Choose any side $\overline{AB}$ of the polygon, and orient the figure so that this side is horizontal and at the bottom of the polygon. At least one parallelogram has a side parallel to $\overline{AB}$, because one such parallelogram overlaps $\overline{AB}$ in a segment. Choose the parallelogram $\mathcal{P}$ with this property whose top edge $\overline{CD}$ (parallel to $\overline{AB}$) is as high up as possible. If this top edge does not overlap with another side of the polygon, then some other parallelogram must lie above $\mathcal{P}$ and overlap with $\overline{CD}$ in a segment, contradicting the extremal definition of $\mathcal{P}$. Hence, some other side of the polygon is parallel to $\overline{AB}$.

In other words, given any side of the polygon, some other side of the polygon is parallel to it. This is clearly impossible if $n = 3$. If some pentagon has this property, then some two of its sides are parallel while the remaining three are pairwise parallel. But some two sides in this triple of parallel sides must be adjacent, implying that they actually *cannot* be parallel, a contradiction. This completes the proof.

## Problem 24

Two pirates divide their loot, consisting of two sacks of coins and one diamond. They decide to use the following rules. On each turn, one pirate chooses a sack and takes $2m$ coins from it, keeping $m$ for himself and putting the rest into the other sack. The pirates alternate taking turns until no more moves are possible. The first pirate unable to make a move loses the diamond, and the other pirate takes it. For what initial numbers of coins can the first pirate guarantee that he will obtain the diamond?

**Solution.**   We claim that if there are $x$ and $y$ coins left in the two sacks, respectively, then the next pirate $P_1$ to move has a winning strategy if and only if $|x - y| > 1$. Otherwise, the other pirate $P_2$ has a winning strategy.

We prove the claim by strong induction on the total number of coins, $x + y$. If $x + y = 0$, then no moves are possible, and the next pirate does not have a winning strategy. Now, assuming that the claim is true when $x + y \le n$ for some nonnegative $n$, we will prove that it is true when $x + y = n + 1$.

First consider the case $|x - y| \le 1$. Assume that a move is possible; otherwise, the next pirate $P_1$ automatically loses, in accordance with our claim. Then $P_1$ must take $2m$ coins from one sack, say the one containing $x$ coins, and put $m$ coins into the sack containing $y$ coins. The new difference between the numbers of coins in the sacks is

$$|(x - 2m) - (y + m)| \ge |-3m| - |y - x| \ge 3 - 1 = 2.$$

At this point, the total number of coins is $x + y - m$, and the difference between the numbers of coins in the two sacks is at least 2, so, by the induction hypothesis, $P_2$ has a winning strategy. This proves the claim when $|x - y| \le 1$.

Now consider the case $|x - y| \ge 2$. Without loss of generality, let $x > y$. $P_1$ would like to find an $m$ such that $2m \le x$, $m \ge 1$, and

$$|(x - 2m) - (y + m)| \le 1.$$

The number $m = \lceil \frac{x - y - 1}{3} \rceil$ satisfies the last two inequalities above, and we claim that $2m \le x$ as well. Indeed, $x - 2m$ is nonnegative because it differs by at most 1 from the positive number $y + m$. After taking $2m$

coins from the sack with $x$ coins, $P_1$ leaves a total of $x + y - m$ coins, and the difference between the numbers of coins in the sacks is at most 1. Hence, by the induction hypothesis, the other pirate $P_2$ has no winning strategy. It follows that $P_1$ has a winning strategy, as desired.

This completes the induction and the proof of the claim. It follows that the first pirate can guarantee that he will obtain the diamond if and only if the numbers of coins initially in the sacks differ by at least 2.

## Problem 25

Do there exist pairwise coprime integers $a, b, c > 1$ such that $2^a + 1$ is divisible by $b$, $2^b + 1$ is divisible by $c$, and $2^c + 1$ is divisible by $a$?

**Solution.** We claim that no such integers exist. Let $\pi(n)$ denote the smallest prime factor of a positive integer $n$.

**Lemma.** *If $p$ is a prime such that $p \mid (2^y + 1)$ and $p < \pi(y)$, then $p = 3$.*

*Proof:* Let $d$ be the order of 2 modulo $p$. Because $2^{p-1} \equiv 1 \pmod{p}$ by Fermat's Little Theorem, we must have $d \leq p - 1$. Hence, each prime factor of $d$ is less than $p < \pi(y)$, implying that $d$ and $y$ are relatively prime.

We are given that $p$ divides $2^y + 1$, so it must divide $2^{2y} - 1 = (2^y + 1)(2^y - 1)$ as well, implying that $d \mid 2y$. Above we showed that $d$ and $y$ are relatively prime, implying that $d \mid 2$. Because $d > 1$, we must have $d = 2$. Hence, $p \mid (2^2 - 1)$, and $p = 3$. ∎

Suppose we did have relatively prime integers $a$, $b$, $c > 1$ such that $b$ divides $2^a + 1$, $c$ divides $2^b + 1$, and $a$ divides $2^c + 1$. Then $a$, $b$, and $c$ are all odd; furthermore, because they are relatively prime, $\pi(a)$, $\pi(b)$, and $\pi(c)$ are distinct. Without loss of generality, assume that $\pi(a) < \pi(b), \pi(c)$. Applying the lemma with $(p, y) = (\pi(a), c)$, we find that $\pi(a) = 3$. Write $a = 3a_0$.

We claim that $3 \nmid a_0$. Otherwise, 9 would divide $2^c + 1$ and hence $2^{2c} - 1$. Because $2^n \equiv 1 \pmod 9$ only if $6 \mid n$, we must have $6 \mid 2c$. Then $3 \mid c$, contradicting the assumption that $a$ and $c$ are coprime. Thus, 3 does not divide $a_0$, $b$, or $c$. Let $q = \pi(a_0 bc)$, so that $\pi(q) = q \leq \min\{\pi(b), \pi(c)\}$.

Suppose, for the sake of contradiction, that $q$ divides $a$. Because $a$ and $c$ are coprime, $q$ cannot divide $c$, implying that $\pi(q) = q$ is not equal to $\pi(c)$. Because $\pi(q) \leq \pi(c)$, we must have $\pi(q) < \pi(c)$. Furthermore, $q$ must divide $2^c + 1$ because it divides a factor of $2^c + 1$ (namely, $a$). Applying our lemma with $(p, y) = (q, c)$, we find that $q = 3$, a contradiction. Hence, our assumption was wrong, and $q$ does not divide $a$. Similarly, $q$ does not divide $c$. It follows that $q$ must divide $b$.

Now, let $e$ be the order of 2 modulo $q$. Then $e \le q - 1$, so $e$ has no prime factors less than $q$. Also, $q$ divides $b$ and hence $2^a + 1$ and $2^{2a} - 1$, implying that $e \mid 2a$. The only prime factors of $2a$ less than $q$ are 2 and 3, so $e \mid 6$. Thus, $q \mid (2^6 - 1)$, and $q = 7$. However, $2^3 \equiv 1 \pmod 7$, so

$$2^a + 1 \equiv (2^3)^{a_0} + 1 \equiv 1^{a_0} + 1 \equiv 2 \pmod 7.$$

Hence, $q$ does not divide $2^a + 1$, a contradiction.


## Problem 26

$2n + 1$ segments are marked on a line. Each of the segments intersects at least $n$ other segments. Prove that one of these segments intersects all the other segments.

**Solution.** Mark the segments on a horizontal number line with the positive direction on the right, so that all endpoints are assigned a coordinate. Of the finitely many segments, some segment $\mathcal{L}$ has a left endpoint whose coordinate is maximal; similarly, some segment $\mathcal{R}$ has a right endpoint whose coordinate is minimal. $\mathcal{L}$ intersects at least $n + 1$ of the given segments (including $\mathcal{L}$ itself), as does $\mathcal{R}$, giving a total count of $2n + 2 > 2n + 1$ segments. Hence, some given segment $\mathcal{S}$ intersects both $\mathcal{L}$ and $\mathcal{R}$.

If any given segment $\mathcal{S}'$ lies entirely to the right of $\mathcal{S}$, then its left endpoint is to the right of $\mathcal{S}$. But the left endpoint of $\mathcal{L}$ cannot lie to the right of $\mathcal{S}$ because $\mathcal{L} \cap \mathcal{S} \ne \emptyset$. Thus, $\mathcal{S}'$ has a left endpoint farther to the right than that of $\mathcal{L}$, a contradiction.

Similarly, no given segment can lie entirely to the left of $\mathcal{S}$. It follows that any given segment intersects $\mathcal{S}$, as required.


## Problem 27

The circles $S_1$ and $S_2$ intersect at points $M$ and $N$. Let $A$ and $D$ be points on $S_1$ and $S_2$, respectively, such that lines $AM$ and $AN$ intersect $S_2$ at $B$ and $C$; lines $DM$ and $DN$ intersect $S_1$ at $E$ and $F$; and $A, E, F$ lie on one side of line $MN$, and $D, B, C$ lie on the other side. Prove that there is a fixed point $O$ such that for any points $A$ and $D$ that satisfy the condition $AB = DE$, $O$ is equidistant from $A, F, C$, and $D$.

**Solution.** Let $P, Q$ be the centers of $S_1, S_2$, respectively. Let $O$ be the reflection of $M$ across the perpendicular bisector of $\overline{PQ}$. We claim that, under the assumption that $AB = DE$, $O$ is equidistant from $A$, $F$, $C$, and $D$.

We use directed angles modulo $\pi$. Note that $\angle MAN = \angle MPQ$ and that $\angle MCN = \angle MQP$, so triangles $MPQ$ and $MAC$ are directly similar. If we let $X, Z$ be the respective midpoints of $\overline{AC}, \overline{PQ}$, then $\angle AXM = \angle PZM$. Now $O$ is the reflection of $M$ across the line through $Z$ perpendicular to $PQ$, so $MZ = OZ$ and $\angle PZM = \angle OMZ = \angle ZOM$. Therefore, as $N, Z, O$ are collinear, $\angle NOM = \angle ZOM = \angle PZM = \angle AXM = \angle NXM$, and the four points $N, X, O, M$ are concyclic. Thus, as $\overline{MN} \perp \overline{MO}$, so $\overline{XN} \perp \overline{XO}$, and, as $X$ is the midpoint of $\overline{AC}$, we have $OA = OC$. For similar reasons, $OD = OF$.

If $\overline{AC}$ coincides with $\overline{FD}$, then the claim is proved. If not, then the lines $AC$ and $DF$ are distinct and not parallel, and it suffices to show that the four points $A$, $F$, $C$, $D$ are concyclic, for then $O$ must be the center of the circle containing them.

Because $\angle NAB = \angle NAM = \angle NEM = \angle NED$, $\angle MBN = \angle MDN$, and by assumption $AB = DE$, triangles $ANB$ and $END$ are congruent. Thus, $AN = NE$ and $DN = NB$. Because $\angle ANE = \angle AME = \angle BMD = \angle BND$, isosceles triangles $ANE$ and $BND$ are similar. Finally, $\angle AEN = \angle NAE = \angle NME = \angle NMD = \angle NBD$. Combining these facts yields

$$\angle FAC = \angle FAN = \angle FEN = \angle AEN - \angle AEF$$

$$= \angle AEN - \angle ANF = \angle NBD - \angle CND$$

$$= \angle NBD - \angle CBD = \angle NBC = \angle NDC = \angle FDC,$$

so that the four points $A$, $F$, $C$, $D$ are indeed concyclic.

## Problem 28

Let the set $M$ consist of the 2000 numbers $10^1 + 1, 10^2 + 1, \ldots, 10^{2000} + 1$. Prove that at least 99% of the elements of $M$ are not prime.

**Solution.** Suppose $n$ is a positive integer and not a power of 2, so that $n$ has an odd factor $s > 1$. Then $10^n + 1$ is composite because $1 < 10^{n/s} + 1 < 10^n + 1$ and

$$10^n + 1 = (10^{n/s} + 1)(10^{s-1} - 10^{s-2} + \cdots - 10 + 1).$$

Among the numbers $1, 2, \ldots, 2000$, there are only 11 powers of 2, namely $1 = 2^0$, $2 = 2^1$, $\ldots$, $1024 = 2^{10}$. Thus, if $n$ is any of the remaining 1989 values, then $10^n + 1$ is not prime. Because $1989 > 1980 = \frac{99}{100} \cdot 2000$, at least 99% of the elements of $M$ are not prime.

## Problem 29

There are 2 counterfeit coins among 5 coins that look identical. Both counterfeit coins have the same weight and the other three real coins have the same weight. The five coins do not all weigh the same, but it is unknown whether the weight of each counterfeit coin is more or less than the weight of each real coin. Find the minimal number of weighings needed to find at least one real coin, and describe how to do so. (The balance scale reports the weight of the objects in the left pan, minus the weight of the objects in the right pan.)

**Solution.** Name the coins $C_1, C_2, \ldots, C_5$. We first show that two weighings suffice. First weigh $C_1$ and $C_2$ against $C_3$ and $C_4$, and record the positive difference $d_1$; then weigh $C_1$ against $C_3$ and record the positive difference $d_2$. At most two of the sets $\{C_1, C_3\}$, $\{C_2, C_4\}$, and $\{C_5\}$ contain a counterfeit coin, so at least one contains only real coins. It is easy to verify that if $C_5$ is real, then (i) $d_1 = 0$ or $d_2 = \frac{1}{2}d_1$. If $C_5$ is counterfeit but $C_2$ and $C_4$ are real, then (ii) $d_1 \neq 0$ and $d_2 = d_1$. And if $C_5$ is counterfeit but $C_1$ and $C_3$ are real, then (iii) $d_1 \neq 0$ and $d_2 = 0$. These exhaust all possible distributions of coins, and the results (i), (ii), (iii) are distinguishable, implying that if we know that (i), (ii), or (iii) holds, then we can identify $C_5$, $C_2$, or $C_3$, respectively, as real. Thus, two weighings suffice.

We now show that one weighing does not suffice. Suppose that we weighed $n_1$ coins on one side ("side A") and $n_2 \leq n_1$ coins on the other side ("side B"). For arbitrary $d > 0$, we show that the weight on side A might be $d$ greater than that on side B, but that it is impossible to identify a real coin given such a difference. First, because it is possible to distinguish one of the five coins, we cannot have $n_1 = n_2 = 0$. Next, pick any coin in our set-up; we claim that we can mark that coin and another such that if there are $m_1$ and $m_2$ marked coins on sides A and B, respectively, then $m_2 \neq m_1$ or $n_2 \neq n_1$. Indeed, if $n_2 = n_1$, then some but not all of the coins are on the scale; hence, we can mark our two coins so that one is on the scale and the other is not. We find $r > 0$ and $\epsilon > -r$ such that

$$(n_1 - n_2)r + (m_1 - m_2)\epsilon = d$$

as follows: if $n_1 = n_2 > 0$, then $m_1 \neq m_2$, so we can we set $\epsilon = \frac{d}{m_1 - m_2}$ and $r > |\epsilon|$; if instead $n_1 > n_2$, then we choose $\epsilon$ sufficiently small and solve for $r$. If the marked coins were counterfeit with weight $r + \epsilon$ and the unmarked coins were real with weight $r$, the balance would report difference $d$.

Repeating this construction but varying the initial marked coin, we find that the scale might report difference $d$ but that each coin is counterfeit in some distribution that reports this difference. Hence, it is not guaranteed that we can determine a real coin after one weighing.

## Problem 30

Let $ABCD$ be a parallelogram with $\angle A = \pi/3$. Let $O$ be the circumcenter of triangle $ABD$. Line $AO$ intersects the external angle bisector of angle $BCD$ at $K$. Find the value $\frac{AO}{OK}$.

**Solution.**   Observe that

$$\angle DCB + \angle BOD = \pi/3 + 2\angle BAD = \pi/3 + 2\pi/3 = \pi,$$

implying that quadrilateral $CBOD$ is cyclic. Let $O'$ be its circumcenter.

Let $P$ be the point diametrically opposite $A$ on the circumcircle of triangle $BAD$, so that $P$ lies on line $AOK$. The reflection across the center of parallelogram $ABCD$ sends $\overline{O'C}$ to $\overline{OA}$, implying that $O'C = OA = OP$ and that lines $O'C$ and $OAP$ are parallel. Hence, $\overline{O'C}$ and $\overline{OP}$ are parallel and congruent, implying that quadrilateral $POO'C$ is a parallelogram; in fact, it must be a rhombus because $O'C = O'O$. Hence, $OP = PC$.

Because $OB = OD$, line $CO$ is the internal angle bisector of angle $BCD$. Hence, line $CO$ is perpendicular to the external angle bisector of angle $BCD$, namely line $CK$. It follows that triangle $OCK$ has a right angle at $C$. $P$ is a point on the hypotenuse $\overline{OK}$ with $OP = PC$, implying that $P$ is actually the midpoint of $\overline{OK}$. Therefore, $OK = 2OP = 2AO$, and the required ratio $AO/OK$ is $\frac{1}{2}$.

## Problem 31

Find the smallest integer $n$ such that an $n \times n$ square can be partitioned into $40 \times 40$ and $49 \times 49$ squares, with both types of squares present in the partition.

**Solution.** We can partition a $2000 \times 2000$ square into $40 \times 40$ and $49 \times 49$ squares: partition one $1960 \times 1960$ corner of the square into $49 \times 49$ squares and then partition the remaining portion into $40 \times 40$ squares.

We now show that $n$ must be at least 2000. Suppose that an $n \times n$ square has been partitioned into $40 \times 40$ and $49 \times 49$ squares, using at least one of each type. Let $\zeta = e^{2\pi i/40}$ and $\xi = e^{2\pi i/49}$. Orient the $n \times n$ square so that two sides are horizontal, and number the rows and columns of unit squares from the top left: $0, 1, 2, \ldots, n-1$. For $0 \le j, k \le n-1$, and write $\zeta^j \xi^k$ in square $(j, k)$. If an $m \times m$ square has its top-left corner at $(x, y)$, then the sum of the numbers written in it is

$$\sum_{j=x}^{x+m-1} \sum_{k=y}^{y+m-1} \zeta^j \xi^k = \zeta^x \xi^y \left( \frac{\zeta^m - 1}{\zeta - 1} \right) \left( \frac{\xi^m - 1}{\xi - 1} \right).$$

The first fraction in parentheses is 0 if $m = 40$, and the second fraction is 0 if $m = 49$. Thus, the sum of the numbers written inside each square in the partition is 0, so the sum of all the numbers must be 0. However, applying the above formula with $(m, x, y) = (n, 0, 0)$, we find that the sum of all the numbers equals 0 only if either $\zeta^n - 1$ or $\xi^n - 1$ equals 0. Thus, $n$ must be either a multiple of 40 or a multiple of 49.

Let $a$ and $b$ be the number of $40 \times 40$ and $49 \times 49$ squares, respectively. The area of the square equals $40^2 \cdot a + 49^2 \cdot b = n^2$. If $40 \mid n$, then $40^2 \mid b$ and hence $b \ge 40^2$. Thus, $n^2 > 49^2 \cdot 40^2 = 1960^2$; because $n$ is a multiple of 40, $n \ge 50 \cdot 40 = 2000$. If instead $49 \mid n$, then $49^2 \mid a$, $a \ge 49^2$, and again $n^2 > 1960^2$. Because $n$ is a multiple of 49, $n \ge 41 \cdot 49 = 2009 > 2000$. In either case, $n \ge 2000$, and 2000 is the minimum possible value of $n$.

## Problem 32

Prove that there exist 10 distinct real numbers $a_1, a_2, \ldots, a_{10}$ such that the equation

$$(x - a_1)(x - a_2) \cdots (x - a_{10}) = (x + a_1)(x + a_2) \cdots (x + a_{10})$$

has exactly 5 different real roots.

**Solution.** Choose distinct real numbers $a_1, a_2, \ldots, a_{10}$ such that $a_6 = a_7 + a_8 = a_9 + a_{10} = 0$ and $a_1, a_2, a_3, a_4, a_5 > 0$. For $6 \le k \le 10$, $x - a_k$ is a factor of both sides of the given equation, so $a_k$ is a real root. Dividing both sides of the given equation by $(x - a_6)(x - a_7) \cdots (x - a_{10})$

and collecting terms onto one side yields

$$(x+a_1)(x+a_2)\cdots(x+a_5)-(x-a_1)(x-a_2)\cdots(x-a_5)=0.$$

For $1\le k\le 5$ and $x>0$, we have

$$|x+a_1|=x+a_1>\max\{x-a_1,a_1-x\}=|x-a_1|.$$

Hence, the left-hand side of the above equation is positive for $x>0$. Because the left-hand side is an even function of $x$, it is also positive for $x<0$. Therefore, the given equation has no other real roots besides the 5 different roots $a_6=0,a_7,a_8,a_9,a_{10}$. This completes the proof.

## Problem 33

We are given a cylindrical region in space, whose altitude is 1 and whose base has radius 1. Find the minimal number of balls of radius 1 needed to cover this region.

**Solution.** First we show that three balls suffice. Let $\mathcal{D}$ be the disc cut off from the cylinder by the plane halfway between the cylinder's two bases, and let $O$ be the center of this disc. Choose three points $A,B,C$ on the boundary of $\mathcal{D}$ such that $\angle AOB=\angle BOC=\angle COA=2\pi/3$, and let $D,E,F$ be the midpoints of $\overline{BC},\overline{CA},\overline{AB}$, respectively. We claim that the unit spheres centered at $D,E,F$ contain the cylinder.

For now, we work in the plane containing $\mathcal{D}$. We show that the disc centered at $D$ with radius $\sqrt{3}/2$ contains all of sector $BOC$; similarly, the discs centered at $E$ and $F$ with radius $\sqrt{3}/2$ contain all of sectors $COA$ and $AOB$, respectively. Introduce coordinates such that $O=(0,0)$ and $D=(1/2,0)$. Given a point $P=(x,y)$ on minor arc $BC$, we have $x\ge 1/2$ and $y^2=1-x^2$. Hence,

$$DP^2=(x-1/2)^2+y^2=(x^2-x+1/4)+(1-x^2)=5/4-x\le 3/4,$$

implying that $P$ lies inside the disc centered at $D$. Hence, this disc contains minor arc $BC$, and it clearly contains $O$. Because the disc is convex, it must contain *all* of sector $BOC$, as claimed.

Therefore, any point in $\mathcal{D}$ is within $\sqrt{3}/2$ of one of $D,E,F$. Now, suppose that we have any point $P$ in the given cylinder; let $Q$ be the foot of the perpendicular from $P$ to $\mathcal{D}$. Without loss of generality, assume that $QD\le\sqrt{3}/2$. Then

$$DP=\sqrt{PQ^2+QD^2}\le\sqrt{(1/2)^2+(\sqrt{3}/2)^2}=1,$$

so that the unit sphere centered at $D$ contains $P$. Therefore, the three described balls cover the cylinder, as desired.

Now we show that two balls are insufficient. Suppose, for the sake of contradiction, that some two unit spheres cover the cylinder. Consider the circular boundary $C_1$ of one base of the cylinder. The plane containing it cuts each ball in a disc, if at all, so each ball can contain at most one continuous arc of the circle. Thus, one ball must contain an arc with measure at least $\pi$, and this ball must then contain two points $X_1, X_2$ on $C_1$ which are diametrically opposite. Hence, the center of the ball must be contained on or within the unit spheres centered at $X_1$ and $X_2$, but the only point with this property is the center of $C_1$.

Hence, one ball $\mathcal{B}_1$ is centered at the center $O_1$ of one base of the cylinder, and similarly another ball $\mathcal{B}_2$ is centered at the center $O_2$ of the other base of the cylinder. But any point on the boundary of $\mathcal{D}$ (the disc halfway between the two bases) is $\sqrt{5}/2$ from $O_1$ and $O_2$ and cannot be in either $\mathcal{B}_1$ or $\mathcal{B}_2$, a contradiction.

## Problem 34

The sequence $a_1, a_2, \ldots, a_{2000}$ of real numbers satisfies the condition

$$a_1^3 + a_2^3 + \cdots + a_n^3 = (a_1 + a_2 + \cdots + a_n)^2$$

for all $n$, $1 \le n \le 2000$. Prove that every element of the sequence is an integer.

**Solution.** We use induction on $n$ to prove that for each $n$, $a_n$ is an integer and

$$a_1 + a_2 + \cdots + a_n = \frac{N_n(N_n + 1)}{2}$$

for a nonnegative integer $N_n$. We extend this sum to the case $n = 0$, for which we use $N_0 = 0$, and we will use this to start our induction.

Assume that our claim holds for $n = k$; we will verify it for $n = k+1$. We are given that

$$\left( \sum_{i=1}^{k+1} a_i \right)^2 = \sum_{i=1}^{k+1} a_i^3,$$

or equivalently

$$\left( \frac{N_k(N_k + 1)}{2} + a_{k+1} \right)^2 = \left( \frac{N_k(N_k + 1)}{2} \right)^2 + a_{k+1}^3.$$

Expanding and factoring, this becomes

$$a_{k+1}(a_{k+1} - (N_k + 1))(a_{k+1} + N_k) = 0.$$

Thus $a_{k+1} \in \{0, N_k + 1, -N_k\}$, so that $a_{k+1}$ is an integer.

Now we determine $N_{k+1}$ to finish the induction. If $a_{k+1} = 0$, then we may set $N_{k+1} = N_k$. If $a_{k+1} = N_k + 1$, then

$$\sum_{i=1}^{k} a_i + a_{k+1} = \frac{N_k(N_k + 1)}{2} + (N_k + 1) = \frac{(N_k + 2)(N_k + 1)}{2},$$

so we may set $N_{k+1} = N_k + 1$. Finally, if $a_{k+1} = -N_k$, then

$$\sum_{i=1}^{k} a_i + a_{k+1} = \frac{N_k(N_k + 1)}{2} - N_k = \frac{N_k(N_k - 1)}{2},$$

so we set $N_{k+1} = N_k - 1$. This completes the inductive step, and the proof.

## Problem 35

The angle bisectors $\overline{AD}$ and $\overline{CE}$ of triangle $ABC$ intersect at $I$. Let $\ell_1$ be the reflection of line $AB$ across line $CE$, and let $\ell_2$ be the reflection of line $BC$ across line $AD$. If lines $\ell_1$ and $\ell_2$ intersect at some point $K$, prove that $\overline{KI} \perp \overline{AC}$.

**Solution.**   If $\angle ABC = \pi/2$, then $\ell_1$ and $\ell_2$ are both perpendicular to $\overline{AC}$, so that the intersection point $K$ could not exist, Hence, $\angle ABC \neq \pi/2$.

Note that $I$ is the incenter of triangle $ABC$, and let $P, Q$, and $R$ be the points of tangency of the incircle of triangle $ABC$ with sides $\overline{BC}$, $\overline{CA}$, and $\overline{AB}$, respectively. Let line $IP$ intersect line $AB$ at $S$, and let line $IR$ intersect line $BC$ at $T$; because angle $ABC$ is not right, these intersection points exist. By equal tangents, $BP = BR$, so right triangles $BPS$ and $BRT$ are congruent.

Note that $Q$ is the reflection of $P$ across line $CE$, as well as the reflection of $R$ across line $AD$. Reflect triangle $BPS$ across line $CE$ to form triangle $UQK_1$, where $U$ is on line $AC$ and $K_1, Q, I$ are collinear and form a line perpendicular to line $AC$. Similarly, reflect triangle $BRT$ across line $AD$ to triangle $VQK_2$, with $V$ on line $AC$ and $K_2, Q, I$ forming a line perpendicular to line $AC$. We have $QK_1 = PS = RT = QK_2$; thus, $K_1 = K_2$. On the other hand, line $UK_1$ is the reflection of line $AB$ across line $CE$, and line $VK_2$ is the reflection of line $BC$ across line $AD$. Therefore, $K = K_1 = K_2$ and $\overline{KI} \perp \overline{AC}$ as desired.

## Problem 36

There are 2000 cities in a country, some pairs of which are connected by a direct airplane flight. For every city $A$ the number of cities connected with $A$ by direct flights equals 1, 2, 4, 8, 16, ..., or 1024. Let $S(A)$ be the number of routes from $A$ to other cities (different from $A$) with at most one intermediate landing. Prove that the sum of $S(A)$ over all 2000 cities $A$ cannot be equal to 10000.

**Solution.** Let $\mathcal{T}$ be the set of cities. For each city $A$, let $d(A)$ be the number of cities *adjacent* to $A$ (i.e., connected to $A$ by a direct flight). We claim that

$$\sum_{A \in \mathcal{T}} d(A)^2 = \sum_{A \in \mathcal{T}} S(A),$$

by showing that each side equals the total number of *short routes*, routes with at most two (distinct) legs.

For each city $A$, exactly $S(A)$ short routes start from $A$. Hence, the right-hand side above, the sum of $S(A)$ over all cities $A$, gives the total number of short routes.

Now we analyze the sum on the left-hand side. Each short flight has either one or two legs. Given any city $A$, exactly $d(A)$ one-leg routes begin at $A$; and exactly $d(A)(d(A)-1)$ two-leg routes make an intermediate stop at $A$. This gives a total of $d(A)^2$ short routes. Hence, the left-hand side above, the sum of $d(A)^2$ over all cities $A$, gives the total number of short routes.

Thus, we see that the two sums are indeed equal. To finish the problem, observe that for any city $A$, we have $d(A) \in \{1, 2, 4, 8, \ldots, 1024\}$ and hence $d(A)^2 \equiv 1 \pmod 3$. Hence,

$$\sum_{A \in \mathcal{T}} S(A) = \sum_{A \in \mathcal{T}} d(A)^2 \equiv 2000 \cdot 1 \equiv 2 \pmod 3.$$

Because $10000 \not\equiv 2 \pmod 3$, it follows that $\sum_{A \in \mathcal{T}} S(A) \neq 10000$.

## Problem 37

A heap of balls consists of one thousand 10-gram balls and one thousand 9.9-gram balls. We wish to pick out two heaps of balls with equal numbers of balls in them but different total weights. What is the minimal number of weighings needed to do this? (The balance scale reports the weight of the objects in the left pan, minus the weight of the objects in the right pan.)

**Solution.** Two heaps of balls with equal numbers of balls in them have the same total weights if and only if they contain the same number of 10-gram balls. Zero weighings cannot suffice; if we pick out two heaps of $n \leq 1000$ balls each, it is possible that each heap contains $\lfloor n/2 \rfloor$ 10-gram balls and $\lceil n/2 \rceil$ 9.9-gram balls, so that both heaps have the same total weight.

However, one weighing does suffice. Split the two thousand balls into three heaps $H_1$, $H_2$, $H_3$ of 667, 667, and 666 balls, respectively. Weigh heaps $H_1$ and $H_2$ against each other. If the total weights are not equal, we are done. Otherwise, discard one ball from $H_1$ to form a new heap $H_1'$ of 666 balls. We claim that $H_1'$ and $H_3$ have different weights. If not, then they have the same number of 10-gram balls, say, $n$. Then $H_1$ and $H_2$ either each had $n$ 10-gram balls or each had $n + 1$ 10-gram balls. This would imply that 1000 equals $3n$ or $3n + 2$, which is impossible.

## Problem 38

Let $D$ be a point on side $\overline{AB}$ of triangle $ABC$. The circumcircle of triangle $BCD$ intersects line $AC$ at $C$ and $M$, and the circumcircle of triangle $ACD$ intersects line $BC$ at $C$ and $N$. Let $O$ be the center of the circumcircle of triangle $CMN$. Prove that $\overline{OD} \perp \overline{AB}$.

**Solution.** All angles are directed modulo $\pi$. We have $\angle NDB = \pi - \angle ADN = \angle NCA = \angle BCA$ and similarly $\angle ADM = \angle BCA$. Thus, $\angle MDN = \pi - 2\angle BCA$. Because $O$ is the circumcenter of triangle $CMN$, $\angle NOM = 2\angle NCM = 2\angle BCA$. Thus, quadrilateral $DMON$ is cyclic. Because angles $MDO$ and $ODN$ cut off equal chords $\overline{OM}$ and $\overline{ON}$ in this circle, they are congruent. Hence,

$$\angle ADO = \angle ADM + \angle MDO = \angle NDB + \angle ODN = \angle ODB,$$

implying that $\angle ADO = \angle ODB = \pi/2$, as desired.

## Problem 39

Every cell of a $200 \times 200$ table is colored black or white. The difference between the numbers of black and white cells is 404. Prove that some $2 \times 2$ square contains an odd number of white cells.

**Solution.** Label the cells in the table with ordered pairs $(r, c)$, where $1 \leq r \leq 200$ is the row number (numbered top to bottom) and $1 \leq c \leq 200$ is the column number (numbered left to right). We begin by stating an obvious subsidiary result. Assume that the $2 \times 2$ square with upper-left corner $(i, j)$ has an even number of white squares. Then $(i, j)$ and $(i+1, j)$ are the same color if and only if $(i, j + 1)$ and $(i + 1, j + 1)$ are the same color.

Now suppose, for the sake of contradiction, that every $2 \times 2$ square contains an even number of white cells, so that each $2 \times 2$ square also contains an even number of black cells. Let $b$ and $w$ be the numbers of black and white cells in the first row of the table. Because $b + w = 200$, $|b - w|$ is an even number, say, equal to $2m \le 200$.

Now consider the next row of the table. If the leftmost cell in the row is the same (resp., different) color than the cell immediately above it, then by applying the subsidiary result we find that *every* cell in the second row is the same (resp., different) color than the cell immediately above it. We can repeat this reasoning for each subsequent row of the table to see that the coloring of each row is either identical to the first row, or directly opposite it. Let the number of rows colored identically to the first row be $x$, and the number of rows that are colored opposite it be $y$. Because $x + y = 200$, $|x - y|$ is even, say, equal to $2n \le 200$. Then the difference between the numbers of black and white cells is $|b - w| \cdot |x - y| = 4mn$. We are given that this difference is 404. Hence, $mn = 101$. But, as 101 is prime, this implies that either $m$ or $n$ is 101, which is impossible because $2m$ and $2n$ are at most 200. Thus, we have a contradiction. This completes the proof.

## Problem 40

Is there a function $f : \mathbb{R} \to \mathbb{R}$ such that

$$|f(x + y) + \sin x + \sin y| < 2$$

for all $x, y \in \mathbb{R}$?

**Solution.** No such function exists. Suppose otherwise, for the sake of contradiction. Setting $x = \pi/2$, $y = \pi/2$ gives $|f(\pi) + 2| < 2$, while setting $x = -\pi/2$, $y = 3\pi/2$ gives $|f(\pi) - 2| < 2$. Hence,

$$4 \le |f(\pi) + 2| + |-f(\pi) + 2|$$
$$< 2 + 2,$$

a contradiction. Thus, no such function exists.

## Problem 41

For any odd integer $a_0 > 5$, consider the sequence $a_0, a_1, a_2, \ldots$, where

$$a_{n+1} = \begin{cases} a_n^2 - 5 & \text{if } a_n \text{ is odd} \\ \frac{a_n}{2} & \text{if } a_n \text{ is even} \end{cases}$$

for all $n \geq 0$. Prove that this sequence is not bounded.

**Solution.** We use induction on $n$ to show that $a_{3n}$ is odd and that $a_{3n} > a_{3n-3} > \cdots > a_0 > 5$ for all $n \geq 1$. The base case $n = 0$ is true by assumption. Now assuming that the claim is true for all $n \leq k$, we prove that it is true for $k+1$. Because $a_{3k}$ is odd, $a_{3k}^2 \equiv 1 \pmod 8$ and hence $a_{3k+1} = a_{3k}^2 - 5 \equiv 4 \pmod 8$. Thus, $a_{3k+1}$ is divisible by 4 but not 8, implying that $a_{3(k+1)} = a_{3k+1}/4$ is indeed odd. Furthermore, $a_{3k} > 5$ by the induction hypothesis, implying that $a_{3k}^2 > 5a_{3k} > 4a_{3k} + 5$. Thus, $a_{3(k+1)} = \frac{1}{4}\left(a_{3k}^2 - 5\right) > a_{3k}$. This completes the induction and shows that the sequence is unbounded.

## Problem 42

Let $ABCD$ be a convex quadrilateral. Let $\ell_a$, $\ell_b$, $\ell_c$, and $\ell_d$ be the external angle bisectors of angles $DAB$, $ABC$, $BCD$, and $CDA$, respectively. The pairs of lines $\ell_a$ and $\ell_b$, $\ell_b$ and $\ell_c$, $\ell_c$ and $\ell_d$, $\ell_d$ and $\ell_a$ intersect at points $K, L, M, N$, respectively. Suppose that the perpendiculars to line $AB$ passing through $K$, to line $BC$ passing through $L$, and to line $CD$ passing through $M$ are concurrent. Prove that $ABCD$ can be inscribed in a circle.

**Solution.** We begin by proving the following lemma.

**Lemma.** *Let $W$, $X$, $Y$, and $Z$ be points in the plane. Suppose $WX^2 + YZ^2 = WZ^2 + XY^2$. Then lines $WY$ and $XZ$ are perpendicular.*

*Proof:* Choose an arbitrary origin in the plane, and let $\mathbf{w}, \mathbf{x}, \mathbf{y}, \mathbf{z}$ denote the vectors from the origin to the points $W, X, Y, Z$, respectively. Using the property that the dot product of a vector with itself is equal to the square of its norm, we can translate the given condition into

$$(\mathbf{w} - \mathbf{x}) \cdot (\mathbf{w} - \mathbf{x}) + (\mathbf{y} - \mathbf{z}) \cdot (\mathbf{y} - \mathbf{z}) = (\mathbf{w} - \mathbf{z}) \cdot (\mathbf{w} - \mathbf{z}) + (\mathbf{x} - \mathbf{y}) \cdot (\mathbf{x} - \mathbf{y}).$$

Expanding and simplifying, we obtain the relation

$$(\mathbf{w} - \mathbf{y}) \cdot (\mathbf{x} - \mathbf{z}) = 0,$$

which proves that $\overline{WY} \perp \overline{XZ}$.                                    ∎

Now we consider the problem at hand. Let the angles of quadrilateral $ABCD$ be $\alpha$, $\beta$, $\gamma$, and $\delta$, and let the three concurrent perpendiculars to lines $AB$, $BC$, and $CD$ meet at $O$. Then

$$\angle AKO = \frac{\alpha}{2}, \ \angle BKO = \frac{\beta}{2}, \ \angle CMO = \frac{\gamma}{2}, \ \text{ and } \ \angle DMO = \frac{\delta}{2}.$$

Because $\alpha + \beta + \gamma + \delta = 2\pi$, it follows that $\angle NKL + \angle LMN = \pi$. Therefore, quadrilateral $KLMN$ is cyclic. Next, observe that $\angle LKO = \frac{\beta}{2} = \angle KLO$, so $OK = OL$. Similarly, $OL = OM$. Thus, $O$ is the circumcenter of $KLMN$. Note that, because $\angle ONA = \angle NKO = \frac{\alpha}{2}$, we have $\overline{ON} \perp \overline{AD}$. Because

$$\angle KNL = \frac{1}{2}\angle KOL = \frac{\pi}{2} - \frac{\beta}{2} = \angle KBA,$$

quadrilateral $ABLN$ is cyclic. By the Power of a Point Theorem, it follows that

$$KL \cdot KB = KN \cdot KA.$$

Repeating this argument for the other three vertices of $KLMN$, we obtain the equalities

$$KL \cdot BL = LM \cdot LC,$$

$$MN \cdot MD = LM \cdot CM,$$

$$MN \cdot DN = KN \cdot AN.$$

Summing these four equations yields

$$KL^2 + MN^2 = KN^2 + LM^2.$$

Applying our lemma, we find that $\overline{KM} \perp \overline{LN}$. It follows that angles $KOL$ and $MON$ are supplementary. Because $\overline{OK} \perp \overline{AB}$ and $\overline{OL} \perp \overline{BC}$, we also know that angles $KOL$ and $ABC$ are supplementary.

We similarly find that angles $MON$ and $CDA$ are supplementary. Therefore, angles $ABC$ and $CDA$ are also supplementary, and quadrilateral $ABCD$ is cyclic.

## Problem 43

There are 2000 cities in a country, and each pair of cities is connected by either no roads or exactly one road. A *cyclic path* is a nonempty, connected path of roads such that each city is at the end of either 0 or 2 roads in the path. For every city, there are at most $N$ cyclic paths which both pass through this city and contain an odd number of roads. Prove that the country can be separated into $2N + 2$ republics such that any two cities from the same republic are not connected by a road.

**Solution.** Equivalently, we show that given a graph in which each vertex belongs to at most $N$ non-self-intersecting odd-length cycles, the vertices can be assigned labels in $\{1, 2, \ldots, 2N+2\}$ so that no two adjacent vertices have the same label. (We say a cycle or path has odd or even length if the number of edges it contains is odd or even, respectively.) It suffices to prove the statement when the graph is connected; otherwise, we can assign labels to each component separately.

Suppose the graph has $n$ vertices. Fix an initial vertex $v_1$, and arrange all the vertices as $v_1, v_2, \ldots, v_n$ in nondecreasing order of distance from $v_1$. Now we successively label vertices $v_1, v_2, \ldots, v_n$ with positive integers as follows: for each $v_i$, assign the smallest positive integer that has not already been assigned to a neighbor of $v_i$. This will ensure that no two adjacent vertices have the same label, and now we simply need to show that no vertex ever receives a label greater than $2N + 2$.

Write $v_i \prec v_j$ or $v_j \succ v_i$ if $v_i, v_j$ are adjacent and $i < j$. Notice that if any $w \neq v_1$ receives the label 1, then there exists $v \prec w$ with a label greater than 1. (Proof: let $v$ be the vertex preceding $w$ on a minimal path from $v_1$ to $w$; because our vertices are arranged in order of distance from $v_1$, $v \prec w$. Also, because $w$ has the label 1, its neighbor $v$ must have a different label.) And, if $w$ receives a label greater than 1, then there exists $v \prec w$ which is labelled 1 (because otherwise our labelling scheme would have assigned the label 1 to $w$). It follows that, for any $w$, there exists a sequence of vertices $w = w_0 \succ w_1 \succ w_2 \succ \cdots \succ w_r = v_1$ whose labels alternate between 1 and numbers greater than 1. We call such a path an *alternating path*.

Now suppose that some $w$ receives a label greater than $2N + 2$. By construction, there exist $u_1, u_2, \ldots, u_{2N+2} \prec w$ such that each $u_m$ is labelled $m$. Let $m$ be any even element of $\{1, 2, \ldots, 2N + 2\}$. Because $u_m$ received a label greater than $m - 1$, there exists $u'_m \prec u_m$ which is labelled $m - 1$. We construct an alternating path $\mathcal{P}_1$ from $u_{m-1}$ to $v_1$ and another alternating path $\mathcal{P}_2$ from $u'_m$ to $v_1$. These paths have even length if $m = 2$, and they have odd length otherwise.

We would like to connect these paths, travelling along $\mathcal{P}_1$ and then backward along $\mathcal{P}_2$, to form an even-length path from $u_{m-1}$ to $u'_m$, but $\mathcal{P}_1$ and $\mathcal{P}_2$ may contain some of the same vertices. To correct this problem, let $v_{\max}$ be the largest vertex occurring in both $\mathcal{P}_1$ and $\mathcal{P}_2$, where by "largest," we mean according to our total ordering $\prec$ of the vertices. (We may have $v_{\max} = u'_m$, if $u_{m-1} = u'_m$.) Then we abridge our alternating paths so that they run from $u_{m-1}$ to $v_{\max}$ and $u'_m$ to $v_{\max}$. If $v_{\max}$ has label 1, this entails trimming an even number of vertices off of both alternating paths; otherwise, we have trimmed an odd number of vertices off of both paths. Either way, we can now append the two to obtain a path $u_{m-1} \succ \cdots \succ v_{\max} \prec \cdots \prec u'_m$, with even length. Finally, adding three edges to this path gives a cycle $w \succ u_{m-1} \succ \cdots \succ v_{\max} \prec \cdots \prec u'_m \prec u_m \prec w$ of odd length. The maximal definition of $v_{\max}$ ensures that this cycle is not self-intersecting.

For each even value of $m = 2, 4, \ldots, 2N + 2$, we obtain such a cycle of odd length passing through $w$. In the cycle corresponding to some $m$, the neighbors of $w$ are $u_{m-1}$ and $u_m$, implying that the $N + 1$ cycles we find are distinct. This is a contradiction, implying that we were mistaken in assuming that some vertex receives a label greater than $2N + 2$. This completes the proof.

## Problem 44

Prove the inequality

$$\frac{1}{\sqrt{1+x^2}} + \frac{1}{\sqrt{1+y^2}} \le \frac{2}{\sqrt{1+xy}}$$

for $0 \le x, y \le 1$.

**Solution.**   If $x = 0$, the problem reduces to the inequality

$$1 + \frac{1}{\sqrt{1+y^2}} \le 2,$$

which is obviously true. Similarly, the inequality is clearly true if $y = 0$.

Now assume $x$ and $y$ are positive, and choose real numbers $u$ and $v$ such that $x = e^{-u}$ and $y = e^{-v}$. Because $x$ and $y$ are both at most 1, $u$ and $v$ must both be nonnegative. Substituting for $x$ and $y$ in terms of $u$ and $v$, we see that it suffices to prove that

$$\frac{1}{\sqrt{1+e^{-2u}}} + \frac{1}{\sqrt{1+e^{-2v}}} \le \frac{2}{\sqrt{1+e^{-(u+v)}}}$$

for nonnegative $u$, $v$. Consider the function

$$f(t) = \frac{1}{\sqrt{1+e^{-2t}}}.$$

Then we are to prove that $f(u) + f(v) \le 2f((u+v)/2)$, for $u, v \ge 0$. To do this, all we need to show is that $f$ is concave on the interval $[0, +\infty)$.

A simple calculation of the derivatives of $f$ does the trick:

$$f'(t) = (1 + e^{-2t})^{-3/2} e^{-2t},$$

and

$$f''(t) = 3(1 + e^{-2t})^{-5/2} e^{-4t} - 2(1 + e^{-2t})^{-3/2} e^{-2t}$$
$$= \frac{3 - 2e^{2t}(1 + e^{-2t})}{(1 + e^{-2t})^{5/2} e^{4t}}$$
$$= \frac{1 - 2e^{2t}}{(1 + e^{-2t})^{5/2} e^{4t}}.$$

The denominator of the last expression is certainly positive, while the numerator is negative because $e^{2t} \geq 1$ for $t \geq 0$. Thus, $f$ is indeed concave for $t \geq 0$. This completes the proof.

## Problem 45

The incircle of triangle $ABC$ touches side $\overline{AC}$ at $K$. A second circle $S$ with the same center intersects each side of the triangle twice. Let $E$ and $F$ be the intersection points on $\overline{AB}$ and $\overline{BC}$ closer to $B$; let $B_1$ and $B_2$ be the intersection points on $\overline{AC}$ with $B_1$ closer to $A$. Finally, let $P$ be the intersection point of $\overline{B_2 E}$ and $\overline{B_1 F}$. Prove that points $B, K, P$ are collinear.

**First Solution.** Let $I$ be the incenter of triangle $ABC$, and let $J$ and $L$ be the points of tangency of the incircle with sides $\overline{BC}$ and $\overline{AB}$. Observe that triangles $ILE$, $IJF$, $IKB_1$, and $IKB_2$ are mutually congruent. Hence $LE = JF = KB_1 = KB_2$. By equal tangents, $AK = AL$ and $CK = CJ$. Adding these to the previous equalities gives $AB_2 = AE$ and $CB_1 = CF$; hence, triangles $AB_2 E$ and $CB_1 F$ are isosceles.

Let $D_1$ and $D_2$ be points on line $AC$ such that $\overline{BD_1} \parallel \overline{B_1 F}$ and $\overline{BD_2} \parallel \overline{B_2 E}$. Then triangles $AD_2 B$ and $CD_1 B$ are also isosceles. Also, triangles $B_1 P B_2$ and $D_1 B D_2$ are homothetic because their sides are parallel.

We now show that the center of the homothety is $K$. We already saw above that $KB_1 = KB_2$. Hence, it suffices to show that $KD_1 = KD_2$. Let $a = BC$, $b = CA$, $c = AB$, and let $s$ denote the semiperimeter $\frac{1}{2}(a + b + c)$ of triangle $ABC$. By a standard calculation using equal tangents, $CK = s - c$ and $AK = s - a$. We also know that $CD_1 = CB = a$ and $AD_2 = AB = c$ because triangles $CD_1 B$ and $AD_2 B$ are isosceles. Therefore,

$$KD_1 = CD_1 - CK = a - (s - c) = c - (s - a) = AD_2 - AK = KD_2,$$

as desired. It follows that a homothety about $K$ takes triangle $B_1PB_2$ to triangle $D_1BD_2$, and hence $K, P$, and $B$ are collinear.

**Second Solution.** Let $X$ be the intersection of $S$ and $\overline{AB}$ closer to $A$, let $Y$ be the intersection of $S$ and $\overline{BC}$ closer to $C$, and let $Z$ be the intersection of lines $XB_1$ and $YB_2$. Applying Pascal's Theorem to the hexagon $YFB_1XEB_2$ shows that $B, P$, and $Z$ are collinear. As in the first solution, $\triangle AB_2E$ is isosceles, so $\angle ZB_1B_2 = \angle XEB_2 = \angle PB_2B_1$, and lines $B_1Z$ and $B_2P$ are parallel. Similarly, lines $B_2Z$ and $B_1P$ are parallel, so $B_1ZB_2P$ is a parallelogram. As in the first solution, $K$ is the midpoint of $\overline{B_1B_2}$, so $K$ lies on line $ZP$. Because $B, P, Z$ are collinear, so are $B, P$, and $K$.

## Problem 46

Each of the numbers $1, 2, \ldots, N$ is colored black or white. We are allowed to simultaneously change the colors of any three numbers in arithmetic progression. For which numbers $N$ can we always make all the numbers white?

**Solution.** Clearly we cannot always make all the numbers white if $N = 1$. Suppose that $2 \le N \le 7$, and suppose that only the number 2 is colored black. Call a number from $\{1, \ldots, N\}$ *heavy* if it is not congruent to 1 modulo 3. Let $X$ be the number of heavy numbers which are black, where $X$ changes as we change the colors. Suppose we change the colors of the numbers in $\{a - d, a, a + d\}$, where $1 \le a - d < a < a + d \le N$. If $d$ is not divisible by 3, then $a - d$, $a$, $a + d$ are all distinct modulo 3, so exactly two of them are heavy. If instead $d$ is divisible by 3, then $a - d$, $a$, $a + d$ must equal 1, 4, 7, none of which are heavy. In either case, changing the colors of these three numbers changes the color of an even number of heavy numbers. Hence, $X$ is always an odd number, and we cannot make all the numbers white.

Now we show that for $N \ge 8$, we can always make all the numbers white. To do this, it suffices to show that we can invert the color of any single number $n$; we prove this by strong induction. If $n \in \{1, 2\}$, then we can invert the color of $n$ by changing the colors of the numbers in $\{n, n + 3, n + 6\}$, $\{n + 3, n + 4, n + 5\}$, and $\{n + 4, n + 5, n + 6\}$.

Now assuming that we can invert the color of $n - 2$ and $n - 1$ (where $3 \leq n \leq N$), we can invert the color of $n$ by first inverting the colors of $n - 2$ and $n - 1$, and then changing the colors of the numbers in $\{n - 2, n - 1, n\}$.

Hence, we can always make all the numbers white if and only if $N \geq 8$.

## 1.18   Taiwan

### Problem 1

In an acute triangle $ABC$, $AC > BC$ and $M$ is the midpoint of $\overline{AB}$. Let altitudes $\overline{AP}$ and $\overline{BQ}$ meet at $H$, and let lines $AB$ and $PQ$ meet at $R$. Prove that the two lines $RH$ and $CM$ are perpendicular.

**Solution.**   Let $S$ be the foot of the altitude from $C$ to line $AB$, and let $X$ be the foot of the perpendicular from $H$ to line $CM$. Because $\angle HPC = \angle HQC = \angle HXC = \pi/2$, the points $H$, $P$, $Q$, $X$, and $C$ are concyclic. Similarly, because $\angle HXM = \angle HSM = \pi/2$, the points $H$, $X$, $S$, and $M$ are concyclic. Furthermore, $P$, $Q$, $S$, and $M$ are concyclic because they all lie on the nine-point circle of triangle $ABC$. By the Radical Axis Theorem, the pairwise radical axes of these three circles — namely, lines $AB$, $PQ$ and $HX$ — must concur. Because $R$ is the intersection of lines $AB$ and $PQ$, it must be collinear with $H$ and $X$. Therefore, line $RH$ (which coincides with line $RX$) is perpendicular to line $CM$.

### Problem 2

Let $\phi(k)$ denote the number of positive integers $n$ satisfying $\gcd(n, k) = 1$ and $n \leq k$. Suppose that $\phi(5^m - 1) = 5^n - 1$ for some positive integers $m$, $n$. Prove that $\gcd(m, n) > 1$.

**Solution.**   In this solution, we use the following well-known facts about $\phi$: it is a multiplicative function (i.e., $\phi(ab) = \phi(a)\phi(b)$ for relatively prime positive integers $a$ and $b$), and $\phi(p^\alpha) = p^\alpha - p^{\alpha-1}$ if $p$ is prime and $\alpha$ is a positive integer.

Suppose, for the sake of contradiction, that $\gcd(m, n) = 1$.

We first show that $m$ is odd. Suppose otherwise for sake of contradiction. Note that $5^x$ is congruent to 1 (resp., 5) modulo 8 if $x$ is even (resp., odd). If $m$ is even, then $5^m - 1$ is divisible by but not equal to 8. Because $5^m - 1$ does not *equal* 8, $5^n - 1 = \phi(5^m - 1)$ is either divisible by $\phi(16) = 8$ or $\phi(8)\phi(p^\alpha) = 4\phi(p^\alpha) \equiv 0 \pmod{8}$ for some odd prime power $p^\alpha > 1$. Therefore, $n$ must be even, contrary to the assumption that $\gcd(m, n) = 1$.

Next suppose, for the sake of contradiction, that $p^2 \mid (5^m - 1)$ for some odd prime $p$. Clearly, $p \nmid 5$, so 5 has some order modulo $p$; let $d$ be this order. Because $p \mid (5^m - 1)$, we have $d \mid m$. Also, $p$ divides $\phi(p^2)$ and hence $p \mid \phi(5^m - 1) = 5^n - 1$, implying that $d \mid n$ as well. But $d > 1$ because $5 \not\equiv 1 \,(\mathrm{mod}\ p)$, implying that $\gcd(m, n) \neq 1$, a contradiction.

Therefore, $5^m - 1 = 4\prod_{p \in S} p$ for some set $S$ of odd primes. Let $p$ be any element of $S$. Because

$$1 = \left(\frac{1}{p}\right) = \left(\frac{5^m}{p}\right) = \left(\frac{5}{p}\right)^m$$

and $m$ is odd, $\left(\frac{5}{p}\right) = 1$. Also, by the law of quadratic reciprocity,

$$\left(\frac{5}{p}\right)\left(\frac{p}{5}\right) = (-1)^{\frac{(5-1)(p-1)}{4}} = 1,$$

implying that $\left(\frac{p}{5}\right) = 1$. Hence, $p$ is congruent to 1 or 4 modulo 5. However, we cannot have $p \equiv 1 \,(\mathrm{mod}\ 5)$, because then 5 would divide $p - 1 = \phi(p)$ and hence $5 \mid \phi(5^m - 1) = 5^n - 1$, which is impossible. Therefore, $p \equiv 4 \,(\mathrm{mod}\ 5)$.

It follows that

$$-1 \equiv 5^m - 1 = 4\prod_{p \in S} p \equiv 4 \cdot 4^{|S|} \,(\mathrm{mod}\ 5)$$

and

$$-1 \equiv 5^n - 1 = \phi(4)\prod_{p \in S}\phi(p) = 2\prod_{p \in S}(p-1) \equiv 2 \cdot 3^{|S|} \,(\mathrm{mod}\ 5).$$

From the first of these equations we have that $|S|$ must be even, but from the second we have that $|S| \equiv 3 \,(\mathrm{mod}\ 4)$, a contradiction. Therefore, our original assumption was false, and $\gcd(m, n) > 1$.

## Problem 3

Let $A = \{1, 2, \ldots, n\}$, where $n$ is a positive integer. A subset of $A$ is *connected* if it is a nonempty set which consists of one element or of consecutive integers. Determine the greatest integer $k$ for which $A$ contains $k$ distinct subsets $A_1, A_2, \ldots, A_k$ such that the intersection of any two distinct sets $A_i$ and $A_j$ is connected.

**Solution.** Let $A_1, \ldots, A_k$ be distinct subsets of $A$ satisfying the required property. Let $m = \max_{1 \leq i \leq n}\{\min\{A_i\}\}$, and suppose that $\min\{A_{i_0}\} = m$.

Every $A_i$ has minimum element less than or equal to $m$, by the definition of $m$. Every $A_i$ also has maximum element greater than or equal to $m$, or else $A_i \cap A_{i_0} = \emptyset$ would not be connected. Therefore, each of the $k$ pairs

$(\min\{A_i\}, \max\{A_i\})$ equals one of the $m(n + 1 - m)$ pairs $(r, s)$ such that $1 \le r \le m \le s \le n$.

We show that at most one $A_i$ has $(\min\{A_i\}, \max\{A_i\}) = (r, s)$ for each such pair $(r, s)$. If there were two such distinct sets, then their intersection would be a connected set containing $r$ and $s$, and hence all of $r, r + 1, \ldots, s$. It would then follow that the two sets both equalled $\{r, r + 1, \ldots, s\}$, a contradiction.

Therefore, $k$ is at most $m(n + 1 - m) \le \lfloor \frac{n+1}{2} \rfloor \lceil \frac{n+1}{2} \rceil$. This maximum is attained when the $A_i$ are the connected subsets of $A$ containing $m_0$, where $m_0$ equals either $\lfloor \frac{n+1}{2} \rfloor$ or $\lceil \frac{n+1}{2} \rceil$.

## Problem 4

Let $f : \mathbb{N} \to \mathbb{N} \cup \{0\}$ be defined recursively by $f(1) = 0$ and

$$f(n) = \max_{1 \le j \le \lfloor \frac{n}{2} \rfloor} \{f(j) + f(n - j) + j\}$$

for all $n \ge 2$. Determine $f(2000)$.

**Solution.** For each positive integer $n$, we consider the binary representation of $n$. Consider the substrings of the representation formed by removing at least one digit from the left side of the representation, such that the substring so formed begins with a 1. We call the decimal values of these substrings the *tail-values of* $n$. Also, for each 1 that appears in the binary representation of $n$, if it represents the number $2^k$, let $2^{k-1}k$ be a *place-value of* $n$.

Let $g(n)$ be the sum of the tail- and place-values of $n$. We prove by induction on $n$ that $f(n) = g(n)$. For convenience, let $g(0) = 0$. It is clear that $g(1) = 0$. It will therefore suffice to show that $g(n)$ satisfies the same recurrence as $f(n)$. First we prove that

$$g(n) \ge g(j) + g(n - j) + j \tag{1}$$

for all $n, j$ such that $0 \le j \le \lfloor \frac{n}{2} \rfloor$. The relation is trivially true for $j = 0$ because we have defined $g(0) = 0$. Now we induct on the number of (binary) digits of $n - j$. For the base case (when $n - j$ has 1 binary digit), we can only have $n - j = 1$. In this case, $(n, j) = (2, 1)$ or $(n, j) = (1, 0)$, in which cases (1) is easily seen to be true. Now we prove the induction step by considering two cases.

*Case 1.* $n - j$ and $j$ have the same number of digits, say $k + 1$. Let $a$ and $b$ be the numbers formed by taking off the leftmost 1's (which represent $2^k$) from $n - j$ and $j$. We want to show that

$$g(n) = g(a + b + 2^{k+1}) \geq g(2^k + a) + g(2^k + b) + (2^k + b).$$

Subtracting the inequality $g(a + b) \geq g(a) + g(b) + b$ (which is true by the induction hypothesis), we see that it suffices to show that

$$\begin{aligned} g(a + b + 2^{k+1}) &- g(a + b) \\ &\geq g(2^k + a) - g(a) + g(2^k + b) - g(b) + 2^k. \end{aligned} \tag{2}$$

On the right-hand side, $g(2^k + a)$ equals $g(a)$ plus the place-value $2^{k-1}k$ and the tail-value $a$. Similarly, $g(2^k + b) = g(b) + 2^{k-1}k + b$. Hence, the right-hand side equals

$$2^{k-1}k + a + 2^{k-1}k + b + 2^k = 2^k(k + 1) + a + b.$$

As for the left-hand side of (2), because $a < 2^k$ and $b < 2^k$, the binary representation of $a + b + 2^{k+1}$ is simply the binary representation of $a + b$, with an additional 1 in the $2^{k+1}$ position. Hence, $g(a + b + 2^{k+1})$ equals $g(a + b)$ plus the additional tail-value $a + b$ and the additional place-value $2^k(k + 1)$. Thus, $g(a + b + 2^{k+1}) - g(a + b)$ equals the right-hand side, proving the inequality in (2).

*Case 2.*   $n - j$ has more digits than $j$. Let $n - j$ have $k + 1$ digits, and — as before — let $a = n - j - 2^k$. We need to prove that

$$g(a + j + 2^k) \geq g(a + 2^k) + g(j) + j.$$

We know by the induction hypothesis that

$$g(a + j) \geq g(a) + g(j) + \min\{a, j\}.$$

Subtracting, we see that it suffices to prove that

$$g(a + j + 2^k) - g(a + j) \geq g(a + 2^k) - g(a) + j - \min\{a, j\}. \tag{3}$$

We find as in Case 1 that on the right-hand side,

$$g(a + 2^k) - g(a) = 2^{k-1}k + a.$$

Hence, the right-hand side equals

$$2^{k-1}k + a + j - \min\{a, j\} = 2^{k-1}k + \max\{a, j\}.$$

On the left-hand side of (3), if $a + j < 2^k$ (i.e., so that the $2^k$ digits do not carry in the sum $(a + j) + 2^k$), then $g(a + j + 2^k)$ equals $g(a + j)$ plus the additional place-value $2^{k-1}k$ and the additional tail-value $a + j$. Hence, the left-hand side of (3) is indeed greater than or equal to the right-hand side. Otherwise, if the $2^k$ digits *do* carry in the sum $(a+j) + 2^k$, then $g(a + j + 2^k)$ equals $g(a+j)$ plus the additional place-value $2^k(k+1)$, minus the original place-value $2^{k-1}k$. Thus, the left-hand side equals

$$2^k(k+1) - 2^{k-1}k = 2^{k-1}k + 2^k > 2^{k-1}k + \max\{a, j\},$$

so again (3) is true. This completes the induction.

Hence, $g(n) \geq \max_{1 \leq j \leq \lfloor n/2 \rfloor}\{g(j) + g(n-j) + j\}$ for all $n$. We now prove that in fact equality holds, by showing that

$$g(n) = g(j) + g(n-j) + j$$

for some $j$. Let $2^k$ be the largest power of 2 less than $n$, and set $j = n - 2^k$. Then $g(n)$ equals $g(n - 2^k)$ plus the additional place-value $g(2^k) = g(n-j)$ and the additional tail-value $n - 2^k = j$.

It follows that $f(n) = g(n)$ for all $n$. Hence, by finding the place- and tail-values of 2000 (with binary representation 11111010000), we may compute that $f(2000) = 10864$.

## 1.19 Turkey

### Problem 1

Find the number of ordered quadruples $(x, y, z, w)$ of integers with $0 \leq x, y, z, w \leq 36$ such that

$$x^2 + y^2 \equiv z^3 + w^3 \pmod{37}.$$

**Solution.** All congruences will be modulo 37. For each $k$ between 0 and 36 inclusive, we find the number of pairs $(x, y)$ of integers with $0 \leq x, y \leq 36$ satisfying $x^2 + y^2 \equiv k$. Notice that this is equivalent to $(x + 6y)(x - 6y) \equiv k$.

First we consider the case $k = 0$. For each $y \in \{0, 1, \ldots, 36\}$, we have $(x + 6y)(x - 6y) \equiv 0$ if and only if $x \equiv \pm 6y$. Thus, there is one pair $(x, y)$ with $y = 0$ such that $x^2 + y^2 \equiv 0$ (namely, $(x, y) = (0, 0)$), and for any other $y$ there are two such pairs $(x, y)$. Hence, there are a total of $2 \cdot 36 + 1 = 73$ pairs $(x, y)$ such that $x^2 + y^2 \equiv 0$.

Now we consider the case when $k \neq 0$. Let $a \equiv x + 6y, b \equiv x - 6y$. For any value $a \in \{1, 2, \ldots, 36\}$, there is exactly one value of $b \in \{1, 2, \ldots, 36\}$ such that $ab \equiv k$. Each of these 36 pairs $(a, b)$ corresponds to a unique solution $(x, y)$, because we must have $x \equiv (a + b)2^{-1}$, $y \equiv (a - b)12^{-1}$. Thus, the equation $(x + 6y)(x - 6y) \equiv k$ has exactly 36 solutions $(x, y)$ whenever $k \neq 0$.

We proceed to count the number of quadruples $(x, y, z, w)$ such that $x^2 + y^2 \equiv z^3 + w^3 \equiv 0$. There are three cube roots $r_1$, $r_2$, $r_3$ of 1 modulo 37 — namely, if we let $g$ be a primitive element modulo 37, then the cube roots are 1, $g^{12}$, and $g^{24}$. Given any $z$, we have $z^3 + w^3 \equiv 0$ if and only if $w$ equals $-r_1 z$, $-r_2 z$, or $-r_3 z$. Hence, there are 109 pairs $(z, w)$ such that $z^3 + w^3 \equiv 0$ — one pair such that $z = 0$, and 3 pairs such that $z = z_0$ for each $z_0 \in \{1, 2, \ldots, 36\}$. Above, we found that there are exactly 73 pairs $(x, y)$ such that $x^2 + y^2 \equiv 0$. Therefore, there are $109 \cdot 73$ quadruples $(x, y, z, w)$ such that $x^2 + y^2 \equiv z^3 + w^3 \equiv 0$.

For each of the $37^2 - 109$ pairs $(z, w)$ such that $z^3 + w^3 \not\equiv 0$, there are exactly 36 pairs $(x, y)$ such that $x^2 + y^2 \equiv z^3 + w^3$. Hence, there are $(37^2 - 109) \cdot 36$ quadruples $(x, y, z, w)$ such that $x^2 + y^2 \equiv z^3 + w^3 \not\equiv 0$.

Therefore, there are

$$109 \cdot 73 + (37^2 - 109) \cdot 36 = 53317$$

quadruples $(x, y, z, w)$ such that $x^2 + y^2 \equiv z^3 + w^3$.

## Problem 2

Given a circle with center $O$, the two tangent lines from a point $S$ outside the circle touch the circle at points $P$ and $Q$. Line $SO$ intersects the circle at $A$ and $B$, with $B$ closer to $S$. Let $X$ be an interior point of minor arc $PB$, and let line $OS$ intersect lines $QX$ and $PX$ at $C$ and $D$, respectively. Prove that

$$\frac{1}{AC} + \frac{1}{AD} = \frac{2}{AB}.$$

**Solution.** Extend ray $PC$ to intersect arc $QB$ at $Y$. By symmetry, arcs $BX$ and $BY$ are congruent, implying that $\angle CPB = \angle YPB = \angle BPX = \angle BPD$. Hence, $\overline{PB}$ is the internal angle bisector of angle $CPD$. Because $\angle APB = \pi/2$, we also have that $\overline{PA}$ is the external angle bisector of angle $CPD$. Applying the internal and external Angle Bisector Theorems, we find that

$$BC/BD = PC/PD = AC/AD.$$

Substituting $BC = AB - AC$ and $BD = AD - AB$ and dividing the left- and right-hand sides by $AB$, we have

$$\frac{AB - AC}{AB \cdot AC} = \frac{AD - AB}{AD \cdot AB}.$$

This implies that

$$\frac{1}{AC} - \frac{1}{AB} = \frac{1}{AB} - \frac{1}{AD},$$

which is equivalent to the desired equality.

## Problem 3

For any two positive integers $n$ and $p$, prove that there are exactly $(p+1)^{n+1} - p^{n+1}$ functions

$$f : \{1, 2, \ldots, n\} \to \{-p, -p+1, \ldots, p\}$$

such that $|f(i) - f(j)| \leq p$ for all $i, j \in \{1, 2, \ldots, n\}$.

**Solution.** Given $m \in \{-p, -p+1, \ldots, p\}$, there are $(\min\{p+1, p-m+1\})^n$ functions satisfying the given conditions which attain values only in $\{m, \ldots, m+p\}$. Of these, $(\min\{p, p-m\})^n$ functions attain values only in $\{m+1, \ldots, m+p\}$. Hence, exactly

$$(\min\{p+1, p+1-m\})^n - (\min\{p, p-m\})^n$$

functions satisfying the given conditions have minimum value $m$.

This expression equals $(p+1)^n - p^n$ for each of the $p+1$ values $m \leq 0$, and it equals $(p+1-m)^n - (p-m)^n$ when $m > 0$. Thus, the sum of the expression over all $m \leq 0$ is $(p+1)((p+1)^n - p^n)$, while the sum of the expression over all $m > 0$ is the telescoping sum $\sum_{m=1}^{p}((p+1-m)^n - (p-m)^n) = p^n$. Adding these two sums, we find that the total number of functions satisfying the given conditions is $(p+1)^{n+1} - p^{n+1}$, as desired.

## Problem 4

In an acute triangle $ABC$ with circumradius $R$, altitudes $\overline{AD}, \overline{BE}, \overline{CF}$ have lengths $h_1, h_2, h_3$, respectively. If $t_1, t_2, t_3$ are the lengths of the tangents from $A, B, C$, respectively, to the circumcircle of triangle $DEF$, prove that

$$\sum_{i=1}^{3} \left( \frac{t_i}{\sqrt{h_i}} \right)^2 \leq \frac{3}{2}R.$$

**Solution.** Let $H$ be the orthocenter of triangle $ABC$, and let $X, Y$, and $Z$ be the respective midpoints of $\overline{AH}, \overline{BH}$, and $\overline{CH}$. Because the circumcircle of triangle $DEF$ is the nine-point circle of triangle $ABC$, it passes through $X, Y$, and $Z$. Hence, $t_1^2 = AX \cdot AD = AX \cdot h_1$, or $(t_1/\sqrt{h_1})^2 = AX$. We can find similar expressions for $BX$ and $CX$. The desired inequality is thus equivalent to $AX + BY + CZ \leq \frac{3}{2}R$, or (multiplying each side by 2)

$$AH + BH + CH \leq 3R.$$

Let $\angle A = \alpha, \angle B = \beta$, and $\angle C = \gamma$. Then,

$$AH = \frac{AF}{\sin \beta} = \frac{AC \cos \alpha}{\sin \beta} = 2R \cos \alpha.$$

Similarly, $BH = 2R \cos \beta$ and $CH = 2R \cos \gamma$, so the required inequality is equivalent to

$$\cos \alpha + \cos \beta + \cos \gamma \leq \frac{3}{2}.$$

Recall that $ABC$ is acute and the function $t \mapsto \cos t$ is concave on the interval $(0, \pi/2)$. Thus, Jensen's Inequality implies that the left-hand side of this last inequality attains its maximum when all three angles are equal to $\pi/3$, in which case the left-hand side equals $3/2$. Thus, this last inequality is true, and the desired inequality is as well.

## Problem 5

(a) Prove that for each positive integer $n$, the number of ordered pairs $(x, y)$ of integers satisfying

$$x^2 - xy + y^2 = n$$

is finite and divisible by 6.

(b) Find all ordered pairs $(x, y)$ of integers satisfying

$$x^2 - xy + y^2 = 727.$$

**Solution.**   (a) Any solution $(x, y)$ must satisfy the inequality

$$n = x^2 - xy + y^2 = \frac{1}{2}(x - y)^2 + \frac{1}{2}(x^2 + y^2) \geq \frac{1}{2}(x^2 + y^2),$$

and only finitely many ordered pairs $(x, y)$ do so. Hence, there are finitely many solutions.

Next we prove that the number of solutions is divisible by 6. If $(x, y)$ is a solution, so is $(y, y - x)$. This linear transformation is invertible, so it permutes the set of all solutions, and we can partition the solution set into orbits. Each such orbit is of the form

$$\big\{(x, y), (y, y - x), (y - x, -x), (-x, -y), (-y, x - y), (x - y, x)\big\}$$

for some initial solution $(x, y)$. It is straightforward to check that no two of the six solutions in each orbit can be equal unless $x = y = 0$, which is impossible. Thus, each orbit has six distinct elements, and the desired result follows.

(b) Given any solution to $x^2 - xy + y^2 = 727$, we can apply the transformations $(x, y) \mapsto (y, y - x)$ (as in part (a)), then possibly $(x, y) \mapsto (y, x)$, to obtain another solution $(x, y)$ with $y \leq 0 \leq x \leq |y|$.

We now find all such solutions with $y \leq 0 \leq x \leq |y|$. Rearranging the required equation gives $y^2 - xy + x^2 - 727 = 0$. Viewing this as a quadratic in $y$, we can apply the quadratic formula to find that

$$y = \frac{x \pm \sqrt{2908 - 3x^2}}{2}.$$

Hence, $2908 - 3x^2$ must be a perfect square, and it is not divisible by 3. Because $3x^2 \leq y^2 - xy + x^2 = 727$, we further know that $2181 \leq 2908 - 3x^2 \leq 2908$, giving $46 < \sqrt{2908 - 3x^2} < 54$. Testing these possibilities, we find that only $\sqrt{2908 - 3x^2} = 49$ has an integer solution $x$, yielding the unique solution $(13, -18)$ of the desired form.

Thus, every solution can be transformed into $(13, -18)$ by applying the two maps described earlier. Hence, any solution is in the orbit of $(13, -18)$ or $(-18, 13)$ under $(x, y) \mapsto (y, y - x)$, implying that all the solutions to $x^2 - xy + y^2 = 727$ are:

$$(13, -18), (-18, -31), (-31, -13), (-13, 18), (18, 31), (31, 13),$$

$$(-18, 13), (13, 31), (31, 18), (18, -13), (-13, -31), (-31, -18).$$

## Problem 6

Given a triangle $ABC$, the internal and external bisectors of angle $A$ intersect line $BC$ at points $D$ and $E$, respectively. Let $F$ be the point (different from $A$) where line $AC$ intersects the circle $\omega$ with diameter $\overline{DE}$. Finally, draw the tangent at $A$ to the circumcircle of triangle $ABF$, and let it hit $\omega$ at $A$ and $G$. Prove that $AF = AG$.

**Solution.**    We give a proof for the case in which $C$, $B$, and $E$ are collinear in that order; the proof for the other case is similar.

Let $O$ be the center of $\omega$. By the Angle Bisector Theorem (for both the internal and exterior angle bisectors),

$$\frac{CD}{DB} = \frac{CA}{AB} = \frac{CE}{BE}.$$

Thus, $CD(CE - CB) = CD \cdot BE = CE \cdot DB = CE(CB - CD)$, or (adding $CD(CB + CE)$ to both sides) $2CD \cdot CE = CB \cdot (CD + CE)$. Because $CD + CE = 2CO$, we have

$$CD \cdot CE = CB \cdot CO.$$

On the other hand, $CD \cdot CE = CA \cdot CF$ by the Power of a Point Theorem applied to $C$ and $\omega$. It follows that $CB \cdot CO = CA \cdot CF$. Hence, by Power of a Point, the points $A, B, O, F$ lie on some circle $\omega_1$.

We perform an inversion about $A$ with radius $AO$. $\omega$ is a circle passing through $A$ which is perpendicular to line $AO$ and contains a point $P$ on ray $AO$ with $AP = 2AO$. Hence, its image $\ell_1$ under the inversion is a line which is perpendicular to line $AO$ and contains a point $P'$ on ray $AO$ with $AP' = AO/2$. In other words, $\omega$'s image $\ell_1$ is the perpendicular bisector of $\overline{AO}$. Next, the inversion takes $\omega_1$ (a circle passing through $A$, passing through $O$, and tangent to line $AG$) to the line $\ell_2$ not passing through $A$, passing through $O$, and parallel to line $AG$.

It follows that the inversion sends $F$, the intersection of $\omega$ and $\omega_1$, to the intersection $F'$ of $\ell_1$ and $\ell_2$; furthermore, the inversion sends $G$, the intersection of $\omega_1$ and line $AG$, to the intersection of $\ell_1$ and line $AG$. The reflection across the midpoint of $\overline{AO}$ sends $\ell_1$ to itself and $\ell_2$ to line $AG$; hence, this reflection sends $\overline{OF'}$ to $\overline{AG'}$, implying that $OF' = AG'$. Because $F'$ lies on the perpendicular bisector of $\overline{AO}$, we also have $OF' = AO$. Therefore, $AF' = AG'$, implying that $AF = AG$. This completes the proof.

## Problem 7

Show that it is possible to cut any triangular prism of infinite length with a plane such that the resulting intersection is an equilateral triangle.

**Solution.** Suppose that a plane perpendicular to the three edges of the prism intersects these edges at $A, B, C$; write $a = BC$, $b = CA$, $c = AB$, and assume without loss of generality that $a \le b \le c$. For $t \ge 0$, define

$$f(t) = \sqrt{a^2 + (t + \sqrt{c^2 - b^2 + t^2})^2} - \sqrt{c^2 + t^2}.$$

Then $f(0) = \sqrt{a^2 + c^2 - b^2} - \sqrt{c^2} \le 0$. On the other hand, we have $f(b) = \sqrt{a^2 + (b + c)^2} - \sqrt{c^2 + b^2} > 0$. Because $f$ is continuous, there exists $t_0$ with $f(t_0) = 0$. Now let $B'$ lie on the same edge as $B$, at a distance of $t$ from $B$. Let $C'$ lie on the same edge of the prism as $C$, at distance $\sqrt{c^2 - b^2 + t_0^2}$ from $C$, and on the opposite side of plane $(ABC)$ from $B'$. Then, by the Pythagorean Theorem,

$$AB' = \sqrt{c^2 + t_0^2}; \quad AC' = \sqrt{b^2 + (c^2 - b^2 + t_0^2)} = \sqrt{c^2 + t_0^2};$$

$$B'C' = \sqrt{a^2 + (t_0 + \sqrt{c^2 - b^2 + t_0^2})^2} = \sqrt{c^2 + t_0^2}.$$

Thus, the plane $(AB'C')$ meets our requirements.

## Problem 8

Given a square $ABCD$, the points $M, N, K, L$ are chosen on the interiors of sides $\overline{AB}, \overline{BC}, \overline{CD}, \overline{DA}$, respectively, such that lines $MN$ and $LK$ are parallel and such that the distance between lines $MN$ and $LK$ equals $AB$. Show that the circumcircles of triangles $ALM$ and $NCK$ intersect each other, while those of triangles $LDK$ and $MBN$ do not.

**Solution.** Orient the square so that $\overline{AB}$ is horizontal and above $\overline{CD}$, where $A$ is due west of $B$. We first claim that $AL > BN$, or in other words, $N$ is north (although not necessarily due north) of $L$. Assume the contrary. Then there is a horizontal segment with left endpoint $L$ and right endpoint on $\overline{MN}$, with length less than or equal to $AB$. On the other hand, the length of this segment is greater than the distance between $\overline{LK}$ and $\overline{MN}$, which is assumed to be $AB$. Thus, we have a contradiction, and $AL > BN$. We may likewise conclude that $AM > DK$.

Construct $P$ and $Q$ so that quadrilaterals $BMPN$ and $DKQL$ are rectangles. We know from above that $P$ is northeast of $Q$. Construct $R$ and $S$ such that $R$ is to the southeast of $Q$ and such that quadrilateral $PRQS$ is a rectangle with sides parallel to those of square $ABCD$.

To show that the circumcircles of triangles $ALM$ and $NCK$ intersect each other, observe that the discs bounded by the circumcircles of triangles $ALM$ and $NCK$ contain rectangles $ALRM$ and $CKSN$, respectively. Hence, these discs both contain rectangle $PRQS$. Because regions inside the circumcircles of triangles $ALM$ and $NCK$ intersect, the circumcircles themselves must also intersect.

We now show that the circumcircle $\omega_1$ of triangle $MBN$ and the circumcircle $\omega_2$ of triangle $LDK$ do not intersect. Notice that they are also the circumcircles of rectangles $BMPN$ and $DKQL$, respectively. Let $l_1$ be the tangent to circle $\omega_1$ at $P$, and let $l_2$ be the tangent to circle $\omega_2$ at $Q$. Because $\overline{MN}$ and $\overline{LK}$ are parallel, so are $\overline{BP}$ and $\overline{QD}$. Because $l_1 \perp \overline{BP}$ and $l_2 \perp \overline{QD}$, we have that $l_1$ and $l_2$ are parallel. Hence, each point of $\omega_1$ lies on or to the right of $l_1$, which in turn lies to the right of $l_2$; on the other hand, each point on $\omega_2$ lies on or to the left of $l_2$. Hence, $\omega_1$ and $\omega_2$ cannot intersect.

## Problem 9

Let $f : \mathbb{R} \to \mathbb{R}$ be a function such that

$$|f(x+y) - f(x) - f(y)| \leq 1$$

for all $x, y \in \mathbb{R}$. Show that there exists a function $g : \mathbb{R} \to \mathbb{R}$ with $|f(x) - g(x)| \leq 1$ for all $x \in \mathbb{R}$, and with $g(x+y) = g(x) + g(y)$ for all $x, y \in \mathbb{R}$.

**Solution.** We claim that the function

$$g(x) = \lim_{n \to \infty} \frac{f(2^n x)}{2^n}$$

satisfies the requirements.

Our first task is to show that the limit exists for all $x$. In fact, we can prove this and prove that $|f(x) - g(x)| \leq 1$ for all $x$ at the same time. First, observe that setting $x = y = 2^m x_0$ in the given inequality for $f$ gives $|f(2^{m+1}x_0) - 2f(2^m x_0)| \leq 1$. Dividing by $2^{m+1}$, we have

$$\left| \frac{f(2^{m+1}x_0)}{2^{m+1}} - \frac{f(2^m x_0)}{2^m} \right| \leq \frac{1}{2^{m+1}}.$$

For any fixed $x$, consider the infinite telescoping sum

$$\sum_{m=0}^{\infty} \left( \frac{f(2^{m+1}x)}{2^{m+1}} - \frac{f(2^m x)}{2^m} \right).$$

Because the absolute values of the terms are bounded by the geometric series $\frac{1}{2}, \frac{1}{4}, \ldots$ which sums to 1, this sum converges absolutely and is bounded by 1 as well. On the other hand, by definition the infinite sum equals

$$\lim_{n \to \infty} \sum_{m=0}^{n} \left( \frac{f(2^{m+1}x)}{2^{m+1}} - \frac{f(2^m x)}{2^m} \right).$$

The telescoping sum inside the limit equals $\left( f(2^{n+1}x)/2^{n+1} \right) - f(x)$, implying that the above limit equals

$$\lim_{n \to \infty} \left( \frac{f(2^{n+1}x)}{2^{n+1}} - f(x) \right).$$

We may now take out the constant $f(x)$ term to obtain

$$\left( \lim_{n \to \infty} \frac{f(2^{n+1}x)}{2^{n+1}} \right) - f(x).$$

It follows that the limit in this last expression converges, and this happens to be exactly the limit we wanted to use to define $g(x)$. Furthermore, we saw above that the last quantity is at most 1, so we also have

$$|g(x) - f(x)| \leq 1.$$

It remains to be shown that $g(x + y) = g(x) + g(y)$ for all $x$ and $y$. Observe that

$$g(x + y) - g(x) - g(y)$$
$$= \lim_{n \to \infty} \frac{f(2^n(x + y))}{2^n} - \lim_{n \to \infty} \frac{f(2^n x)}{2^n} - \lim_{n \to \infty} \frac{f(2^n y)}{2^n}$$
$$= \lim_{n \to \infty} \frac{f(2^n(x + y)) - f(2^n x) - f(2^n y)}{2^n}.$$

From the given,

$$\left| f(2^n(x+y)) - f(2^n x) - f(2^n y) \right| \le 1$$

for any $n$, implying that the term inside the limit of the last expression above is between

$$-\frac{1}{2^n} \quad \text{and} \quad \frac{1}{2^n}.$$

Because

$$\lim_{n\to\infty} \frac{1}{2^n} = 0,$$

it follows that the limit in the last expression above is 0. Hence,

$$g(x+y) = g(x) + g(y),$$

as wanted.

## 1.20 United Kingdom

### Problem 1

Two intersecting circles $C_1$ and $C_2$ have a common tangent which touches $C_1$ at $P$ and $C_2$ at $Q$. The two circles intersect at $M$ and $N$. Prove that the triangles $MNP$ and $MNQ$ have equal areas.

**Solution.** Let $X$ be the intersection of lines $MN$ and $PQ$. Because line $MN$ is the radical axis of $C_1$ and $C_2$, $X$ has equal power with respect to these two circles. Thus, $XP^2 = XQ^2$, or $XP = XQ$. Also, because $\angle PXM + \angle MXQ = \pi$, we have $\sin \angle PXM = \sin \angle MXQ$. Therefore,

$$[MNP] = \frac{1}{2} MN (XP \sin \angle PXM)$$
$$= \frac{1}{2} MN (XQ \sin \angle MXQ)$$
$$= [MNQ],$$

as desired.

### Problem 2

Given that $x$, $y$, $z$ are positive real numbers satisfying $xyz = 32$, find the minimum value of

$$x^2 + 4xy + 4y^2 + 2z^2.$$

**Solution.** Applying the arithmetic mean-geometric mean inequality twice, we find that

$$x^2 + 4xy + 4y^2 + 2z^2 = (x^2 + 4y^2) + 4xy + 2z^2$$
$$\geq 2\sqrt{x^2 \cdot 4y^2} + 4xy + 2z^2$$
$$= 4xy + 4xy + 2z^2$$
$$\geq 3\sqrt[3]{4xy \cdot 4xy \cdot 2z^2}$$
$$= 3\sqrt[3]{32(xyz)^2}$$
$$= 96.$$

Equality holds when $x^2 = 4y^2$ and $4xy = 2z^2$, i.e., when $(x, y, z) = (4, 2, 4)$.

## Problem 3

(a) Find a set $A$ of ten positive integers such that no six distinct elements of $A$ have a sum which is divisible by 6.

(b) Is it possible to find such a set if "ten" is replaced by "eleven"?

**Solution.** (a) An example of such a set is

$$A = \{6j + k \mid 1 \leq j \leq 5, 1 \leq k \leq 2\}.$$

In any six-element subset of $A$, if there are $t$ numbers congruent to 1 modulo 6, then $t \in \{1, 2, \ldots, 5\}$. The others in the subset are congruent to 0 modulo 6. Thus, the sum of the elements in the subset is congruent to $t \not\equiv 0 \pmod{6}$.

(b) It is not possible. Given any set of eleven positive integers, we find six distinct elements of this set whose sum is divisible by 6. Because there are more than two integers in this set, we may choose two whose sum is even. Similarly, among the other ten integers, we may choose two more whose sum is even. Continuing in a similar manner, we can find five disjoint two-element subsets whose sums are congruent to either 0, 2, or 4 modulo 6. If all three types of sums occur, the six elements in the corresponding subsets have sum congruent to $0 + 2 + 4 \equiv 0 \pmod{6}$. Otherwise, only two types of sums occur. By the Pigeonhole Principle, three subsets have sums of the same type. Then the elements in these three pairs will have sum divisible by 6.

## 1.21   United States of America

### Problem 1

Call a real-valued function $f$ *very convex* if

$$\frac{f(x) + f(y)}{2} \geq f\left(\frac{x+y}{2}\right) + |x - y|$$

holds for all real numbers $x$ and $y$. Prove that no very convex function exists.

**First Solution.**   Fix $n \geq 1$. For each integer $i$, define

$$\Delta_i = f\left(\frac{i+1}{n}\right) - f\left(\frac{i}{n}\right).$$

The given inequality with $x = (i+2)/n$ and $y = i/n$ implies

$$\frac{f\left(\frac{i+2}{n}\right) + f\left(\frac{i}{n}\right)}{2} \geq f\left(\frac{i+1}{n}\right) + \frac{2}{n},$$

or

$$f\left(\frac{i+2}{n}\right) - f\left(\frac{i+1}{n}\right) \geq f\left(\frac{i+1}{n}\right) - f\left(\frac{i}{n}\right) + \frac{4}{n}.$$

In other words, $\Delta_{i+1} \geq \Delta_i + 4/n$. Combining this for $n$ consecutive values of $i$ gives

$$\Delta_{i+n} \geq \Delta_i + 4.$$

Summing this inequality for $i = 0$ to $i = n-1$ and cancelling terms yields

$$f(2) - f(1) \geq f(1) - f(0) + 4n.$$

This cannot hold for all $n \geq 1$. Hence, there are no very convex functions.

**Second Solution.**   We show by induction that the given inequality implies

$$\frac{f(x) + f(y)}{2} - f\left(\frac{x+y}{2}\right) \geq 2^n |x - y|$$

for all nonnegative integers $n$. This will yield a contradiction, because for fixed $x$ and $y$ such that $x \neq y$ the right side gets arbitrarily large, while the left side remains fixed.

We are given the base case $n = 0$. Now, if the inequality holds for a given $n$, then for $a, b$ real,

$$\frac{f(a) + f(a + 2b)}{2} \geq f(a + b) + 2^{n+1}|b|,$$

$$f(a + b) + f(a + 3b) \geq 2(f(a + 2b) + 2^{n+1}|b|),$$

and

$$\frac{f(a + 2b) + f(a + 4b)}{2} \geq f(a + 3b) + 2^{n+1}|b|.$$

Adding these three inequalities and cancelling terms yields

$$\frac{f(a) + f(a + 4b)}{2} \geq f(a + 2b) + 2^{n+3}|b|.$$

Setting $x = a$, $y = a + 4b$, we obtain

$$\frac{f(x) + f(y)}{2} \geq f\left(\frac{x + y}{2}\right) + 2^{n+1}|x - y|,$$

and the induction is complete.


## Problem 2

Let $S$ be the set of all triangles $ABC$ for which

$$5\left(\frac{1}{AP} + \frac{1}{BQ} + \frac{1}{CR}\right) - \frac{3}{\min\{AP, BQ, CR\}} = \frac{6}{r},$$

where $r$ is the inradius and $P$, $Q$, $R$ are the points of tangency of the incircle with sides $AB$, $BC$, $CA$, respectively. Prove that all triangles in $S$ are isosceles and similar to one another.

**Solution.** We start with the following lemma.

**Lemma.** *Let $A$, $B$, $C$ be the angles of triangle $ABC$. Then*

$$\tan\frac{A}{2}\tan\frac{B}{2} + \tan\frac{B}{2}\tan\frac{C}{2} + \tan\frac{C}{2}\tan\frac{A}{2} = 1.$$

*Proof.* Note that $A/2 + B/2 + C/2 = \pi/2$. With this fact, and using the trigonometric identities

$$\tan\alpha + \tan\beta = \tan(\alpha + \beta)[1 - \tan\alpha\tan\beta],$$

$$\tan(\pi/2 - \alpha) = \cot\alpha = 1/\tan\alpha,$$

we find that

$$\tan\frac{A}{2}\tan\frac{B}{2} + \tan\frac{B}{2}\tan\frac{C}{2}$$

$$= \tan\frac{B}{2}\left(\tan\frac{A}{2} + \tan\frac{C}{2}\right)$$

$$= \tan\frac{B}{2}\tan\left(\frac{A}{2} + \frac{C}{2}\right)\left[1 - \tan\frac{A}{2}\tan\frac{C}{2}\right]$$

$$= \tan\frac{B}{2}\tan\left(\pi/2 - \frac{B}{2}\right)\left[1 - \tan\frac{A}{2}\tan\frac{C}{2}\right]$$

$$= 1 - \tan\frac{A}{2}\tan\frac{C}{2}.$$

∎

Without loss of generality, assume that $AP = \min\{AP, BQ, CR\}$. Let $x = \tan(\angle A/2)$, $y = \tan(\angle B/2)$, and $z = \tan(\angle C/2)$. Then $AP = r/x, BQ = r/y$, and $CR = r/z$. Then the equation given in the problem statement becomes

$$2x + 5y + 5z = 6, \tag{1}$$

and the lemma implies that

$$xy + yz + zx = 1. \tag{2}$$

Eliminating $x$ from (1) and (2) yields

$$5y^2 + 5z^2 + 8yz - 6y - 6z + 2 = 0,$$

or

$$(2y + z - 1)^2 + (y + 2z - 1)^2 = 0.$$

The only real solution to this equation is $y = z = 1/3$. Thus, there is only one possible value for $(x, y, z)$, namely $(4/3, 1/3, 1/3)$. Thus, all the triangles in $S$ are isosceles and similar to one another.

## Problem 3

A game of solitaire is played with $R$ red cards, $W$ white cards, and $B$ blue cards. A player plays all the cards one at a time. With each play he accumulates a penalty. If he plays a blue card, then he is charged a penalty which is the number of white cards still in his hand. If he plays a white card, then he is charged a penalty which is twice the number of red cards still in his hand. If he plays a red card, then he is charged a penalty which is three times the number of blue cards still in his hand. Find, as a function of $R$, $W$, and $B$, the minimal total penalty a player can amass and all the ways in which this minimum can be achieved.

**Solution.** The minimum achievable penalty is

$$\min\{BW, 2WR, 3RB\}.$$

This penalty is achievable because the three penalties $BW, 2WR$, and $3RB$ can be obtained by playing cards in one of the three orders

- $\mathbf{bb}\cdots\mathbf{brr}\cdots\mathbf{rww}\cdots\mathbf{w},$
- $\mathbf{rr}\cdots\mathbf{rww}\cdots\mathbf{wbb}\cdots\mathbf{b},$
- $\mathbf{ww}\cdots\mathbf{wbb}\cdots\mathbf{brr}\cdots\mathbf{r}.$

Given an order of play, let a "run" of some color denote a set of cards of that color played consecutively in a row. Then the optimality of one of the three above orders follows immediately from the following lemma, along with the analogous observations for blue and white cards.

**Lemma 1.** *For any given order of play, we may combine any two runs of red cards without increasing the penalty.*

*Proof:* Suppose that $w$ white cards and $b$ blue cards are played between the two red runs. If we move a red card from the first run to the second, we increase the penalty of our order of play by $2w$ because we now have one more red card in our hand when we play the $w$ white cards. However, the penalty decreases by $3b$ because this red card is now after the $b$ blue cards. If the net gain $3b - 2w$ is non-negative, then we can move all the red cards in the first run to the second run without increasing the penalty. If the net gain $3b - 2w$ is negative, then we can move all the red cards in the second run to the first run without increasing the penalty. In either case, we may combine any two runs of red cards without increasing the penalty. ∎

Thus, there must be an optimal game where cards are played in one of the three given orders. To determine whether there are other optimal orders, first observe that $\mathbf{wr}$ can never appear during an optimal game; otherwise, if we instead play these two cards in the order $\mathbf{rw}$, then we accrue a smaller penalty. Similarly, $\mathbf{bw}$ and $\mathbf{rb}$ can never appear. Now we prove the following lemma.

**Lemma 2.** *Any optimal order of play must have less than 5 runs.*

*Proof:* Suppose that some optimal order of play had at least five runs. Assume the first card played is red; the proof is similar in the other cases. Say we first play $r_1, w_1, b_1, r_2, w_2$ cards of each color, where each $r_i, w_i, b_i$ is positive and where we cycle through red, white, and blue runs. From the proof of our first lemma, we must have both $3b_1 - 2w_1 = 0$ and $b_1 - 2r_2 = 0$. Hence, the game starting with playing $r_1, w_1 + w_2, b_1, r_2, 0$ cards is optimal as well, so we must also have $3b_1 - 2(w_1 + w_2) = 0$, a contradiction. ∎

Thus, any optimal game has at most 4 runs. Now from our initial observations and the proof of lemma 1, we see that any order of play of the form

$$\mathbf{rr} \cdots \mathbf{rww} \cdots \mathbf{wbb} \cdots \mathbf{brr} \cdots \mathbf{r},$$

is optimal if and only if $2W = 3B$ and $2WR = 3RB \le WB$. Similar conditions hold for 4-run games that start with **w** or **b**.

## Problem 4

Find the smallest positive integer $n$ such that if $n$ unit squares of a $1000 \times 1000$ unit-square board are colored, then there will exist three colored unit squares whose centers form a right triangle with legs parallel to the edges of the board.

**Solution.** We show that the minimum such $n$ is 1999. Indeed, $n \ge 1999$ because we can color 1998 squares without producing a right triangle: color every square in the first row and the first column, except for the one square at their intersection.

Now assume that some squares have been colored so that there is no right triangle of the described type. Call a row or column *heavy* if it contains more than one colored square, and *light* otherwise. Our assumption then states that no heavy row and heavy column intersect in a colored square.

If there are no heavy rows, then each row contains at most one colored square, so there are at most 1000 colored squares. We reach the same conclusion if there are no heavy columns. If there is a heavy row and a heavy column, then by the initial observation, each colored square in the heavy row or column must lie in a light column or row, and no two can lie in the same light column or row. Thus, the number of colored squares is at most the number of light rows and columns, which is at most $2 \cdot (1000 - 1) = 1998$.

We conclude that in fact 1999 is the minimum number of colored squares needed to force the existence of a right triangle of the type described.

## Problem 5

Let $A_1 A_2 A_3$ be a triangle and let $\omega_1$ be a circle in its plane passing through $A_1$ and $A_2$. Suppose there exist circles $\omega_2, \omega_3, \ldots, \omega_7$ such that for $k = 2, 3, \ldots, 7$, $\omega_k$ is externally tangent to $\omega_{k-1}$ and passes through $A_k$ and $A_{k+1}$, where $A_{n+3} = A_n$ for all $n \geq 1$. Prove that $\omega_7 = \omega_1$.

**Solution.** Without loss of generality, we may assume that in counterclockwise order, the vertices of the triangle are $A_1, A_2, A_3$. Let $\theta_1$ be the measure of the arc from $A_1$ to $A_2$ along $\omega_1$, taken in the counterclockwise direction. Define $\theta_2, \ldots, \theta_7$ analogously.

Let $\ell$ be the line through $A_2$ tangent to $\omega_1$ and $\omega_2$. Then the angle from the line $A_1 A_2$ to $\ell$, again measured counterclockwise, is $\theta_1/2$. Similarly, the angle from $\ell$ to $A_2 A_3$ is $\theta_2/2$. Therefore, writing $\angle A_1 A_2 A_3$ for the counterclockwise angle from the line $A_1 A_2$ to the line $A_2 A_3$, we have

$$\theta_1 + \theta_2 = 2\angle A_1 A_2 A_3.$$

By similar reasoning we obtain the system of six equations:

$$\theta_1 + \theta_2 = 2\angle A_1 A_2 A_3, \qquad \theta_2 + \theta_3 = 2\angle A_2 A_3 A_1,$$
$$\theta_3 + \theta_4 = 2\angle A_3 A_1 A_2, \qquad \theta_4 + \theta_5 = 2\angle A_1 A_2 A_3,$$
$$\theta_5 + \theta_6 = 2\angle A_2 A_3 A_1, \qquad \theta_6 + \theta_7 = 2\angle A_3 A_1 A_2.$$

Adding the equations on the left column, and subtracting the equations on the right yields $\theta_1 = \theta_7$.

To see that this last equality implies $\omega_1 = \omega_7$, simply note that as the center $O$ of a circle passing through $A_1$ and $A_2$ moves along the perpendicular bisector of $\overline{A_1 A_2}$, the angle $\theta_1$ goes monotonically from 0 to $2\pi$. Thus the angle determines the circle.

## Problem 6

Let $a_1, b_1, a_2, b_2, \ldots, a_n, b_n$ be nonnegative real numbers. Prove that

$$\sum_{i,j=1}^{n} \min\{a_i a_j, b_i b_j\} \leq \sum_{i,j=1}^{n} \min\{a_i b_j, a_j b_i\}.$$

**Solution.** Define

$$L(a_1, b_1, \ldots, a_n, b_n) = \sum_{i,j} (\min\{a_i b_j, a_j b_i\} - \min\{a_i a_j, b_i b_j\}).$$

Our goal is to show that

$$L(a_1, b_1, \ldots, a_n, b_n) \geq 0$$

for $a_1, b_1, \ldots, a_n, b_n \geq 0$. Our proof is by induction on $n$, the case $n = 1$ being evident. Using the obvious identities

- $L(a_1, 0, a_2, b_2, \ldots) = L(0, b_1, a_2, b_2, \ldots) = L(a_2, b_2, \ldots)$,
- $L(x, x, a_2, b_2, \ldots) = L(a_2, b_2, \ldots)$,

and the less obvious but easily verified identities

- $L(a_1, b_1, a_2, b_2, a_3, b_3, \ldots) = L(a_1 + a_2, b_1 + b_2, a_3, b_3, \ldots)$ if $a_1/b_1 = a_2/b_2$,
- $L(a_1, b_1, a_2, b_2, a_3, b_3, \ldots) = L(a_2 - b_1, b_2 - a_1, a_3, b_3, \ldots)$ if $a_1/b_1 = b_2/a_2$ and $a_1 \leq b_2$,

we may deduce the result from the induction hypothesis unless we are in the following situation:

1. all of the $a_i$ and $b_i$ are nonzero;
2. for $i = 1, \ldots, n$, $a_i \neq b_i$;
3. for $i \neq j$, $a_i/b_i \neq a_j/b_j$ and $a_i/b_i \neq b_j/a_j$.

For $i = 1, \ldots, n$, let $r_i = \max\{a_i/b_i, b_i/a_i\}$. Without loss of generality, we may assume $1 < r_1 < \cdots < r_n$, and that $a_1 < b_1$. Now notice that $f(x) = L(a_1, x, a_2, b_2, \ldots, a_n, b_n)$ is a *linear* function of $x$ in the interval $[a_1, r_2 a_1]$. Explicitly,

$$f(x) = \min\{a_1 x, x a_1\} - \min\{a_1^2, x^2\} + L(a_2, b_2, \ldots, a_n, b_n)$$

$$+ 2 \sum_{j=2}^{n} (\min\{a_1 b_j, x a_j\} - \min\{a_1 a_j, x b_j\})$$

$$= (x - a_1)(a_1 + 2 \sum_{j=2}^{n} c_j) + L(a_2, b_2, \ldots, a_n, b_n),$$

where $c_j = -b_j$ if $a_j > b_j$ and $c_j = a_j$ if $a_j < b_j$.

In particular, because $f$ is linear, we have

$$f(x) \geq \min\{f(a_1), f(r_2 a_1)\}.$$

Note that $f(a_1) = L(a_1, a_1, a_2, b_2, \ldots) = L(a_2, b_2, \ldots)$ and

$f(r_2 a_1) = L(a_1, r_2 a_1, a_2, b_2, \ldots)$

$$= \begin{cases} L(a_1 + a_2, r_2 a_1 + b_2, a_3, b_3, \ldots) & \text{if } r_2 = b_2/a_2, \\ L(a_2 - r_2 a_1, b_2 - a_1, a_3, b_3, \ldots) & \text{if } r_2 = a_2/b_2, \ b_2 \geq a_1, \\ L(r_2 a_1 - a_2, a_1 - b_2, a_3, b_3, \ldots) & \text{if } r_2 = a_2/b_2, \ b_2 < a_1, \end{cases}$$

Thus, we deduce the desired inequality from the induction hypothesis in all cases.

**Note.**   More precisely, it can be shown that for $a_i, b_i > 0$, equality holds if and only if, for each $r > 1$, the set $S_r$ of indices $i$ in $\{1, \ldots, n\}$ such that $a_i/b_i \in \{r, 1/r\}$ has the property that

$$\sum_{i \in S_r} a_i = \sum_{i \in S_r} b_i.$$

## 1.22 Vietnam

### Problem 1

Two circles $\omega_1$ and $\omega_2$ are given in the plane, with centers $O_1$ and $O_2$, respectively. Let $M_1'$ and $M_2'$ be two points on $\omega_1$ and $\omega_2$, respectively, such that the lines $O_1 M_1'$ and $O_2 M_2'$ intersect. Let $M_1$ and $M_2$ be points on $\omega_1$ and $\omega_2$, respectively, such that when measured clockwise the angles $\angle M_1' O_1 M_1$ and $\angle M_2' O_2 M_2$ are equal.

(a) Determine the locus of the midpoint of $\overline{M_1 M_2}$.

(b) Let $P$ be the point of intersection of lines $O_1 M_1$ and $O_2 M_2$. The circumcircle of triangle $M_1 P M_2$ intersects the circumcircle of triangle $O_1 P O_2$ at $P$ and another point $Q$. Prove that $Q$ is fixed, independent of the locations of $M_1$ and $M_2$.

**Solution.** (a) We use complex numbers. Let a lowercase letter denote the complex number associated with the point with the corresponding uppercase label. Let $M'$, $M$, and $O$ denote the midpoints of segments $\overline{M_1' M_2'}$, $\overline{M_1 M_2}$, and $\overline{O_1 O_2}$, respectively. Also let $z = \frac{m_1 - o_1}{m_1' - o_1} = \frac{m_2 - o_2}{m_2' - o_2}$, so that multiplication by $z$ is a rotation about the origin through some angle. Then $m = \frac{m_1 + m_2}{2}$ equals

$$\frac{1}{2}\left(o_1 + z(m_1' - o_1)\right) + \frac{1}{2}\left(o_2 + z(m_2' - o_2)\right) = o + z(m' - o),$$

that is, the locus of $M$ is the circle centered at $O$ with radius $OM'$.

(b) We shall use directed angles modulo $\pi$. Observe that

$$\angle Q M_1 M_2 = \angle Q P M_2 = \angle Q P O_2 = \angle Q O_1 O_2.$$

Similarly, $\angle Q M_2 M_1 = \angle Q O_2 O_1$, implying that triangles $Q M_1 M_2$ and $Q O_1 O_2$ are similar with the same orientations. Hence,

$$\frac{q - o_1}{q - o_2} = \frac{q - m_1}{q - m_2},$$

or equivalently

$$\frac{q - o_1}{q - o_2} = \frac{(q - m_1) - (q - o_1)}{(q - m_2) - (q - o_2)} = \frac{o_1 - m_1}{o_2 - m_2} = \frac{o_1 - m_1'}{o_2 - m_2'}.$$

Because lines $O_1 M_1'$ and $O_2 M_2'$ meet, $o_1 - m_1' \neq o_2 - m_2'$ and we can solve this equation to find a unique value for $q$.

## Problem 2

Suppose that all circumcircles of the four faces of a tetrahedron have congruent radii. Show that any two opposite edges of the tetrahedron are congruent.

**Solution.** We first prove the following claim ($*$): if $\overline{XZ}$ and $\overline{YW}$ are opposite edges of tetrahedron $ABCD$ such that $\angle XYZ + \angle XWZ = \pi$, then $\angle YXW + \angle YZW < \pi$. By the triangle inequality for solid angles, we find that

$$\angle XYZ + \angle YZW + \angle ZWX + \angle WXY$$

$$< (\angle ZYW + \angle WYX) + \angle YZW + (\angle XWY + \angle YWZ) + \angle WXY$$

$$= (\angle ZYW + \angle YWZ + \angle YZW) + (\angle XWY + \angle WYX + \angle WXY)$$

$$= \pi + \pi = 2\pi.$$

The claim ($*$) follows immediately.

We next prove the following claim (†): any two angles of tetrahedron $ABCD$ opposite the same side are either congruent or supplementary. Let $R$ be the common circumradius of tetrahedron $ABCD$'s four faces. Note that given any two angles of tetrahedron $ABCD$ opposite the same side, say angles $ABC$ and $ADC$, we have

$$\sin \angle ABC = \frac{AC}{2R} = \sin \angle ADC$$

by the Extended Law of Sines. The claim (†) follows immediately.

Let us now assume for the sake of contradiction that some pair of angles opposite the same side, say angles $ABC$ and $CDA$, are supplementary. If all of the other pairs of opposite angles were congruent, then we would have

$$\angle BCD + \angle DAB = (\pi - \angle CDB - \angle DBC) + (\pi - \angle ADB - \angle DBA)$$

$$= (\pi - \angle CAB - \angle DAC) + (\pi - \angle ACB - \angle DCA)$$

$$= (\pi - \angle CAB - \angle ACB) + (\pi - \angle DAC - \angle DCA)$$

$$= \angle ABC + \angle CDA = \pi,$$

a contradiction. Therefore, besides the angles opposite $\overline{AC}$, some other two angles opposite the same edge are supplementary. From claim ($*$), angles $BAD$ and $BCD$ are not supplementary. Hence, the angles opposite one of the remaining edges $\overline{AB}$, $\overline{AD}$, $\overline{CB}$, $\overline{CD}$ are supplementary. Without loss of generality, assume that $\angle ACB = \angle ADB$. By claims (∗) and (†), we have $\angle CAD = \angle CBD$.

Furthermore,

$$\angle CDB = \pi - \angle DCB - \angle DBC$$
$$= \pi - \angle DAB - \angle DAC$$
$$= \pi - (\pi - \angle ABD - \angle ADB) - (\pi - \angle ACD - \angle ADC)$$
$$= \angle ABD + \angle ACD + \angle ADB + \angle ADC - \pi$$
$$= \angle ABD + \angle ACD + (\pi - \angle ACB) + (\pi - \angle ABC) - \pi$$
$$= \angle ABD + \angle ACD + (\pi - \angle ACB - \angle ABC)$$
$$= \angle ABD + \angle ACD + \angle CAB,$$

implying that $\angle CDB > \angle CAB$. Because angles $CDB$ and $CAB$ are not congruent, they must be supplementary (by claim (†)). Now because angles $ADB$, $BDC$ and $CDA$ form a convex solid angle,

$$(\angle ADB + \angle BDC) + \angle CDA < \angle ADC + \angle CDA < 2\pi.$$

But

$$\angle ADB + \angle BDC + \angle CDA$$
$$= (\pi - \angle ACB) + (\pi - \angle BAC) + (\pi - \angle CBA)$$
$$= 3\pi - \pi = 2\pi,$$

a contradiction.

Therefore, the angles opposite the same side of the tetrahedron are congruent. As we argued before, in this case we have

$$\angle BCD + \angle DAB = \angle ABC + \angle CDA,$$

implying that $2\angle DAB = 2\angle ABC$ or $\angle DAB = \angle ABC$. Therefore, $DB = 2R\sin\angle DAB = 2R\sin\angle ABC = AC$. Similarly, $DA = BC$ and $DC = BA$. This completes the proof.

## Problem 3

Two circles $C_1$ and $C_2$ intersect at two points $P$ and $Q$. The common tangent of $C_1$ and $C_2$ closer to $P$ than to $Q$ touches $C_1$ and $C_2$ at $A$ and $B$, respectively. The tangent to $C_1$ at $P$ intersects $C_2$ at $E$ (distinct from $P$) and the tangent to $C_2$ at $P$ intersects $C_1$ at $F$ (distinct from $P$). Let

$H$ and $K$ be two points on the rays $AF$ and $BE$, respectively, such that $AH = AP$, $BK = BP$. Prove that the five points $A$, $H$, $Q$, $K$, $B$ lie on the same circle.

**Solution.** Because the given conditions are symmetric, we need only prove that $ABKQ$ is cyclic.

We use directed angles modulo $\pi$. Line $AP$ intersects ray $BE$, say at the point $R$. Let lines $AB$ and $PE$ intersect at $T$. Using tangents and cyclic quadrilaterals, we have $\angle QAR = \angle QAP = \angle QPE = \angle QBE = \angle QBR$, so $ABRQ$ is cyclic. We claim that $K = R$, from which our desired result follows.

Using the properties of the exterior angles of triangles $ABP$ and $EPR$, tangents $\overline{AB}$ and $\overline{PT}$, and cyclic quadrilaterals, we obtain

$$\angle BPR = \angle BAP + \angle PBA = \angle AQP + \angle PQB$$

$$= \angle APT + \angle PEB = \angle RPE + \angle PER = \angle PRB.$$

Hence, triangle $BPR$ is isosceles with $BP = BR$, implying that $R = K$. Our proof is complete.

## Problem 4

Let $a, b, c$ be pairwise coprime positive integers. An integer $n \geq 1$ is said to be *stubborn* if it cannot be written in the form

$$n = bcx + cay + abz$$

for any positive integers $x$, $y$, $z$. Determine, as a function of $a$, $b$, and $c$, the number of stubborn integers.

**Solution.** We claim that any integer $n$ can be written in the form $bcx + cay + abz$ where $x$, $y$, $z$ are integers with $0 < y \leq b$ and $0 < z \leq c$, where $x$ is possibly negative. Because $a$ and $bc$ are coprime, we can write $n = an' + bcx_0$ for some integers $n'$, $x_0$. Because $b$, $c$ are coprime, $n' = cy_0 + bz_0$ for some integers $y_0$, $z_0$. Hence, $n = bcx_0 + cay_0 + abz_0$. Choosing integers $\beta, \gamma$ such that $0 < y_0 + \beta b \leq b$ and $0 < z_0 + \gamma c \leq c$, we find that

$$n = bc(x_0 - \beta a - \gamma a) + ca(y_0 + \beta b) + ab(z_0 + \gamma c),$$

of the desired form.

Observe that any positive integer less than $bc + ca + ab$ is clearly stubborn. On the other hand, we claim that every integer $n > 2abc$ is not stubborn.

Given such an integer, write $n = bcx + cay + abz$ with $0 < y \le b$ and $0 < z \le c$. Then

$$2abc < bcx + cay + abz \le bcx + cab + abc = bcx + 2abc,$$

implying that $x > 0$, as needed.

Next, we prove that exactly half the positive integers in $S = [bc + ca + ab, 2abc]$ are stubborn. To do so, it suffices to prove that $n \in S$ is stubborn if and only if $f(n) = (2abc + bc + ca + ab) - n$ is not stubborn. For the "only if" direction, suppose that $n$ is stubborn and write $f(n) = bcx + cay + abz$ with $0 < y \le b$ and $0 < z \le c$. If $x$ were not positive, then we could write $n = bc(1-x) + ca(b+1-y) + ab(c+1-z)$, with $1 - x_0$, $b + 1 - y_0$, and $c + 1 - z_0$ positive — but this is impossible because $n$ is stubborn. Therefore, $x > 0$ and $f(n)$ is not stubborn.

To prove the "if" direction, suppose for the sake of contradiction that $f(n)$ is not stubborn and that $n$ is not stubborn as well. Write $f(n) = bcx_0 + cay_0 + abz_0$ and $n = bcx_1 + cay_1 + abz_1$ for positive integers $x_i, y_i, z_i$. Then

$$2abc = bc(x_0 + x_1 - 1) + ca(y_0 + y_1 - 1) + ab(z_0 + z_1 - 1).$$

Write $x = x_0 + x_1 - 1$ and define $y$ and $z$ similarly. Taking the above equation modulo $a$ shows that $0 \equiv bcx \pmod{a}$. Because $bc$ is relatively prime to $a$, $x$ must be divisible by $a$, implying that $x \ge a$. Similarly, $y \ge b$ and $z \ge c$. Thus, $2abc = bcx + cay + abz \ge 3abc$, a contradiction.

To summarize: the $bc + ca + ab - 1$ positive integers less than $bc + ca + ab$ are stubborn, every integer greater than $2abc$ is not stubborn, and half of the $2abc - (bc + ca + ab) + 1$ remaining positive integers are stubborn. This yields a total of

$$bc + ca + ab - 1 + \frac{2abc - (bc + ca + ab) + 1}{2} = \frac{2abc + bc + ca + ab - 1}{2}$$

stubborn positive integers.

## Problem 5

Let $\mathbb{R}^+$ denote the set of positive real numbers, and let $a, r > 1$ be real numbers. Suppose that $f : \mathbb{R}^+ \to \mathbb{R}$ is a function such that $(f(x))^2 \le ax^r f\left(\frac{x}{a}\right)$ for all $x > 0$.

(a) If $f(x) < 2^{2000}$ for all $x < \dfrac{1}{2^{2000}}$, prove that $f(x) \le x^r a^{1-r}$ for all $x > 0$.

(b) Construct such a function $f : \mathbb{R}^+ \to \mathbb{R}$ (not satisfying the condition given in (a)) such that $f(x) > x^r a^{1-r}$ for all $x > 0$.

**Solution.**   Observe that we can rewrite the given inequality in the form

$$\left( \frac{f(x)}{x^r a^{1-r}} \right)^2 \le \frac{f(x/a)}{(x/a)^r a^{1-r}}. \tag{$*$}$$

(a) Assume, for the sake of contradiction, that there exists $x_0$ such that $f(x_0) > x_0^r a^{1-r}$. Define $x_n = x_0/a^n$ and $\lambda_n = f(x_n)/(x_n^r a^{1-r})$ for $n \ge 0$, so that $\lambda_0 > 1$. From $(*)$, we have that $\lambda_{n+1} \ge \lambda_n^2$ for $n \ge 0$, and a straightforward proof by induction shows that $\lambda_n \ge \lambda_0^{2^n}$ for $n \ge 0$. We use this fact again soon; for now, observe that it implies that each $\lambda_n \ge \lambda_0^{2^n}$ is positive and hence that each $f(x_n)$ is positive as well. We may then set $x = x_n$ into the given inequality and rearrange the inequality to yield

$$\frac{f(x_{n+1})}{f(x_n)} \ge \frac{f(x_n)}{a x_n^r} = \frac{\lambda_n x_n^r a^{1-r}}{a x_n^r} = \frac{\lambda_n}{a^r}$$

for all $n \ge 0$.

There exists $N$ such that $2a^r < \lambda_0^{2^n} \le \lambda_n$ for all $n > N$. For all such $n$, we have $f(x_{n+1})/f(x_n) \ge 2$ or equivalently (because $f(x_n)$ is positive) $f(x_{n+1}) \ge 2f(x_n)$. Therefore, $f(x_n) \ge 2^{2000}$ for all sufficiently large $n$, but at the same time

$$x_n = \frac{x_0}{a^n} < \frac{1}{2^{2000}}$$

for all sufficiently large $n$. This contradicts the condition in (i), implying that our original assumption was false. Therefore, $f(x) \le x^r a^{1-r}$ for all $x$.

(b) For each real $x$, there exists a unique value $x_0 \in (1, a]$ such that $x_0/x = a^n$ for some integer $n$; let $\lambda(x) = x_0^{2^n}$, and set $f(x) = \lambda(x) x^r a^{1-r}$. By construction, we have $\lambda(x)^2 = \lambda(x/a)$ for all $x$; in other words, $(*)$ holds for all $x$. We also have that $\lambda(x) > 1$ for all $x$; in other words, $f(x) > x^r a^{1-r}$ for all $x$. This completes the proof.

# 2

# 2000 Regional Contests: Problems and Solutions

## 2.1   Asian Pacific Mathematical Olympiad

### Problem 1

Compute the sum

$$S = \sum_{i=0}^{101} \frac{x_i^3}{1 - 3x_i + 3x_i^2},$$

where $x_i = \frac{i}{101}$ for $i = 0, 1, \ldots, 101$.

**Solution.**   Because $1 - 3x + 3x^2 = x^3 + (1-x)^3 = x^3 - (x-1)^3$ is nonzero for all $x$, we can let

$$f(x) = \frac{x^3}{1 - 3x + 3x^2} = \frac{x^3}{x^3 + (1-x)^3}$$

for all $x$. Setting $x = x_i$ and $x = 1 - x_i = x_{101-i}$ above, and adding the two resulting equations, we find that $f(x_i) + f(x_{101-i}) = 1$. Therefore,

$$S = \sum_{i=0}^{101} f(x_i) = \sum_{i=0}^{50} (f(x_i) + f(1 - x_i)) = 51.$$

### Problem 2

We are given an arrangement of nine circular slots along three sides of a triangle: one slot at each corner, and two more along each side. Each of the numbers $1, 2, \ldots, 9$ is to be written into exactly one of these circles, so that

(i) the sums of the four numbers on each side of the triangle are equal;

(ii) the sums of the squares of the four numbers on each side of the triangle are equal.

Find all ways in which this can be done.

**Solution.** Take any such arrangement of the numbers. Let $x$, $y$, $z$ be the numbers in the corner slots, and let $S_1$ (resp., $S_2$) denote the sum of the four numbers (resp., of their squares) on any side. By the given conditions, we have

$$3S_1 = x + y + z + \sum_{k=1}^{9} k = x + y + z + 45,$$

$$3S_2 = x^2 + y^2 + z^2 + \sum_{k=1}^{9} k^2 = x^2 + y^2 + z^2 + 285.$$

From the second equation, we find that $x$, $y$, $z$ are either all divisible by three or all not divisible by three. By the Pigeonhole Principle, some two are congruent modulo 3. Taking the first equation modulo 3, we also find that $3 \mid (x + y + z)$. Hence, $x \equiv y \equiv z \pmod 3$.

If $(x, y, z) = (3, 6, 9)$ or $(1, 4, 7)$, then $S_2$ equals either 137 or 117. In either case, $S_2$ is congruent to 1 modulo 4, implying that exactly one number on each of the three sides of the triangle is odd. This is impossible because there are $5 > 3$ odd numbers to be written in the slots.

Hence, $(x, y, z) = (2, 5, 8)$, and $S_2 = 126$. Because $9^2 + 8^2 > 126$, the number 9 cannot lie on the same side as $8$ — i.e., it lies on the side containing the numbers 2 and 5. Because $\min\{7^2 + 9^2, 7^2 + 5^2 + 8^2\} > 126$, the number 7 must lie on the side containing 2 and 8. Given this information, the quadruples of numbers on the three sides must be $(2, 4, 9, 5)$, $(5, 1, 6, 8)$, and $(8, 7, 3, 2)$ in order for the sum of squares of the numbers on each side to equal 126. Indeed, all such arrangements satisfy the given conditions.

## Problem 3

Let $ABC$ be a triangle with median $\overline{AM}$ and angle bisector $\overline{AN}$. Draw the perpendicular to line $NA$ through $N$, hitting lines $MA$ and $BA$ at $Q$ and $P$, respectively. Also let $O$ be the point where the perpendicular to line $BA$ through $P$ meets line $AN$. Prove that $\overline{QO} \perp \overline{BC}$.

**Solution.** Let the line through $Q$ perpendicular to $OQ$ intersect lines $AB$ and $AC$ at $B'$ and $C'$, respectively, and let $R$ be the reflection of $P$ across line $AO$. Because $P, N, Q, R$ are collinear, $\angle OPQ = \angle QRO$. Because $\angle OPB' = \angle B'QO = \pi/2$, $O, B', P, Q$ are concyclic, and $\angle OPQ = \angle OB'Q$. Similarly, $\angle QRO = \angle QC'O$. Therefore, $\angle OB'Q = \angle QC'O$, and $OB'C'$ is an isosceles triangle, implying $B'Q = QC'$. Now I claim that lines $B'C'$ and $BC$ are parallel. If they were not, then let the line through $Q$ parallel to $BC$ intersect lines $AB$ and $AC$ at $B_1$ and $C_1$, respectively. Because $B_1 Q = QC_1$ and $B'Q = QC'$, we would have $\triangle B_1 B'Q \sim \triangle C_1 C'Q$. But this would imply that lines $B'B_1$ and $C'C_1$ were parallel, which could not be the case. Therefore, $B'C' \parallel BC$, and because $\overline{QO} \perp \overline{B'C'}$, we have $\overline{QO} \perp \overline{BC}$.

## Problem 4

Let $n, k$ be positive integers with $n > k$. Prove that

$$\frac{1}{n+1} \cdot \frac{n^n}{k^k(n-k)^{n-k}} < \frac{n!}{k!(n-k)!} < \frac{n^n}{k^k(n-k)^{n-k}}.$$

**Solution.** We use the Binomial Theorem to write $n^n = (k + (n - k))^n$ in the form $\sum_{m=0}^n a_m$, where

$$a_m = \binom{n}{m} k^m (n-k)^{n-m} > 0$$

for each $m$. The desired chain of inequalities is then equivalent to

$$\frac{n^n}{n+1} < a_k < n^n.$$

The right inequality holds because $n^n = \sum_{m=0}^n a_m > a_k$. To prove the left inequality, it suffices to prove that $a_k$ is larger than each of $a_0, \ldots, a_{k-1}, a_{k+1}, \ldots, a_n$, because then

$$n^n = \sum_{m=0}^n a_m < \sum_{m=0}^n a_k = (n+1)a_k.$$

Indeed, we prove that $a_m$ is increasing for $m \leq k$ and decreasing for $m \geq k$. Observe that

$$\binom{n}{m} = \frac{m+1}{n-m}\binom{n}{m+1}.$$

Hence,

$$\frac{a_m}{a_{m+1}} = \frac{\binom{n}{m} k^m (n-k)^{n-m}}{\binom{n}{m+1} k^{m+1}(n-k)^{n-m-1}} = \frac{n-k}{n-m} \cdot \frac{m+1}{k}.$$

This expression is less than 1 when $m < k$, and it is greater than 1 when $m \geq k$. In other words, $a_0 < \cdots < a_k$ and $a_k > \cdots > a_n$, as desired.

## Problem 5

Given a permutation $(a_0, a_1, \ldots, a_n)$ of the sequence $0, 1, \ldots, n$, a transposition of $a_i$ with $a_j$ is called *legal* if $a_i = 0$, $i > 0$, and $a_{i-1} + 1 = a_j$. The permutation $(a_0, a_1, \ldots, a_n)$ is called *regular* if after finitely many legal transpositions it becomes $(1, 2, \ldots, n, 0)$. For which numbers $n$ is the permutation $(1, n, n - 1, \ldots, 3, 2, 0)$ regular?

**Solution.** Fix $n$, and let $\pi_0$ and $\pi_1$ denote the permutations

$$(1, n, n - 1, \ldots, 3, 2, 0) \quad \text{and} \quad (1, 2, \ldots, n, 0),$$

respectively. We say that $\pi_0$ *terminates* in a permutation $\pi_1'$ if applying some legal transpositions to $\pi_0$ eventually yields $\pi_1'$, and if no legal transpositions may be applied to $\pi_1'$. Because no legal transpositions may be applied to $\pi_1$, if $\pi_0$ is regular then it terminates in $\pi_1$. As we apply legal transformations to $\pi_0$, at most one legal transposition may be applied to each resulting permutation. Hence, $\pi_0$ terminates in at most one permutation.

If $n$ equals 1 or 2, it is easy to check that $(1, n, n - 1, \ldots, 3, 2, 0)$ is regular. If $n$ is instead greater than 2 and even, we claim that $\pi_0$ does not terminate in $\pi_1$ and hence is not regular. For $k \in [0, (n-2)/2]$, applying $k$ legal transpositions to $\pi_0$ yields a permutation that begins with the entries $1, n, n-1, \ldots, 2k+2, 0$. Hence, $\pi_0$ terminates in a permutation beginning with $1, n, 0$, obtained after $(n - 2)/2$ legal transpositions.

Now suppose that $n > 2$ and $n$ is odd. In order to consider this case, we introduce some notation: For all integers $s > 0$, $t \geq 0$ such that $s + t$ divides $n + 1$, we construct a permutation called the $(s, t)$-*staircase* one entry at a time as follows, applying (1) once and then repeatedly applying (2) and (3) in alternating fashion:

(1) Let the first $s$ entries be $1, 2, \ldots, s - 1, 0$.

(2) Let the next $t$ entries be the largest $t$ numbers in $\{1, 2, \ldots, n\}$ not yet assigned to an entry, arranged in increasing order.

(3) Let the next $s$ entries be the largest $s$ numbers in $\{1, 2, \ldots, n\}$ not yet assigned to an entry, arranged in increasing order.

If $(s+t) \mid (n+1)$ and $t > 0$, then applying $n/(s+t)$ legal transpositions to the $(s, t)$-staircase yields the $(s + 1, t - 1)$-staircase. Repeatedly performing legal transpositions thus eventually yields the $(s + t, 0)$-staircase, a process we refer to as *collecting a staircase*.

Next, suppose that $s \mid (n + 1)$. If $2s \nmid (n + 1)$, then applying $n/s - 2$ legal transpositions to the $(s, 0)$-staircase yields a permutation different from $\pi_1$ to which no further legal transpositions may be applied. If instead $2s \mid (n + 1)$, then the $(s, 0)$-staircase is actually the $(s, s)$-staircase, which can be collected into the $(2s, 0)$-staircase.

We now prove that if $n > 2$ and $n$ is odd, then $\pi_0$ is regular if and only if $n + 1$ is a power of 2. Because $n + 1$ is even, we may write $n + 1 = 2^q r$ where $q$ is a positive integer and $r$ is an odd integer. Applying $(n - 1)/2$ legal transpositions to $\pi_0$ yields the $(2, 0)$-staircase. If $2^q > 2$, then because $2s$ divides $n + 1$ for $s = 2^1, \ldots, 2^{q-1}$, we can repeatedly collect staircases to eventually yield the $(2^q, 0)$-staircase. If $2^q = 2$, then we already have the $(2^q, 0)$-staircase.

If $r = 1$, then we have obtained $\pi_1$, and $\pi_0$ is regular. Otherwise, applying $r - 2$ additional legal transposition yields a permutation in which 0's left neighbor is $n$. Hence, no more legal transpositions are possible. However, this final permutation begins $1, 2, \ldots, 2^q$ rather than $1, n$ — implying that $\pi_0$ does not terminate in $\pi_1$ and hence that $\pi_0$ is not regular.

Therefore, $\pi_0$ is regular if and only if $n$ equals 2 or $n + 1$ is a power of 2.

## 2.2  Austrian-Polish Mathematics Competition

### Problem 1

Find all positive integers $N$ whose only prime divisors are 2 and 5, such that the number $N + 25$ is a perfect square.

**Solution.**  We are given that $N$ is of the form $2^a \cdot 5^b$, where $a$ and $b$ are nonnegative integers. For some integer $x > 5$, we have $x^2 = N + 25$ or equivalently $(x + 5)(x - 5) = N$. Thus, $N$ is expressible as the product of two natural numbers differing by 10. We consider two cases:

*Case 1.* $b = 0$. Then $2^a = (x + 5)(x - 5)$, so $x + 5$ and $x - 5$ are powers of 2. But no two powers of 2 differ by 10, so this case yields no solutions.

*Case 2.* $b \geq 1$. In this case, $x^2$ is divisible by 5, so it must be divisible by 25. It follows that $b \geq 2$. Let $x = 5y$, giving $y > 1$ and $(y - 1)(y + 1) = 2^a \cdot 5^{b-2}$. If $y - 1$ and $y + 1$ are odd, then $2^a \cdot 5^{b-2}$ is odd and must be a power of 5 — implying that $y - 1$ and $y + 1$ are powers of 5 that differ by 2, which is impossible. Thus, $y - 1$ and $y + 1$ are even, and $p = \frac{1}{2}(y - 1)$ and $q = \frac{1}{2}(y + 1)$ are consecutive positive integers whose product is $2^{a-2}5^{b-2}$. Hence, $p$ and $q$ equal $2^m$ and $5^n$ in some order, for some nonnegative integers $m$ and $n$. We consider two possible subcases:

*Subcase 1.* $5^n - 2^m = 1$. Because $5^n, 2^m \not\equiv 0 \,(\mathrm{mod}\ 3)$, we must have $2^m \equiv 1 \,(\mathrm{mod}\ 3)$ and $5^n \equiv 2 \,(\mathrm{mod}\ 3)$. Thus, $n$ is odd and $5^n \equiv 5 \,(\mathrm{mod}\ 8)$. It follows that $2^m \equiv 5^m - 1 \equiv 4 \,(\mathrm{mod}\ 8)$, implying that $m = 2$. This yields the solution $N = 2000$.

*Subcase 2.* $2^m - 5^n = 1$. Because any power of 5 is congruent to 1 modulo 4, we have $2^m \equiv 5^n + 1 \equiv 2 \,(\mathrm{mod}\ 4)$, implying that $m = 1$ and $n = 0$. This yields the solution $N = 200$.

Thus, the only solutions are $N = 200, 2000$.

### Problem 2

For which integers $n \geq 5$ is it possible to color the vertices of a regular $n$-gon using at most 6 colors such that any 5 consecutive vertices have different colors?

**Solution.**  Let the colors be $a$, $b$, $c$, $d$, $e$, $f$. Denote by $S_1$ the sequence $a$, $b$, $c$, $d$, $e$, and by $S_2$ the sequence $a$, $b$, $c$, $d$, $e$, $f$. If $n > 0$ is representable in the form $5x + 6y$, for $x, y \geq 0$, then $n$ satisfies the conditions of the problem: we may place $x$ consecutive $S_1$ sequences, followed by $y$ consecutive $S_2$ sequences, around the polygon. Setting $y$ equal to 0, 1, 2, 3, or 4, we find that $n$ may equal any number of the form $5x$, $5x + 6$, $5x + 12$, $5x + 18$, or $5x + 24$. The only numbers greater than 4 not of this form are 7, 8, 9, 13, 14, and 19. We show that none of these numbers has the required property.

Assume for a contradiction that a coloring exists for $n$ equal to one of 7, 8, 9, 13, 14, and 19. There exists a number $k$ such that $6k < n < 6(k+1)$. By the Pigeonhole Principle, at least $k + 1$ vertices of the $n$-gon have the same color. Between any two of these vertices are at least 4 others, because any 5 consecutive vertices have different colors. Hence, there are at least $5k + 5$ vertices, and $n \geq 5k + 5$. However, this inequality fails for $n = 7, 8, 9, 13, 14, 19$, a contradiction.

Hence, a coloring is possible for all $n \geq 5$ except $7, 8, 9, 13, 14$, and 19.

## Problem 3

Let the *3-cross* be the solid made up of one central unit cube with six other unit cubes attached to its faces, such as the solid made of the seven unit cubes centered at $(0,0,0)$, $(\pm 1, 0, 0)$, $(0, \pm 1, 0)$, and $(0, 0, \pm 1)$. Prove or disprove that (three-dimensional) space can be tiled with 3-crosses in such a way that no two of them share any interior points.

**Solution.**    We give a tiling of space with 3-crosses, letting the centers of the unit cubes comprising the solids coincide with the lattice points. To each lattice point $(x, y, z)$, assign the index number $x + 2y + 3z$, modulo 7. We call two lattice points *adjacent* if and only if they differ by 1 in exactly one coordinate. It is clear by inspection that any 3-cross contains 7 cubes whose centers have precisely the indices 0 through 6. From this, it is also clear that any lattice point with index not equal to 0 is adjacent to a unique lattice point with index 0. Therefore, space may be tiled with the 3-crosses whose centers are those lattice points with index 0.

## Problem 4

In the plane the triangle $A_0 B_0 C_0$ is given. Consider all triangles $ABC$ satisfying the following conditions: (i) $C_0$, $A_0$, and $B_0$ lie on $\overline{AB}$, $\overline{BC}$, and $\overline{CA}$, respectively; (ii) $\angle ABC = \angle A_0 B_0 C_0$, $\angle BCA = \angle B_0 C_0 A_0$, and $\angle CAB = \angle C_0 A_0 B_0$. Find the locus of the circumcenter of all such triangles $ABC$.

**First Solution.** Note that at least one such triangle $ABC$ exists: for instance, we could let triangle $A_0B_0C_0$ be the medial triangle of $ABC$. Hence, the desired locus is nonempty.

Let triangle $A_0B_0C_0$ have circumcircle $\omega$ with center $O$. We use a vector coordinate system with origin at $O$, where the vector labelled by a lowercase letter corresponding to the point with the corresponding uppercase letter.

Suppose that triangle $ABC$ satisfies the given conditions. Because $A$ and $A_0$ are on opposite sides of line $BC$, and $\angle C_0A_0B_0 = \angle CAB$, $A$ must lie on the reflection of $\omega$ across line $B_0C_0$. Hence, the circumcenter $O_1$ of triangle $B_0C_0A$ must be the reflection of $O$ across line $B_0C_0$. We can similarly locate the circumcenters $O_2$ and $O_3$ of triangles $A_0C_0B$ and $A_0B_0C$. Then quadrilateral $OB_0O_1C_0$ is a rhombus so that $\mathbf{o}_1 = \mathbf{b}_0 + \mathbf{c}_0$. Similarly, $\mathbf{o}_2 = \mathbf{a}_0 + \mathbf{c}_0$.

Let $M$ be the midpoint of $\overline{O_1O_2}$. For the rest of this paragraph, we introduce a temporary Cartesian coordinate system so that line $AB$ is the $x$-axis, where $A = (a,0)$, $C_0 = (c_0,0)$, and $B = (b,0)$. Because $O_1$ and $O_2$ lie on the perpendicular bisectors of $\overline{AC_0}$ and $\overline{C_0B}$, respectively, their $x$-coordinates are $\frac{1}{2}(a+c_0)$ and $\frac{1}{2}(b+c_0)$, respectively. Hence, their midpoint $M$ has $x$-coordinate $\frac{1}{4}(a+b) + \frac{1}{2}c_0$. Thus, the reflection $H'$ of $C_0$ across $M$ has $x$-coordinate $\frac{1}{2}(a+b)$, implying that it lies on the perpendicular bisector of $\overline{AB}$. Note that $\mathbf{h}' = \mathbf{m} + (\mathbf{m} - \mathbf{c}_0) = \mathbf{a}_0 + \mathbf{b}_0 + \mathbf{c}_0$. In other words, $H'$ lies on the ray from $O$ through the centroid $G$ of triangle $A_0B_0C_0$, where $OH' = 3OG$; it follows that $H'$ is the orthocenter $H$ of triangle $A_0B_0C_0$.

Likewise, the perpendicular bisectors of $\overline{BC}$ and $\overline{CA}$ also pass through the orthocenter $H$ of triangle $A_0B_0C_0$. Therefore, the desired locus (which we already showed is nonempty) must consist of one single point, the orthocenter of triangle $A_0B_0C_0$.

**Second Solution.** We observe as in the first solution that $A, B, C$ are on the reflections of the circumcircle $\omega$ of triangle $A_0B_0C_0$ over the lines $B_0C_0, C_0A_0, A_0B_0$, respectively. Let those reflections of $\omega$ be $\omega_1, \omega_2, \omega_3$, respectively. We claim that $\omega_1, \omega_2, \omega_3$ all pass through the orthocenter $H$ of triangle $A_0B_0C_0$. We prove this claim for the case in which $A_0B_0C_0$ is acute; the proof for the obtuse case is similar. Consider triangle $HB_0C_0$. We have $\angle HB_0C_0 = \pi/2 - \angle C_0$ and $\angle HC_0B_0 = \pi/2 - \angle B_0$, implying that $\angle B_0HC_0 = \angle B_0 + \angle C_0$. Because we are given that $\angle B_0AC_0 = \angle A_0$, we see that $\angle B_0HC_0$ and $\angle B_0AC_0$ are supplementary.

Therefore, quadrilateral $B_0HC_0A$ is cyclic. It follows that $H$ lies on $\omega_1$, and similarly it lies on $\omega_2$ and $\omega_3$ as well.

Now observe that angles $A_0BH$ and $A_0CH$ both intercept segment $A_0H$. Because $\omega_2$ and $\omega_3$ are congruent, it follows that $\angle A_0BH = \angle A_0CH$. Therefore, triangle $HBC$ is isosceles with $HB = HC$. Similarly, we have $HA = HB$. Thus, $H$ is the circumcenter of triangle $ABC$. It follows that the desired locus cannot contain any points except $H$. As shown in the first solution, this locus is nonempty, implying that it consists of the single point $H$.

## Problem 5

We are given a set of 27 distinct points in the plane, no three collinear. Four points from this set are vertices of a unit square; the other 23 points lie inside this square. Prove that there exist three distinct points $X, Y, Z$ in this set such that $[XYZ] \leq \frac{1}{48}$.

**Solution.** We prove by induction on $n$ that, given $n \geq 1$ points inside the square (with no three collinear), the square may be partitioned into $2n + 2$ triangles, where each vertex of these triangles is either one of the $n$ points or one of the vertices of the square. For the base case $n = 1$, because the square is convex, we may partition the square into 4 triangles by drawing line segments from the interior point to the vertices of the square. For the inductive step, assume the claim holds when $n$ equals some value $k \geq 1$. Then for the case $n = k + 1$, take $k$ of the points, and partition the square into $2k + 2$ triangles whose vertices are either vertices of the square or are among the $k$ chosen points. Call the remaining point $P$. Because no three of the points in the set are collinear, $P$ lies inside one of the $2n + 2$ partitioned triangles $ABC$. We may further divide this triangle into the triangles $APB$, $BPC$, and $CPA$. This yields a partition of the square into $2(n + 1) + 2 = 2n + 4$ triangles, completing the induction.

For the special case $n = 23$, we may divide the square into 48 triangles with total area 1. One of the triangles has area at most $\frac{1}{48}$, as desired.

## Problem 6

For all real numbers $a, b, c \geq 0$ such that $a + b + c = 1$, prove that

$$2 \leq (1 - a^2)^2 + (1 - b^2)^2 + (1 - c^2)^2 \leq (1 + a)(1 + b)(1 + c)$$

and determine when equality occurs for each of the two inequalities.

**Solution.**   Let the left, middle, and right portions of the inequality be denoted by $L$, $M$, and $R$, respectively. Also, given a function $f$ in three variables, let the *symmetric sum* $\sum_{\text{sym}} f(a, b, c)$ denote the sum $f(x, y, z)$ over all permutations $(x, y, z)$ of $(a, b, c)$.

Because $a + b + c = 1$,

$$L = 2(a + b + c)^4 = \sum_{\text{sym}}(a^4 + 8a^3b + 6a^2b^2 + 12a^2bc),$$

$$M = \frac{1}{2}\sum_{\text{sym}}((a + b + c)^2 - a^2)^2$$

$$= \sum_{\text{sym}}(a^4 + 8a^3b + 7a^2b^2 + 16a^2bc),$$

$$R = (2a + b + c)(a + 2b + c)(a + b + 2c)(a + b + c)$$

$$= \sum_{\text{sym}}(a^4 + 9a^3b + 7a^2b^2 + 15a^2bc).$$

From these computations, we have

$$M - L = \sum_{\text{sym}}(a^2b^2 + 4a^2bc),$$

which is nonnegative because $a, b, c \geq 0$. Equality holds if and only if at least two of $a$, $b$, $c$ are 0; that is, if and only if $(a, b, c) = (0, 0, 1)$, $(0, 1, 0)$, or $(1, 0, 0)$. Also,

$$R - M = \sum_{\text{sym}}(a^3b - a^2bc) = \sum_{\text{sym}}\left(\frac{1}{3}(a^3b + a^3b + bc^3) - a^2bc\right),$$

which is nonnegative by the arithmetic mean-geometric mean inequality. Equality holds if and only if $a^3b = bc^3$, $b^3c = ca^3$, and $c^3a = ab^3$ hold simultaneously. This implies that either two of $a$, $b$, $c$ equal zero, or $a = b = c$. So equality holds if and only if $(a, b, c) = \left(\frac{1}{3}, \frac{1}{3}, \frac{1}{3}\right)$, $(1, 0, 0)$, $(0, 1, 0)$, or $(0, 0, 1)$.

## 2.3  Balkan Mathematical Olympiad

### Problem 1

Let $E$ be a point inside nonisosceles acute triangle $ABC$ lying on median $\overline{AD}$, and drop perpendicular $\overline{EF}$ to line $BC$. Let $M$ be an arbitrary point on $\overline{EF}$, and let $N$ and $P$ be the orthogonal projections of $M$ onto lines $AC$ and $AB$, respectively. Prove that the angle bisectors of angles $PMN$ and $PEN$ are parallel.

**Solution.**

**Lemma.** *In a (possibly concave) quadrilateral $TUVW$ in which angles $UVW$ and $WTU$ are less than $\pi$, the bisectors of angles $TUV$ and $VWT$ are parallel if $\angle UVW = \angle WTU$.*

*Proof:* We use directed angles modulo $\pi$ in the proof of this lemma. Let $U'$, $V'$, $W'$, and $T'$ be points on the angle bisectors of angles $TUV$, $UVW$, $VWT$, and $WTU$, respectively. Let line $WW'$ intersect line $TU$ at $Q$. Note that

$$\pi = \angle TQW + \angle QWT + \angle WTQ$$
$$= \angle TQW + \angle W'WT + \angle WTU$$

and

$$\pi = \angle T'TU + \angle U'UV + \angle V'VW + \angle W'WT.$$

Setting these two expressions for $\pi$ equal to each other, we find that

$$\angle TQW + \angle T'TU = \angle U'UV + \angle V'VW.$$

Because angles $WTU$ and $UVW$ of the quadrilateral measure less than $\pi$ and are congruent, we have $\angle T'TU = \angle V'VW$. Hence,

$$\angle TQW = \angle U'UV = \angle TUU',$$

implying that the angle bisectors of $\angle TUV$ and $\angle VWT$ are parallel.  ∎

Let the line through $E$ parallel to $\overline{BC}$ intersect $\overline{AB}$ at $X$ and $\overline{AC}$ at $Y$. Let the line through $M$ parallel to $\overline{BC}$ intersect $\overline{AB}$ at $G$ and $\overline{AC}$ at $H$. If $X$ coincides with $P$, then $\angle EPM = \angle EXM$. Otherwise, because $\angle XPM = \pi/2 = \angle XEM$, quadrilateral $XEPM$ is cyclic with $X$ and $P$ on the same side of line $EM$; hence, we again find that $\angle EPM = \angle EXM$. Similarly, $\angle ENM = \angle EYM$.

The homothety about $A$ that sends $D$ to $X$ sends $\overline{BC}$ to $\overline{XY}$, implying that $XE = EY$. Hence, right triangles $XEM$ and $YEM$ are congruent, implying that $\angle EXM = \angle EYM$. Therefore, $\angle EPM = \angle ENM$. Because $P$ and $N$ lie on opposite sides of line $EF$, the interior angles $EPM$ and $ENM$ of quadrilateral $NEPM$ measure less than $\pi$. Applying the lemma to quadrilateral $NEPM$, we find that the angle bisectors of angles $PMN$ and $PEN$ are parallel.

## Problem 2

Find the maximum number of $1 \times 10\sqrt{2}$ rectangles one can remove from a $50 \times 90$ rectangle by using cuts parallel to the edges of the original rectangle.

**Solution.**    We begin by proving that it is possible to remove 315 rectangles. Place the $50 \times 90$ rectangle in the first quadrant of the coordinate plane so that its vertices are $(0, 0)$, $(90, 0)$, $(90, 50)$, and $(0, 50)$. We first remove the rectangular region with opposite vertices $(0, 0)$ and $(6 \cdot 10\sqrt{2}, 50 \cdot 1)$ by dividing it into 50 rows and 6 columns of $1 \times 10\sqrt{2}$ rectangles (oriented with their longer sides parallel to the $x$-axis). We may then also remove the rectangular region with opposite vertices $(60\sqrt{2}, 0)$ and $(60\sqrt{2}+5, 30\sqrt{2})$ by dividing it into 3 rows and 5 columns of $10\sqrt{2} \times 1$ rectangles (oriented with their shorter sides parallel to the $x$-axis). In total, we remove a total of 315 rectangles.

Now we prove that 315 is the maximum number of rectangles that can be removed. Partition the rectangle into square and rectangular regions by drawing the lines of the form $x = 5\sqrt{2}n$ and $y = 5\sqrt{2}n$ for each nonnegative integer $n$. Color the resulting regions chessboard-style so that the colors of the regions alternate between black and white. Without loss of generality, assume that the total black area is at least as large as the total white area.

Let $\mathcal{R}_1$ and $\mathcal{R}_2$ be the uppermost regions in the right column of our partition, so that $\mathcal{R}_1$ is a $(50 - 35\sqrt{2}) \times (90 - 60\sqrt{2})$ rectangle and $\mathcal{R}_2$ is a $5\sqrt{2} \times (90 - 60\sqrt{2})$ rectangle. There are six rectangles immediately below them in the partition; these can be divided into three pairs of adjacent, congruent rectangles. The remaining rectangles in the partition — forming twelve columns of width $5\sqrt{2}$ — can also be divided into pairs of (horizontally) adjacent, congruent rectangles. Because any two adjacent rectangles in the partition are of different colors, it follows that outside $\mathcal{R}_1$ and $\mathcal{R}_2$, the total black area equals the total white area of the partition.

Therefore, the total black area in the partition must exceed the total white area by

$$|\text{Area}(\mathcal{R}_2) - \text{Area}(\mathcal{R}_1)|$$
$$= |5\sqrt{2} \cdot (90 - 60\sqrt{2}) - (50 - 35\sqrt{2}) \cdot (90 - 60\sqrt{2})|$$
$$= (40\sqrt{2} - 50)(90 - 60\sqrt{2}).$$

Hence, the total white area in the partition is

$$\frac{1}{2}(90 \cdot 50) - \frac{1}{2}(40\sqrt{2} - 50)(90 - 60\sqrt{2}) = 6900 - 3300\sqrt{2}.$$

Any $1 \times 10\sqrt{2}$ rectangle we remove contains $5\sqrt{2}$ units of white area, implying that we remove at most

$$\frac{6900 - 3300\sqrt{2}}{5\sqrt{2}} < 316$$

rectangles. This proves that the maximum number of rectangles removed is indeed 315.

## Problem 3

Call a positive integer $r$ a *perfect power* if it is of the form $r = t^s$ for some integers $s$, $t$ greater than 1. Show that for any positive integer $n$, there exists a set $S$ of $n$ distinct positive integers with the following property: given any nonempty subset $T$ of $S$, the arithmetic mean of the elements in $T$ is a perfect power.

**Solution.** Given a set $Z$ of positive integers and a positive integer $m$, let $mZ = \{mz \mid z \in Z\}$ and let $\mu(Z)$ denote the arithmetic mean of the elements in $Z$. Because $\mu(Z)$ is a linear function of the elements in $Z$, $\mu(mZ) = m\mu(Z)$.

**Lemma.** *Let $A$ be a nonempty set of positive integers. There exists $m \in \mathbb{N}$ such that $mA$ contains only perfect powers.*

*Proof:* Let $k = |A|$ and write $A = \{a_1, a_2, \ldots, a_k\}$. Let $p_1, p_2, \ldots, p_N$ be all the prime factors of $\prod_{i=1}^{k} a_i$. For $i = 1, 2, \ldots, k$, there exist nonnegative integers $\alpha_{i,j}$ such that $a_i = \prod_{j=1}^{N} p_j^{\alpha_{i,j}}$.

Let $q_1, q_2, \ldots, q_k$ be distinct primes. For $j = 1, 2, \ldots, N$, by the Chinese Remainder Theorem, there exists $\beta_j$ such that $\beta_j \equiv -\alpha_{i,j} \pmod{q_i}$

for $i = 1, 2, \ldots, k$. Let $m = \prod_{j=1}^{N} p_j^{\beta_j}$. Then for $i = 1, 2, \ldots, k$,

$$ma_i = \prod_{j=1}^{N} p_j^{\alpha_{i,j} + \beta_j} = \left( \prod_{j=1}^{N} p_j^{\frac{\alpha_{i,j} + \beta_j}{q_i}} \right)^{q_i}$$

is a perfect power, as desired.                                    ∎

Fix a positive integer $n$. Let $\tilde{S}$ be a set of $n$ distinct positive multiples of $\mathrm{lcm}(1, 2, \ldots, n)$, so that $\mu(\tilde{T})$ is a positive integer for all nonempty subsets $\tilde{T}$ of $\tilde{S}$. Let $A$ equal the set of all such values $\mu(\tilde{T})$. By the lemma, there exists $m \in \mathbb{N}$ such that $mA$ contains only perfect powers; we claim that $S = m\tilde{S}$ has the required property.

Indeed, suppose that $T$ is an arbitrary nonempty subset of $S$. Then $T$ equals $m\tilde{T}$ for some nonempty subset $\tilde{T}$ of $\tilde{S}$, implying that $\mu(T) = m\mu(\tilde{T}) \in mA$ is a perfect power, as needed.

## 2.4 Mediterranean Mathematical Competition

### Problem 1

We are given $n$ different positive numbers $a_1, a_2, \ldots, a_n$ and the set $\{\sigma_1, \sigma_2, \ldots, \sigma_n\}$, where each $\sigma_i \in \{-1, 1\}$. Prove that there exist a permutation $(b_1, b_2, \ldots, b_n)$ of $a_1, a_2, \ldots, a_n$ and a set $\{\beta_1, \beta_2, \ldots, \beta_n\}$ where each $\beta_i \in \{-1, 1\}$, such that the sign of $\sum_{j=1}^{i} \beta_j b_j$ equals the sign of $\sigma_i$ for all $1 \leq i \leq n$.

**Solution.** We construct a sequence of nonzero numbers $x_1, x_2, \ldots, x_n$ with the following properties: (i) for $1 \leq i \leq n$, $x_1, x_2, \ldots, x_i$ have distinct absolute values; (ii) when sorted in order of increasing absolute value, their signs alternate; and (iii) for $1 \leq i \leq n$, the sign of the number in $x_1, x_2, \ldots, x_i$ with largest absolute value equals $\sigma_i$. To do so, we simply construct $x_1, x_2, \ldots, x_n$ in that order, at each step choosing $x_{i_0}$ with the proper sign so that property (ii) holds for $i = i_0$, and either setting $|x_{i_0}| > \max\{|x_1|, |x_2|, \ldots, |x_{i_0-1}|\}$ or $|x_{i_0}| < \min\{|x_1|, |x_2|, \ldots, |x_{i_0-1}|\}$ so that property (iii) holds for $i = i_0$.

Choose the $b_i$ and $\beta_i$ such that $b_{j_1} < b_{j_2} \iff |x_{j_1}| < |x_{j_2}|$ and $\beta_j x_j > 0$ for all $j, j_1, j_2$. Suppose that $1 \leq i \leq n$. Arrange $b_1, b_2, \ldots, b_i$ in increasing order to obtain $b_{k_1}, b_{k_2}, \ldots, b_{k_i}$. By construction, the sequence of signs $\beta_{k_1}, \beta_{k_2}, \ldots, \beta_{k_i}$ alternates, and $\beta_{k_i} = \sigma_i$. Therefore,

$$\sum_{j=1}^{i} \beta_j b_j = \sigma_i \left( b_{k_i} - b_{k_{i-1}} + b_{k_{i-2}} - b_{k_{i-3}} + \cdots \pm b_{k_1} \right).$$

The expression in parentheses is the sum of $\lfloor k/2 \rfloor$ positive expressions of the form $b_{k_{j+1}} - b_{k_j}$ and perhaps an additional positive term $b_{k_1}$. Therefore, $\sum_{j=1}^{i} \beta_j b_j$ has the same sign as $\sigma_i$ for each $i$, as desired.

### Problem 2

Outwards along the sides of convex quadrilateral $ABCD$ are constructed equilateral triangles $WAB$, $XBC$, $YCD$, $ZDA$ with centroids $S_1$, $S_2$, $S_3$, $S_4$, respectively. Prove that $\overline{S_1 S_3} \perp \overline{S_2 S_4}$ if and only if $AC = BD$.

**Solution.** Choose an arbitrary point $O$ as the origin. Let $\mathbf{a}$, $\mathbf{b}$, $\mathbf{c}$, and $\mathbf{d}$ denote the vectors from $O$ to $A$, $B$, $C$, and $D$, respectively. Let

$M_1, M_2, M_3, M_4$ denote the midpoints of $\overline{AB}$, $\overline{BC}$, $\overline{CD}$, $\overline{DA}$, respectively, and let $\mathbf{s}_i'$ denote the vector from the $M_i$ to $S_i$ for $i = 1, 2, 3, 4$.

Given two vectors $\mathbf{x}$ and $\mathbf{y}$, let $\angle(\mathbf{x}, \mathbf{y})$ denote the clockwise angle between them. (All angles are directed modulo $2\pi$.) Without loss of generality, assume that $ABCD$ is oriented clockwise, and let $\varphi$ be the transformation that rotates any vector $\pi/2$ counterclockwise and multiplies its magnitude by $1/2\sqrt{3}$. Then

$$\varphi(\mathbf{x}) \cdot \varphi(\mathbf{y}) = |\varphi(\mathbf{x})||\varphi(\mathbf{y})| \cos \angle(\varphi(\mathbf{x}), \varphi(\mathbf{y}))$$

$$= \left(\frac{|\mathbf{x}|}{2\sqrt{3}}\right)\left(\frac{|\mathbf{y}|}{2\sqrt{3}}\right) \cos \angle(\mathbf{x}, \mathbf{y}) = \frac{1}{12}\mathbf{x} \cdot \mathbf{y}.$$

The dot product of the vector from $S_3$ to $S_1$ with the vector from $S_4$ to $S_2$ equals

$$\left(\frac{(\mathbf{b} - \mathbf{d}) + (\mathbf{a} - \mathbf{c})}{2} + \mathbf{s}_1' - \mathbf{s}_3'\right) \cdot \left(\frac{(\mathbf{b} - \mathbf{d}) - (\mathbf{a} - \mathbf{c})}{2} + \mathbf{s}_2' - \mathbf{s}_4'\right),$$

which equals the sum of the following four expressions:

$$\left(\frac{|\mathbf{b} - \mathbf{d}|^2 - |\mathbf{a} - \mathbf{c}|^2}{4}\right), \qquad (\mathbf{s}_1' - \mathbf{s}_3') \cdot (\mathbf{s}_2' - \mathbf{s}_4'),$$

$$\frac{1}{2}\left[\mathbf{s}_1' \cdot (\mathbf{b} - \mathbf{a}) - \mathbf{s}_3' \cdot (\mathbf{c} - \mathbf{d}) + (\mathbf{b} - \mathbf{c}) \cdot \mathbf{s}_2' - (\mathbf{a} - \mathbf{d}) \cdot \mathbf{s}_4'\right]$$

$$\frac{1}{2}\left[\mathbf{s}_1' \cdot (\mathbf{c} - \mathbf{d}) - \mathbf{s}_3' \cdot (\mathbf{b} - \mathbf{a}) + (\mathbf{a} - \mathbf{d}) \cdot \mathbf{s}_2' - (\mathbf{b} - \mathbf{c}) \cdot \mathbf{s}_4'\right].$$

The first expression equals $\frac{1}{4}(BD^2 - AC^2)$. The four terms in the third expression all equal zero: $\overline{MS_1} \perp \overline{AB}$ implies that $\mathbf{s}_1' \cdot (\mathbf{b} - \mathbf{a}) = 0$, and so on.

In the second expression, observe that

$$\mathbf{s}_1' - \mathbf{s}_3' = \varphi\big((\mathbf{b} - \mathbf{a}) - (\mathbf{d} - \mathbf{c})\big) = \varphi\big((\mathbf{c} - \mathbf{a}) + (\mathbf{b} - \mathbf{d})\big)$$

and

$$\mathbf{s}_2' - \mathbf{s}_4' = \varphi\big((\mathbf{c} - \mathbf{b}) - (\mathbf{a} - \mathbf{d})\big) = \varphi\big((\mathbf{c} - \mathbf{a}) - (\mathbf{b} - \mathbf{d})\big).$$

Thus, their dot product is one-twelfth of

$$\big((\mathbf{c} - \mathbf{a}) + (\mathbf{b} - \mathbf{d})\big) \cdot \big((\mathbf{c} - \mathbf{a}) - (\mathbf{b} - \mathbf{d})\big) = |\mathbf{c} - \mathbf{a}|^2 - |\mathbf{b} - \mathbf{d}|^2,$$

or $\frac{1}{12}(CA^2 - BD^2)$.

As for the fourth expression,

$$\mathbf{s}_1' \cdot (\mathbf{c} - \mathbf{d}) = (AB/2\sqrt{3})(CD) \cos\big(\pi/2 + \angle(\mathbf{a} - \mathbf{b}, \mathbf{c} - \mathbf{d})\big)$$

while

$$-s_3'\cdot(\mathbf{b}-\mathbf{a}) = s_3'\cdot(\mathbf{a}-\mathbf{b}) = (CD/2\sqrt{3})(AB)\cos\left(\pi/2+\angle(\mathbf{c}-\mathbf{d},\mathbf{a}-\mathbf{b})\right).$$

The sum of the arguments of the two cosines is

$$\pi + \left(\angle(\mathbf{a}-\mathbf{b},\mathbf{c}-\mathbf{d}) + \angle(\mathbf{c}-\mathbf{d},\mathbf{a}-\mathbf{b})\right) = 3\pi,$$

implying that the value of each cosine is the negative of the other. Thus, the $s_1'$ and $s_3'$ terms in the fourth expression cancel each other out. Similarly, so do the $s_2'$ and $s_4'$ terms.

Hence, the entire dot product equals $\left(\frac{1}{4}-\frac{1}{12}\right)(BD^2-AC^2)$. Because $\overline{S_1S_3}\perp\overline{S_2S_4}$ if and only if this dot product equals 0, $\overline{S_1S_3}\perp\overline{S_2S_4} \iff BD = AC$, as desired.

## Problem 3

$P$, $Q$, $R$, $S$ are the midpoints of sides $BC$, $CD$, $DA$, $AB$, respectively, of convex quadrilateral $ABCD$. Prove that

$$4(AP^2 + BQ^2 + CR^2 + DS^2) \le 5(AB^2 + BC^2 + CD^2 + DA^2).$$

**Solution.** It is a well known formula that if $\overline{XM}$ is a median in triangle $XYZ$, then $XM^2 = \frac{1}{2}XY^2 + \frac{1}{2}XZ^2 - \frac{1}{4}YZ^2$. This can be proven, for example, by applying Stewart's Theorem to the cevian $\overline{XM}$ in triangle $XYZ$. We set $(X,Y,Z,M)$ equal to $(A,B,C,P)$, $(B,C,D,Q)$, $(C,D,A,R)$, and $(D,A,B,S)$ into this formula and add the four resulting equations to obtain a fifth equation. Multiplying both sides of the fifth equation by 4, we find that the left-hand side of the desired inequality equals

$$AB^2 + BC^2 + CD^2 + DA^2 + 4(AC^2 + BD^2).$$

Thus, it suffices to prove that

$$AC^2 + BD^2 \le AB^2 + BC^2 + CD^2 + DA^2.$$

This is the well-known "parallelogram inequality." To prove it, let $O$ be an arbitrary point in the plane, and for each point $X$, let $\mathbf{x}$ denote the vector from $O$ to $X$. We may expand each of the terms in $AB^2 + BC^2 + CD^2 + DA^2 - AC^2 - BD^2$ — for instance, writing

$$AB^2 = |\mathbf{a}-\mathbf{b}|^2 = |\mathbf{a}|^2 - 2\mathbf{a}\cdot\mathbf{b} + |\mathbf{b}|^2$$

— to find that this expression equals

$$|\mathbf{a}|^2 + |\mathbf{b}|^2 + |\mathbf{c}|^2 + |\mathbf{d}|^2 - 2(\mathbf{a}\cdot\mathbf{b} + \mathbf{b}\cdot\mathbf{c} + \mathbf{c}\cdot\mathbf{d} + \mathbf{d}\cdot\mathbf{a} - \mathbf{a}\cdot\mathbf{c} - \mathbf{b}\cdot\mathbf{d})$$
$$= |\mathbf{a} + \mathbf{c} - \mathbf{b} - \mathbf{d}|^2 \geq 0,$$

with equality if and only if $\mathbf{a} + \mathbf{c} = \mathbf{b} + \mathbf{d}$ (i.e., if and only if quadrilateral $ABCD$ is a parallelogram). This completes the proof.

## 2.5 St. Petersburg City Mathematical Olympiad (Russia)

### Problem 1

Let $\overline{AA_1}, \overline{BB_1}, \overline{CC_1}$ be the altitudes of an acute triangle $ABC$. The points $A_2$ and $C_2$ on line $A_1C_1$ are such that line $CC_1$ bisects $\overline{A_2B_1}$ and line $AA_1$ bisects $\overline{C_2B_1}$. Lines $A_2B_1$ and $AA_1$ meet at $K$, and lines $C_2B_1$ and $CC_1$ meet at $L$. Prove that lines $KL$ and $AC$ are parallel.

**Solution.**  Let $K_1$ and $L_1$ be the midpoints of $\overline{C_2B_1}$ and $\overline{A_2B_1}$, so that $K_1$ lies on line $AA_1$ and $L_1$ lies on line $CC_1$. It is well known (and not difficult to prove) that the altitude $\overline{AA_1}$ of triangle $ABC$ is an angle bisector of triangle $A_1B_1C_1$. From this it follows that $\overline{A_1K_1}$ is both an angle bisector and a median in triangle $A_1C_2B_1$. Thus, $A_1C_2 = A_1B_1$, and $\overline{A_1K_1}$ is also an altitude in triangle $A_1C_2B_1$. That is, $\overline{A_1K_1} \perp \overline{B_1C_2}$. Similarly, $\overline{C_1L_1} \perp \overline{A_2B_1}$.

Hence, lines $KK_1$ and $LL_1$ are altitudes of the triangle $KLB_1$, implying that they concur with the altitude $\ell$ from $B_1$ in this triangle. Because lines $KK_1$ and $LL_1$ meet at the orthocenter $H$ of triangle $ABC$, $\ell$ must pass through $B_1$ and $H$ as well. Hence, $\ell$ is perpendicular to $\overline{AC}$. Because $\ell$ is the altitude in triangle $KLB_1$ passing through $B_1$, it is also perpendicular to $\overline{KL}$. We conclude that $\overline{KL} \parallel \overline{AC}$, as desired.

### Problem 2

One hundred points are chosen in the coordinate plane. Show that at most $2025 = 45^2$ rectangles with vertices among these points have sides parallel to the axes.

**First Solution.**  Let $O$ be one of the 100 points, and call a rectangle good if (i) its vertices are $O$ and three other chosen points, and (ii) its sides are parallel to the axes. We claim that there are at most 81 good rectangles. Draw through $O$ the lines $\ell_1$ and $\ell_2$ parallel to the coordinate axes, where $m$ chosen points lie on $\ell_1 - \{O\}$ and $n$ chosen points lie on $\ell_2 - \{O\}$. Given any fixed chosen point $P$ not on $\ell_1$ or $\ell_2$, at most one good rectangle has $P$ as a vertex; furthermore, every good rectangle is of this form for some $P$. Because there are $99 - m - n$ such points $P$, there are at most this many good rectangles. If $m + n > 17$, we are done. Otherwise, given a pair $(P, Q)$ of chosen points, where $P \in \ell_1 - \{O\}$ and $Q \in \ell_2 - \{O\}$, at most one good rectangle has $P$ and $Q$ as vertices; furthermore, every good rectangle is of this form for some such pair $(P, Q)$. Because there are $mn \leq m(17 - m) \leq 8 \cdot 9 = 72$ such pairs, there are at most $72 < 81$ good rectangles.

We conclude that in any case, there are at most 81 rectangles whose vertices are $O$ and three other chosen points. Vary $O$ over all 100 points, counting the number of such rectangles for each $O$. The sum of the tallies is at most 8100, and we count any rectangle whose vertices are chosen points 4 times. Therefore, there are at most $8100/4 = 2025$ rectangles, as desired.

**Second Solution.** Call a rectangle *proper* if its four vertices are chosen points and if its sides are parallel to the axes. Draw all of the vertical lines $\ell_1, \ldots, \ell_n$ passing through at least one of the chosen points. Suppose that $\ell_i$ contains $x_i$ chosen points, so that $s := \sum_{i=1}^{n} x_i = 100$. The number of proper rectangles with sides on the $i$th and $j$th lines is at most $\min\{\binom{x_i}{2}, \binom{x_j}{2}\}$. Observe that

$$\min\left\{\binom{x}{2}, \binom{y}{2}\right\} \leq \frac{2xy - x - y}{4}$$

for positive integers $x$ and $y$, because if $x \leq y$ then the left-hand side is at most

$$\frac{x(x-1)}{2} \leq \frac{1}{4}\left[x(y-1) + y(x-1)\right].$$

Hence, the number of proper rectangles is at most

$$\sum_{1 \leq i < j \leq n} \frac{2x_i x_j}{4} - \sum_{1 \leq i < j \leq n} \frac{x_i + x_j}{4} = \frac{1}{4}\left(s^2 - \sum_{i=1}^{n} x_i^2\right) - \frac{1}{4}(n-1)s$$

$$= 2525 - \frac{1}{4}\left(\sum_{i=1}^{n} x_i^2 + 100n\right).$$

Applying the root mean square inequality and the arithmetic mean-geometric mean inequality, we find that this final expression is at most

$$2525 - \frac{1}{4}(s^2/n + 100n) = 2525 - 25(100/n + n)$$

$$\geq 2525 - 25 \cdot 2\sqrt{(100/n)(n)}$$

$$= 2025,$$

as desired.

## Problem 3

(a) Find all pairs of distinct positive integers $a$, $b$ such that $(b^2 + a) \mid (a^2 + b)$ and $b^2 + a$ is a power of a prime.

(b) Let $a$ and $b$ be distinct positive integers greater than 1 such that $(b^2 + a - 1) \mid (a^2 + b - 1)$. Prove that $b^2 + a - 1$ has at least two distinct prime factors.

**Solution.**    (a) We prove that the only such pair is $(a, b) = (5, 2)$. If $b = 1$, then $(a + 1) \mid (a^2 + 1)$ implies $(a + 1) \mid [a(a + 1) - (a^2 + 1)] = a - 1$. Hence, $a = 1$, which does not give a solution as $a$ and $b$ were required to be distinct. So assume $b > 1$, and write $b^2 + a = p^m$ where $p$ is prime and $m \geq 1$.

Observe that $b(b^3 + 1) \equiv (b^2)^2 + b \equiv a^2 + b \equiv 0 \pmod{b^2 + a}$, so that $b^2 + a$ divides $b(b^3 + 1)$. But $\gcd(b, b^3 + 1) = 1$ and $b^2 + a$ is a power of a prime, so one of $b$ or $b^3 + 1$ is divisible by $b^2 + a$. The first case is clearly impossible; in the second case, we have $(b^2 + a) \mid (b + 1)(b^2 - b + 1)$.

Each of $b + 1$ and $b^2 - b + 1$ is less than $b^2 + a$, so neither can be divisible by $b^2 + a$. Because $b^2 + a = p^m$ is a power of $p$, we conclude that $p$ divides both $b + 1$ and $b^2 - b + 1$. It must then also divide $(b^2 - b + 1) - (b + 1)(b - 2) = 3$, implying that $p = 3$.

There is no solution for $m = 1$. If $m = 2$, then we have $b^2 + a = 9$, yielding the solution $(a, b) = (5, 2)$. Otherwise, suppose that $m \geq 3$.

One of $b + 1$ and $b^2 - b + 1$ is divisible by 3 and the other is divisible by $3^{m-1}$. But $b^2 < b^2 + a = 3^m$, implying that $b + 1$ is at most $3^{m/2} + 1 < 3^{m-1}$ and so cannot be divisible by $3^{m-1}$. We conclude that $3^{m-1}$ divides $b^2 - b + 1$ and hence that 9 divides $4(b^2 - b + 1) = (2b - 1)^2 + 3$. This is impossible because no square is congruent to 6 modulo 9. Thus, $(5, 2)$ is the only solution.

(b) Assume, for the sake of contradiction, that $b^2 - 1 + a$ is a prime power. Because $(b^2 - 1)^2 - a^2$ is divisible by $b^2 - 1 + a$, as is $a^2 + b - 1$ by hypothesis, so then is their sum $(b^2 - 1)^2 + b - 1 = b(b - 1)(b^2 + b - 1)$. Observe that $b$, $b - 1$, and $b^2 + b - 1 = b(b + 1) - 1 = (b + 2)(b - 1) + 1$ are pairwise relatively prime. Hence, one of $b, b - 1, b^2 + b - 1$ must be divisible by the prime power $b^2 + a - 1$.

Because $b$ and $b - 1$ are smaller than $b^2 + a - 1$, we must have that $b^2 + a - 1$ divides $b^2 + b - 1$ and hence that $a \leq b$; because $a \neq b$ by hypothesis, $a < b$. On the other hand, because $(b^2 + a - 1) \mid (a^2 + b - 1)$, we must have $0 \leq (a^2 + b - 1) - (b^2 + a - 1) = (a - b)(a + b - 1)$. Therefore, $a \geq b$ as well, a contradiction.

## Problem 4

In a country of 2000 airports, there are initially no airline flights. Two airlines take turns introducing new roundtrip nonstop flights. (Between any two cities, only one nonstop flight can be introduced.) The transport authority would like to achieve the goal that if any airport is shut down, one can still travel between any two other airports, possibly with transfers. The airline that causes the goal to be achieved loses. Which airline wins with perfect play?

**Solution.** The second airline wins. Consider a situation where the goal is not achieved, but adding any single flight causes the goal to be achieved. Because the goal is not achieved, there is some airport $A$ which when shut down splits the cities into two disconnected groups $G_1$ and $G_2$. Then any two cities within $G_1$ or within $G_2$ must already be joined, because otherwise adding the flight between those two cities would not cause the goal to be achieved. Similarly, every city must be joined to $A$, but no city in $G_1$ can be joined to any city of $G_2$. Hence, if there are $k$ cities in $G_1$, then there are $k(k-1)/2$ flights between cities in $G_1$, $(1999-k)(1998-k)/2$ flights between cities in $G_2$, and 1999 flights between $A$ and another city. The total number of flights is thus $k(k-1999)+1999000$, which is even. In particular, it is never the second airline's turn to add a new flight when such a situation occurs. Therefore, the second airline can always avoid losing.

## Problem 5

We are given several monic quadratic polynomials, all with the same discriminant. The sum of any two of the polynomials has distinct real roots. Show that the sum of all of the polynomials also has distinct real roots.

**Solution.** The common discriminant must be positive, because otherwise each of the polynomials would take only positive values, so the sum of any two of them would not have real roots. Let this common discriminant be $4D$, so that each polynomial is of the form $(x-c)^2-D$ for some $c$. For each of the polynomials, consider the interval on which that polynomial takes negative values; that interval has length $2\sqrt{D}$. If two of these intervals $(c_1 - \sqrt{D}, c_1 + \sqrt{D})$ and $(c_2 - \sqrt{D}, c_2 + \sqrt{D})$ did not intersect, then $|c_2 - c_1| > \sqrt{D}$ and $\frac{1}{2}(c_1 + c_2)$ does not lie in either interval. Both polynomials are thus nonnegative at $\frac{1}{2}(c_1+c_2)$, but this point is where the polynomials' sum $p$ attains its minimum — contradicting the assumption that $p$ has two distinct real roots.

Hence, any two of the intervals intersect. Choose an interval $(c - \sqrt{D}, c + \sqrt{D})$ such that $c$ is minimal. Because every other interval intersects this one, we find that every interval contains $c + \sqrt{D} - \epsilon$ for some $\epsilon$. At this point, the sum of all of the polynomials takes a negative value, implying that this sum must have distinct real roots.

## Problem 6

On an infinite checkerboard are placed 111 nonoverlapping *corners*, L-shaped figures made of 3 unit squares. The collection has the following property: for any corner, the $2 \times 2$ square containing it is entirely covered by the corners. Prove that one can remove between 1 and 110 of the corners so that the property will be preserved.

**Solution.** If some $2 \times 3$ rectangle is covered by two corners, then we may remove all of the corners except those two. Thus, we may assume that no such rectangle exists.

We construct a directed graph whose vertices are the corners, as follows: for each corner, draw the $2 \times 2$ square containing that corner, and add an edge from this corner to the other corner covering the remainder of the $2 \times 2$ square. If one corner has no edge pointing toward it, we may remove that corner, so we may assume that no such corner exists. Hence, each edge of the graph is in some cycle. If there is more than one cycle, then we may remove all the corners except those in a cycle of minimal length, and the required property is preserved. Thus, it suffices to show that there cannot exist a single cycle consisting of all 111 vertices.

By the *center* of a corner we refer to the point at the center of the $2 \times 2$ square containing that corner. Recalling that we assumed that no two corners cover a $2 \times 3$ rectangle, one easily checks that if there is an edge pointing from one corner to another, then these corners' centers differ by 1 in both their $x$- and $y$- coordinates. Hence, in any cycle, the parities of the $x$-coordinates of the vertices in that cycle alternate, implying that the number of vertices in the cycle is even. Therefore, there cannot be a cycle containing all 111 vertices, as desired.

## Problem 7

We are given distinct positive integers $a_1, a_2, \ldots, a_{20}$. The set of pairwise sums $\{a_i + a_j \mid 1 \leq i \leq j \leq 20\}$ contains 201 elements. What is the smallest possible number of elements in the set $\{|a_i - a_j| \mid 1 \leq i < j \leq 20\}$, the set of positive differences between the integers?

**Solution.** The smallest possible number of differences is 100. This number can be attained: set $a_i = 10^{11} + 10^i$ and $a_{10+i} = 10^{11} - 10^i$ for $i = 1, \ldots, 10$. Then $\{|a_i - a_j| \mid 1 \leq i < j \leq 20\}$ equals

$$\{2 \cdot 10^i \mid 1 \leq i \leq 10\} \cup \{|10^i \pm 10^j| \mid 1 \leq i < j \leq 10\},$$

with the desired total of $10 + 2 \cdot 45 = 100$ differences.

Now suppose, for the sake of contradiction, that there are fewer than 100 distinct differences. Let $S = \{a_1, \ldots, a_{20}\}$. We obtain two contradictory bounds on the number of multisets of the form $\{x, y, z, w\}$ with $x, y, z, w \in S$, $x < y \leq z < w$, and $x + w = y + z$.

For each of the 190 pairs $(b, c)$ of elements in $S$ with $b > c$, consider the difference $b - c$. Because there are fewer than 100 distinct differences, there are more than 90 pairs $(b, c)$ for which $b - c = b' - c'$ for some values $b' < b$, $c' < c$ in $S$. For each of these more than 90 pairs $(b, c)$ with corresponding values $b'$, $c'$, we form the multiset $\{b, c, b', c'\}$. Each such multiset $\{x, y, z, w\}$ with $x < y \leq z < w$ corresponds to at most two pairs $(b, c)$, namely $(b, c) = (w, z)$ and $(b, c) = (w, y)$. Hence, there are more than 45 such multisets.

On the other hand, the number of such multisets equals $\sum s_i(s_i - 1)/2$, where for each integer $i$ we define $s_i$ to be the number of pairs $(b, c)$ such that $b, c \in S$, $b \leq c$, and $b + c = i$. For each $i$, any element $s$ of $S$ can appear in at most one such pair $(b, c)$ — namely, $(s, i - s)$ if $s \leq i - s$, or $(i - s, s)$ otherwise. Thus, $s_i \leq \frac{20}{2} = 10$ for all $i$. Hence, if $s_i \neq 0$, then $1 \leq s_i \leq 10$ and $s_i(s_i - 1)/2 \leq 5s_i - 5$.

The given information implies that there are 201 integers $i$ such that $s_i \geq 1$; let $T$ be the set of these integers. There are $\binom{20}{2}$ pairs $(b, c)$ of elements in $S$ such that $b < c$, and 20 such pairs with $b = c$, implying that $\sum_{i \in T} s_i = \binom{20}{2} + 20 = 210$. Therefore, the number of multisets of the described form is

$$\sum_{i \in T} s_i(s_i - 1)/2 \leq \sum_{i \in T}(5s_i - 5) = 5\sum_{i \in T} s_i - 5 \cdot 201 = 45,$$

a contradiction. This completes the proof.

## Problem 8

Let $ABCD$ be an isosceles trapezoid with $\overline{AD} \parallel \overline{BC}$. An arbitrary circle tangent to $\overline{AB}$ and $\overline{AC}$ intersects $\overline{BC}$ at $M$ and $N$. Let $X$ and $Y$ be the intersections closer to $D$ of the incircle of triangle $BCD$ with $\overline{DM}$ and $\overline{DN}$, respectively. Show that $\overline{XY} \parallel \overline{AD}$.

**Solution.** Denote the first circle by $S_1$, and suppose it is tangent to line $AB$ at $P$ and to line $AC$ at $Q$. Choose points $P_1$ and $Q_1$ on the extensions of rays $DB$ and $DC$ beyond $B$ and $C$, respectively, so that $BP_1 = BP$ and $CQ_1 = CQ$. Then

$$DP_1 - DQ_1 = (DB + BP_1) - (DC + CQ_1)$$
$$= (AC + BP) - (AB + CQ)$$
$$= (AC - CQ) - (AB - BP) = AQ - AP = 0.$$

Thus, there exists a circle $S_2$ tangent to lines $DB$ and $DC$ at $P_1$ and $Q_1$, respectively. Because $BP^2 = BP_1^2$ (resp., $CQ^2 = CQ_1^2$), the powers of the point $B$ (resp., $C$) with respect to the circles $S_1$ and $S_2$ are equal. Thus, the line $BC$ is the radical axis of the two circles; because $M$ and $N$ lie on this axis and lie on $S_1$, they also lie on $S_2$. Furthermore, because $\overline{MN}$ lies within triangle $DP_1Q_1$, $M$ and $N$ are the intersections closer to $D$ of $S_2$ with lines $DM$ and $DN$, respectively. Consider the homothety with positive ratio centered at $D$ taking $S_2$ to the incircle of triangle $DBC$. This homothety must carry $M$ and $N$ to $X$ and $Y$, implying that $\overline{XY}$ is parallel to the base of the trapezoid, as desired.

## Problem 9

In each square of a chessboard is written a positive real number such that the sum of the numbers in each row is 1. It is known that for any eight squares, no two in the same row or column, the product of the numbers in these squares is no greater than the product of the numbers on the main diagonal. Prove that the sum of the numbers on the main diagonal is at least 1.

**Solution.** Suppose, for the sake of contradiction that the sum of the numbers on the main diagonal is less than 1. Call a square *good* if its number is greater than the number in the square in the same column that lies on the main diagonal. Each row must contain a good square, because otherwise the numbers in that row would have sum less than 1.

For each square on the main diagonal, draw a horizontal arrow from that square to a good square in its row, and then draw a vertical arrow from that good square back to the main diagonal. Among these arrows, some must form a loop. We consider the following squares: the squares on the main diagonal which are not in the loop, and the good squares which are in the loop. Each row and column contains exactly one of these squares. However, the product of the numbers in these squares is greater than the product of the numbers on the main diagonal, a contradiction.

## Problem 10

Is it possible to draw finitely many segments in three-dimensional space such that any two segments either share an endpoint or do not intersect, any endpoint of a segment is the endpoint of exactly two other segments, and any closed polygon made from these segments has at least 30 sides?

**Solution.** Yes, this is possible. We prove by induction that for $n \geq 3$, there exists a graph $G_n$ in which each vertex has degree 3 and no cycle contains fewer than $n$ vertices. We may then embed $G_{30}$ in three-dimensional space by letting its vertices be any points in general position (that is, such that no four points are coplanar) and letting each edge between two vertices be a segment. No two segments intersect unless they share an endpoint, and any closed polygon made from these segments has at least 30 sides.

Proving the base case $n = 3$ is easy — we may simply take a complete graph on three vertices. Now suppose we are given such a graph $G_n$, consisting of $m$ edges labelled $1, \ldots, m$ in some fashion. Choose an integer $M > n2^m$. We construct a new graph $G_{n+1}$ whose vertices are pairs $(v, k)$, where $v$ is a vertex of $G_n$ and $k \in \{0, \ldots, M-1\}$ (or rather, the integers modulo $M$) as follows. For $i = 1, \ldots, m$, suppose that edge $i$ has endpoints $a$ and $b$. In $G_{n+1}$, join $(a, j)$ to $(b, j + 2^i)$ with an edge for $j = 0, \ldots, M-1$.

We claim that every vertex in $G_n$ has degree 3. Indeed, if $a \in G_n$ is adjacent to $b_1, b_2, b_3 \in G_n$, then any $(a, j) \in G_{n+1}$ is adjacent to exactly three vertices, of the form $(b_1, j_1)$, $(b_2, j_2)$, $(b_3, j_3)$. Furthermore, note that $b_1, b_2, b_3$ are distinct; in other words, given a fixed $a$ and $b$, at most one edge connects two vertices of the form $(a, j)$ and $(b, j')$.

We now show that $G_{n+1}$ contains no cycles of length less than $n + 1$. Suppose on the contrary that $(a_1, j_1), \ldots, (a_t, j_t)$ is such a cycle with $t \leq n$ distinct vertices. If the edges in the path $a_1, a_2, \ldots, a_t$ form a tree, then there exists some leaf $a_k$ in this tree besides $a_1 = a_t$. Hence, $a_{k-1}$ and $a_{k+1}$ are equal, say to some value $b$. Then $(a_k, j_k)$ is adjacent to $(a_{k-1}, j_{k-1})$ and $(a_{k+1}, j_{k+1})$, two distinct vertices of the form $(b, j')$ — a contradiction.

Therefore, the edges in the path $a_1, a_2, \ldots, a_t$ contain a cycle. By the induction hypothesis, this cycle must contain at least $n$ edges, implying that $t \geq n$ and that the edges in our path are distinct. Let $i_1, \ldots, i_n$ be the labels of the edges of this cycle in $G_n$. Then the second coordinates of $(a_1, j_1), \ldots, (a_t, j_t)$ change by $\pm 2^{i_1}, \ldots, \pm 2^{i_n}$ modulo $M$ as we travel along the cycle. Because $j_1 = j_t$, we must have

$$\pm 2^{i_1} \pm \cdots \pm 2^{i_n} \equiv 0 \pmod{M}.$$

However, $\pm 2^{i_1} \pm \cdots \pm 2^{i_n}$ is a nonzero integer whose absolute value is less than $n2^m < M$, a contradiction. Thus, $G_{n+1}$ has no cycles of length less than $n + 1$. This completes the inductive step and the proof.

## Problem 11

What is the smallest number of weighings on a balance scale needed to identify the individual weights of a set of objects known to weigh $1, 3, 3^2, \ldots, 3^{26}$ in some order? (The balance scale reports the weight of the objects in the left pan, minus the weight of the objects in the right pan.)

**Solution.**   At least three weighings are necessary: each of the first two weighings divides the weights into three categories (the weights in the left pan, the weights in the right pan, and the weights remaining off the scale). Because $27 > 3 \cdot 3$, some two weights must fall into the same category on both weighings, implying that these weights cannot be distinguished. We now show that three weighings indeed suffice.

Label the 27 weights using the three-letter words made up of the letters $L, R, O$. In the $i$th weighing, put the weights whose $i$th letter is $L$ on the left pan and the weights whose $i$th letter is $R$ on the right pan. The difference between the total weight of the objects in the left pan and the total weight of the objects in the right pan equals

$$\epsilon_0 3^0 + \epsilon_1 3^1 + \cdots + \epsilon_{26} 3^{26},$$

where $\epsilon_j$ equals $1$, $-1$, or $0$ if $3^j$ is in the left pan, in the right pan, or off the scale, respectively. The value of the above sum uniquely determines all of the $\epsilon_j$: the value of the sum modulo 3 determines $\epsilon_0$, then the value of the sum modulo 9 determines $\epsilon_1$, and so on. (This is a case of a more general result, that each integer has a unique representation in base 3 using the digits $-1$, $0$, $1$.)

Thus, for $j = 0, \ldots, 26$, the $i$th weighing determines the $i$th letter of the weight that measures $3^j$. After three weighings, we thus know exactly which weight measures $3^j$, as desired.

## Problem 12

The line $\ell$ is tangent to the circumcircle of acute triangle $ABC$ at $B$. Let $K$ be the projection of the orthocenter of triangle $ABC$ onto line $\ell$, and let $L$ be the midpoint of side $\overline{AC}$. Show that triangle $BKL$ is isosceles.

**First Solution.** All angles are directed modulo $\pi$. If $B = K$, then clearly $LB = LK$; we now assume that $B \neq K$. Let $H$ be the orthocenter of triangle $ABC$, and let $A_1$ and $C_1$ be the feet of the altitudes in triangle $ABC$ from $A$ and $C$, respectively. Because $\angle BKH = \angle BC_1H = \angle BA_1H = \pi/2$, the points $A_1, B, C_1, H, K$ lie on a circle with diameter $\overline{BH}$. Note that

$$\angle KBC_1 = \angle KBA = \angle BCA = \frac{\pi}{2} - \angle HBC = \angle A_1HB = \angle A_1C_1B,$$

where the last equality holds because quadrilateral $BA_1HC_1$ is cyclic. Hence, $\overline{BK}$ and $\overline{A_1C_1}$ are parallel, implying that these segments' perpendicular bisectors (which both pass through the center of the circle in which quadrilateral $C_1KBA_1$ is inscribed) coincide.

Next, because $\angle AC_1C = \angle AA_1C = \pi/2$, quadrilateral $AC_1A_1C$ is cyclic with center $L$. Thus, $L$ lies on the perpendicular bisector of $\overline{A_1C_1}$, which is also the perpendicular bisector of $\overline{BK}$. Hence, $LB = LK$, and triangle $BKL$ is isosceles.

**Second Solution.** Again, let $H$ be the orthocenter of triangle $ABC$. Reflect $H$ across the point $L$ to obtain $H'$; then quadrilateral $AHCH'$ is a parallelogram, implying that

$$\angle BAH' = \angle BAC + \angle LAH' = \angle BAC + \angle LCH = \frac{\pi}{2}.$$

Likewise, $\angle BCH' = \pi/2$. Hence, $H'$ is the point on the circumcircle of $ABC$ diametrically opposite $B$. Thus, the projection of $H'$ onto the tangent line $\ell$ is $B$, while we are given that the projection of $H$ onto $\ell$ is $K$. It follows that the projection of $L$ (the midpoint of $\overline{H'H}$) onto $\ell$ is the midpoint of $\overline{BK}$, implying that $LB = LK$. Hence, triangle $BKL$ is isosceles.

## Problem 13

Two balls of negligible size move within a vertical $1 \times 1$ square at the same constant speed. Each travels in a straight path except when it hits a wall, in which case it reflects off the wall so that its angle of incidence equals its angle of reflection. Show that a spider (also of negligible size), moving at the same speed as the balls, can descend straight down on a string from the top edge of the square to the bottom so that while the spider is within the square, neither the spider nor its string is touching one of the balls.

**Solution.** Suppose the square has side length one meter, and that the spider and the balls move at one meter per minute. Then the horizontal projection of each ball moves at a speed no greater than one meter per minute, and thus completes a full circuit in no less than two minutes.

If both balls are moving vertically, then the spider can descend along some vertical line at any time. Otherwise, at some time $t_0$ (measured in minutes) one ball touches the left wall. If the second ball touches the wall between time $t_0$ and $t_0 + 1$, then it does not touch the wall between time $t_0 + 1$ and $t_0 + 2$; hence, the spider may descend safely if it begins at time $t_0 + 1$. Otherwise, if the second ball does *not* touch the wall between time $t_0$ and $t_0 + 1$, then the spider may descend safely if it begins at time $t_0$.

## Problem 14

Let $n \geq 3$ be an integer. Prove that for positive numbers $x_1 \leq x_2 \leq \cdots \leq x_n$,

$$\frac{x_n x_1}{x_2} + \frac{x_1 x_2}{x_3} + \cdots + \frac{x_{n-1} x_n}{x_1} \geq x_1 + x_2 + \cdots + x_n.$$

**Solution.** Suppose that $0 \leq x \leq y$ and $0 < a \leq 1$. We have $1 \geq a$ and $y \geq x \geq ax$, implying that $(1 - a)(y - ax) \geq 0$ or $ax + ay \leq a^2 x + y$. Dividing both sides of this final inequality by $a$, we find that

$$x + y \leq ax + \frac{y}{a}. \qquad (*)$$

We may set

$$(x, y, a) = \left( x_n, x_n \cdot \frac{x_{n-1}}{x_2}, \frac{x_1}{x_2} \right)$$

in $(*)$ to find that

$$x_n + \frac{x_{n-1} x_n}{x_2} \leq \frac{x_n x_1}{x_2} + \frac{x_{n-1} x_n}{x_1}. \qquad (\dagger)$$

Furthermore, for $i = 1, 2, \ldots, n-2$, we may set

$$(x, y, a) = \left( x_i, x_{n-1} \cdot \frac{x_{i+1}}{x_2}, \frac{x_{i+1}}{x_{i+2}} \right)$$

in $(*)$ to find that

$$x_i + x_{n-1} \cdot \frac{x_{i+1}}{x_2} \leq x_i \cdot \frac{x_{i+1}}{x_{i+2}} + x_{n-1} \frac{x_{i+1}}{x_2} \cdot \frac{x_{i+2}}{x_{i+1}}$$

$$= x_i \cdot \frac{x_{i+1}}{x_{i+2}} + x_{n-1} \cdot \frac{x_{i+2}}{x_2}.$$

Summing this inequality for $i = 1, 2, \ldots, n-2$ yields

$$(x_1 + \cdots + x_{n-2}) + x_{n-1} \leq \left( \frac{x_1 x_2}{x_3} + \cdots + \frac{x_{n-2} x_{n-1}}{x_n} \right) + \frac{x_{n-1} x_n}{x_2}.$$

Summing this last inequality with (†) yields the desired inequality.

## Problem 15

In the plane is given a convex $n$-sided polygon $\mathcal{P}$ with area less than 1. For each point $X$ in the plane, let $F(X)$ denote the area of the union of all segments joining $X$ to points of $\mathcal{P}$. Show that the set of points $X$ such that $F(X) = 1$ is a convex polygon with at most $2n$ sides. (Of course, by "polygon" here we refer to a 1-dimensional border, not a closed 2-dimensional region.)

**Solution.** For each point $X$ in the plane, we let $\mathbf{F}_X$ denote the union of all segments joining $X$ to points of $\mathcal{P}$. Also, we let $\mathbf{P}$ denote the closed region bounded by $\mathcal{P}$. Let $\mathcal{Q}$ be the set of points $X$ such that $F(X) = 1$. In addition, let the vertices of $\mathcal{P}$ be $A_1, A_2, \ldots, A_n$ in clockwise order, with indices taken modulo $n$. Finally, let $O$ be the centroid of $\mathbf{P}$.

**Lemma.** *Any ray whose endpoint is $O$ intersects $\mathcal{Q}$ in at most one point.*

*Proof:* Suppose, for the sake of contradiction, that some ray with endpoint $O$ intersected $\mathcal{Q}$ at two distinct points $R_1$ and $R_2$. Observe that $R_1$ and $R_2$ must lie outside $\mathbf{P}$, because otherwise

$$F(R_1) = F(R_2) = \text{Area}(\mathbf{P}) < 1.$$

Without loss of generality, assume that $R_2$ is closer to $O$ than $R_1$, so that $R_2 \in \mathbf{F}_{R_1}$. Then $\{R_2\} \cup \mathcal{P}$ lies in the convex set $\mathbf{F}_{R_1}$. It follows that the convex hull of $\{R_2\} \cup \mathcal{P}$ lies in $\mathbf{F}_{R_1}$ as well. But this convex hull is precisely $\mathbf{F}_{R_2}$, implying that $\mathbf{F}_{R_2} \subseteq \mathbf{F}_{R_1}$.

Suppose, for the sake of contradiction, that $R_1 \in \mathbf{F}_{R_2}$. We already know that $R_2 \in \mathbf{F}_{R_1}$. Hence, by the definitions of these regions, we have that $R_2 \in \overline{R_1 S_1}$ and $R_1 \in \overline{R_2 S_2}$ for some $S_1, S_2 \in \mathcal{P}$. In this case, $R_1$ and $R_2$ lie on $\overline{S_1 S_2}$. Because $\overline{S_1 S_2}$ is contained in the convex set $\mathbf{P}$, it follows that $R_1$ and $R_2$ are as well — a contradiction. Thus, our assumption at the beginning of this paragraph was false, and in fact $R_1 \notin \mathbf{F}_{R_2}$.

Hence, $\mathbf{F}_{R_2} \subseteq \mathbf{F}_{R_1}$, but equality does not hold because $R_1$ is contained in the right hand side but not the left hand side. Because $\mathbf{F}_{R_2}$ and $\mathbf{F}_{R_1}$ are closed polygonal regions, we must have $1 = \text{Area}(\mathbf{F}_{R_2}) < \text{Area}(\mathbf{F}_{R_1}) = 1$, a contradiction. Therefore, our assumption at the beginning of the proof of this lemma was false, and in fact any ray with endpoint $O$ intersects $\mathcal{Q}$ in at most one point. ∎

The rays $A_1 A_2, A_2 A_3, \ldots, A_n A_1$ partition the exterior of $\mathbf{P}$ into $n$ regions. Let $\mathbf{M}_j$ be the region bounded by rays $A_{j-1} A_j$ and $A_j A_{j+1}$. Observe that it consists of all the points $X$ such that $\mathbf{P}$ lies to the left of ray $X A_j$. (By this, we mean that $\mathbf{P}$ lies entirely in one of the two half-planes formed by line $X A_j$, and that if we walk from $X$ to $A_j$ then $\mathbf{P}$ lies in the half-plane to our left.)

Similarly, the rays $A_2 A_1, A_3 A_2, \ldots, A_1 A_n$ partition the exterior of $\mathbf{P}$ into $n$ regions. Let $\mathbf{N}_j$ be the region bounded by rays $A_j A_{j-1}$ and $A_{j+1} A_j$. Observe that it consists of all the points $X$ such that $\mathbf{P}$ lies to the right of ray $X A_j$.

Now, choose an arbitrary point $Y_0$ on $\mathcal{Q}$, and suppose that it lies in $\mathbf{M}_j \cap \mathbf{N}_k$. Observe that for each point $Y \in \mathbf{M}_j \cap \mathbf{N}_k$, the region $\mathbf{F}_Y$ can be partitioned into two polygons: triangle $A_j Y A_k$ and the polygon $A_k A_{k+1} \cdots A_j$. Hence, $F(Y)$ equals

$$\text{Area}(A_k A_{k+1} \cdots A_j) + \text{Area}(A_j Y A_k)$$

$$= \text{Area}(A_k A_{k+1} \cdots A_j) + \frac{1}{2} A_j A_k \cdot \text{Distance}(Y, \text{ line } A_j A_k).$$

So, if $Y \in \mathbf{M}_j \cap \mathbf{N}_k$ has the property that $\overline{Y Y_0} \parallel \overline{A_j A_k}$, then $F(Y) = F(Y_0) = 1$. Starting from $Y_0$, we travel along this line clockwise around $O$ until we reach a border of $\mathbf{M}_j \cap \mathbf{N}_k$. Each point $Y$ on our path so far satisfies $F(Y) = 1$.

At this point, we are in the region $\mathbf{M}_{j'} \cap \mathbf{N}_{k'}$, where $(j', k')$ equals either $(j+1, k)$ or $(j, k+1)$. An argument similar to the one above shows that we may now travel clockwise around $O$ and parallel to $\overline{A_{j'} A_{k'}}$ until we reach a border of $\mathbf{M}_{j'} \cap \mathbf{N}_{k'}$. Each point $Y$ on our path so far continues to satisfy $F(Y) = 1$.

We can continue similarly until we have travelled completely around $\mathbf{P}$. Rigorously, letting $O$ be the centroid of $\mathbf{P}$, if $Y$ is our (changing) position as we travel then we can continue travelling until $\angle Y_0 O Y = 2\pi$. Suppose we stop at $Y = Y_1$.

We claim that the path we travel must be *all* of $Q$. If not, then there exists some $Z_1$ not on $Q$ such that $F(Z_1) = 1$. Then ray $O Z_1$ intersects $Q$ at some point $Z_2 \neq Z_1$, but this contradicts lemma 1.

It similarly follows that $Y_0 = Y_1$, because otherwise ray $O Y_0$ contains two points (namely, $Y_0$ and $Y_1$) in $Q$ — again contradicting lemma 1.

Therefore, $Q$ consists of a closed, simple polygonal path. Also, we change direction at most $2n$ times, because we can only change directions when we cross any of the $2n$ rays $A_j A_{j+1}$ and $A_j A_{j-1}$. Hence, $Q$ is a polygon with at most $2n$ sides.

Note that as we travel, at times we switch from travelling parallel to $\overline{A_j A_k}$ to travelling parallel to $\overline{A_{j+1} A_k}$; in these cases, our direction of travel rotates clockwise by $\pi - \angle A_j A_k A_{j+1}$. At other times, we switch from travelling parallel to $\overline{A_j A_k}$ to travelling parallel to $\overline{A_j A_{k+1}}$; in these cases, our direction of travel rotates clockwise by $\pi - \angle A_{k+1} A_j A_k$. Hence, whenever our direction of travel changes, it rotates by some clockwise angle measuring less than $\pi$. It follows that $Q$ is actually a *convex* polygon.

In summary: $Q$ is a convex polygon with at most $2n$ sides. This completes the proof.

## Problem 16

What is the smallest number of unit segments that can be erased from the interior of a $2000 \times 3000$ rectangular grid so that no smaller rectangle remains intact?

**Solution.**  The minimum number is 4 million. Suppose some segments have been removed so that no smaller rectangle remains intact. We construct a graph whose vertices are the $1 \times 1$ squares contained in the $2000 \times 3000$ grid, where two vertices are adjacent if the corresponding $1 \times 1$ squares are adjacent and the segment between then has been erased. Suppose that there are $t$ connected components in this graph.

Because no $1 \times 1$ or $1 \times 2$ rectangle remains intact, each vertex of the graph is in a connected component with at least three vertices. Because there are six million vertices in total, $t$ must be at most two million. For any of these $t$ components $C$, let $v(C)$ and $e(C)$ denote the number of vertices and edges, respectively, in that component. (Of course, the term "edge" here refers to edges of the graph, not to segments on the original grid.) We have $e(C) \geq v(C) - 1$ for each component; summing over all components, we find that

$$\sum_C e(C) \geq \sum_C (v(C) - 1) = 6 \cdot 10^6 - t \geq 4 \cdot 10^6. \qquad (*)$$

To see that it in fact suffices to erase only four million segments, we first build up a tiling of the $3000 \times 2000$ rectangle by translations of the following $4 \times 6$ rectangular figure:

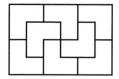

Note that the only rectangles formed by the segments in this tiling are the $4 \times 6$ rectangles themselves. Each of these $4 \times 6$ rectangles has at least one vertex within the interior of the $3000 \times 2000$ rectangle, and this vertex is at the center of a figure arranged as follows:

We can thus ensure that the $4 \times 6$ rectangles do not remain intact by rearranging the above pattern in the following form instead:

Although this construction removes the $4 \times 6$ rectangles, we must ensure that we do not reintroduce any new rectangles. When we modify the above figures, we do not alter any of the segments outside their interior; so any new rectangles we reintroduce must intersect the interior of at least one of these figures. Such a rectangle would intersect the figure either in a straight line segment cutting all the way through the above figure, or in a broken line segment consisting of two segments at right angles to each other. However, no such segment or broken line segment intersects the interior of the above figure, so in fact no new rectangles are introduced.

Therefore, if we erase segments to obtain the tiling found with our procedure above, no rectangles are formed by the remaining segments. Also, if we construct the graph described at the beginning of the solution, then because each connected component has exactly 3 vertices we know that equality holds in $(*)$. Hence, the number of erased edges is exactly four million, as needed. This completes the proof.

## Problem 17

Let $\overline{AA_1}$ and $\overline{CC_1}$ be altitudes of acute triangle $ABC$. The line through the incenters of triangles $AA_1C$ and $AC_1C$ meets lines $AB$ and $BC$ at $X$ and $Y$, respectively. Prove that $BX = BY$.

**Solution.**   Let $P$ and $Q$ be the incenters of triangles $AC_1C$ and $AA_1C$, respectively. Then

$$\angle APC = \pi - \frac{1}{2}(\angle C_1CA + \angle CAC_1) = 3\pi/4,$$

Similarly, $\angle AQC = 3\pi/4$. Hence, quadrilateral $APQC$ is cyclic.

We now use directed angles modulo $\pi$. Observe that

$$\angle BXY = \angle BAP + \angle APX$$

$$= \angle BAP + (\pi - \angle APQ) = \angle BAP + \angle ACQ.$$

Similarly, $\angle BYX = \angle BCQ + \angle CAP$. Therefore,

$$\angle BXY + \angle BYX = (\angle BAP + \angle CAP) + (\angle ACQ + \angle BCQ) = 0,$$

implying that $BX = BY$.

## Problem 18

Does there exist a 30-digit number such that the number obtained by taking any five of its consecutive digits is divisible by 13?

**Solution.**   We claim that no such number exists. Suppose on the contrary that such a number exists, and let $\overline{a_1 \cdots a_{30}}$ be its decimal expansion. Then for $1 \le i \le 25$, we have

$$4a_i - a_{i+5} \equiv 10^5 a_i - a_{i+5}$$

$$= 10 \cdot \overline{a_i \cdots a_{i+4}} - \overline{a_{i+1} \cdots a_{i+5}}$$

$$\equiv 0 \pmod{13}.$$

Hence, each term in the sequence $a_1$, $a_6$, $a_{11}$, $a_{16}$, $a_{21}$, $a_{26}$ is congruent modulo 13 to four times the previous term. In other words, these six terms are consecutive terms of one of the following sequences:

$$1, 4, 3, 12, 9, 10, 1, 4, 3, \ldots;$$

$$2, 8, 6, 11, 5, 7, 2, 8, 6, \ldots;$$

$$0, 0, \ldots.$$

Any six adjacent terms of the first two sequence contains one of 10, 11, or 12, but each $a_i$ is a single digit. Hence, $a_1, a_6, \ldots, a_{26}$ must be adjacent terms of the third sequence above. However, this is impossible because $a_1 \neq 0$. Therefore, our original assumption was false, and no 30-digit number has the required property.

## Problem 19

Let $ABCD$ be a convex quadrilateral, and let $M$ and $N$ be the midpoints of $\overline{AD}$ and $\overline{BC}$, respectively. Suppose $A$, $B$, $M$, $N$ lie on a circle and that $\overline{AB}$ is tangent to the circumcircle of triangle $BMC$. Prove that $\overline{AB}$ is also tangent to the circumcircle of triangle $AND$.

**Solution.** Because quadrilateral $ABNM$ is cyclic, $\angle MAB = \pi - \angle BNM = \angle MNC$. Because $\overline{AB}$ is tangent to the circumcircle of triangle $BMC$, we have $\angle ABM = \angle BCM = \angle NCM$. Thus, triangles $ABM$ and $NCM$ are similar, implying that $AM/AB = NM/NC$; equivalently, $MD/AB = MN/BN$.

Moreover, $\angle DMN = \angle ABN$ because quadrilateral $ABNM$ is cyclic. Thus, triangles $DMN$ and $ABN$ are similar, so $\angle MDN = \angle BAN$. We conclude that $\overline{AB}$ is also tangent to the circumcircle of triangle $AND$, as desired.

## Problem 20

Let $n \geq 3$ be a positive integer. For all positive numbers $a_1, a_2, \ldots, a_n$, show that

$$\frac{a_1 + a_2}{2} \frac{a_2 + a_3}{2} \cdots \frac{a_n + a_1}{2} \leq \frac{a_1 + a_2 + a_3}{2\sqrt{2}} \cdots \frac{a_n + a_1 + a_2}{2\sqrt{2}}.$$

**Solution.** We take the indices of the $a_i$ modulo $n$. Observe that

$$4(a_{i-1} + a_i + a_{i+1})^2 = \left[ (2a_{i-1} + a_i) + (a_i + 2a_{i+1}) \right]^2$$

$$\geq 4(2a_{i-1} + a_i)(a_i + 2a_{i+1})$$

by the arithmetic mean-geometric mean inequality. Equivalently,

$$(a_{i-1} + a_i + a_{i+1})^2 \geq (2a_{i-1} + a_i)(a_i + 2a_{i+1}).$$

In addition to this inequality, note that

$$(2a_{i-1} + a_i)(2a_i + a_{i-1}) \geq 2(a_{i-1} + a_i)^2$$

by the arithmetic mean-geometric mean inequality. Multiplying these two inequalities together for $i = 1, 2, \ldots, n$, and then taking the square root of each side of the resulting inequality, gives the desired result.

## Problem 21

A connected graph is said to be *2-connected* if after removing any single vertex, the graph remains connected. Prove that given any 2-connected graph in which the degree of every vertex is greater than 2, it is possible to remove a vertex (and all edges adjacent to that vertex) so that the remaining graph is still 2-connected.

**Solution.** Let $G$ be the given graph. Given two vertices $v, w$, we write "there exists $v \rightsquigarrow w$ (possibly specifying some condition)" if there exists a path from $v$ to $w$ (satisfying the condition, if one is specified). Given any vertex $v$, we are told that the graph obtained by removing $v$ from $G$ remains connected. In other words, we are told:

($*$) For any distinct vertices $v_1, v_2, v$, there exists $v_1 \rightsquigarrow v_2$ not passing through $v$.

Later we will prove that unless the desired result holds trivially, we can partition the vertices in $G$ into three nonempty sets $S, \{a, b\}, T$ such that the following conditions hold:

(i) $S$ is not adjacent to $T$ (that is, no vertex in $S$ is adjacent to a vertex in $T$); and

(ii) given any two vertices $w_1, w_2 \in S \cup \{a, b\}$, there exists a *pseudopath* between any other vertices in $S \cup \{a, b\}$ which does not pass through $v_1$ or $v_2$. (Here, a pseudopath is a path which contains only vertices in $S \cup \{a, b\}$, and contains no edges besides $\overline{ab}$ and the edges in $G$. In other words, a pseudopath is like a path in $S \cup \{a, b\}$, except it is allowed to contain one additional type of edge.)

Given such a partition, let $x$ be any vertex in $S$. We claim that the graph obtained by removing $x$ from $G$ is 2-connected. In other words, we claim that given any distinct vertices $y, v_1, v_2$ distinct from $x$, there exists $v_1 \rightsquigarrow v_2$ not passing through the *forbidden points* $x$ and $y$. We say that we can *legally travel* between two points, or that a path between two points is *legal*, if it does not pass through either $x$ or $y$.

**Lemma 1.** *If $y \in T$, then (1) one can legally travel from any nonforbidden point of $S$ to any nonforbidden point of $\{a, b\}$, and (2) one can legally travel from any nonforbidden point of $T$ to some nonforbidden point of $\{a, b\}$.*

*Proof:* We are given that $x \in S$ and $y \in T$. Because one forbidden point lies in $S \cup \{a, b\}$ and this forbidden point is $x$, $a$ and $b$ are not forbidden points. We prove that we may legally travel from any nonforbidden point $s \in S$ to $a$; a similar proof shows that we may travel from $s$ to $b$. By (ii), there exists a pseudopath from $s$ to $a$ not passing through $x$ or $b$. Because it does not pass through $b$, this pseudopath does not contain $\overline{ab}$, so it is actually a path. By construction, it does not contain $x$; nor can it contain $y \in T$, because no pseudopath contains vertices in $T$. This proves (1).

Suppose that we have $t \in T$ distinct from $y$. By $(*)$, there exists $t \rightsquigarrow a$ that does not pass through $y$. We truncate this path as soon as it passes through $\{a, b\}$. Because the path starts in $T$, and $T$ is not adjacent to $\{x\} \subseteq S$, the path cannot pass through $x$ without first passing through $\{a, b\}$. But because our path stops as soon as it passes through $\{a, b\}$, it does not pass through $x$. Hence, our path from $t$ to $\{a, b\}$ contains neither $x$ nor $y$, as desired. ∎

**Lemma 2.** *If $y$ lies in $S \cup \{a, b\}$ instead of $T$, then (1) one can legally travel from any nonforbidden point of $S$ to some nonforbidden point of $\{a, b\}$, and (2) one can legally travel from any nonforbidden point of $T$ to any nonforbidden point of $\{a, b\}$.*

*Proof:* To prove (1), assume without loss of generality that $a$ is not forbidden. Let $s$ be any nonforbidden point in $S$. By (ii), there exists a pseudopath from $s$ to $a$ that does not pass through $x$ or $y$. Truncate this path as soon is it passes through $\{a, b\}$; because the truncated path cannot contain $\overline{ab}$, it must be a path. This proves (1).

To prove (2), let $t$ be a point in $T$. It suffices to prove that if $a$ is nonforbidden, then we may legally travel from $t$ to $a$. (A similar proof holds if $b$ is nonforbidden and we wish to travel from $t$ to $b$.) By $(*)$, there exists $t \rightsquigarrow a$ that does not pass through $b$. Truncate this path as soon as it passes through $a$. It does not pass through $S$ because to do so it would first have to pass through $a$ (since $T$ is not adjacent to $S$). Nor does the path pass through $b$. Because the forbidden points lie in $S \cup \{b\}$, it follows that our path from $t$ to $a$ is legal. This proves (2).                    ∎

**Lemma 3.** *One can legally travel between any two nonforbidden points of* $\{a, b\}$.

*Proof:*   Of course, if there is only one nonforbidden point in the pair $\{a, b\}$, the claim is obvious. Otherwise, if $a$ and $b$ are not forbidden, we wish to travel between them without passing through any forbidden points. Either the conditions of lemma 1 hold, or else the conditions of lemma 2 hold. In the first case, because $S$ is nonempty, it must contain at least one vertex; because this vertex has degree at least 3, $S$ must contain an additional vertex. Hence, $S$ contains at least 2 vertices, and one of these vertices $s$ is not forbidden. By lemma 1, we can legally travel from $a$ to $s$ and then from $s$ to $b$. Combining these two paths allows us to legally travel between $a$ and $b$, as desired.

If instead the conditions of lemma 2 hold, then a similar proof shows that we may legally travel from $a$ to some $t \in T$ and then to $b$.        ∎

With these three lemmas, we show that one can legally travel between any two nonforbidden points $v$ and $w$. From lemmas 1 and 2, we can legally travel from $v$ to some point $v' \in \{a, b\}$; similarly, we can legally travel from $w$ to some point $w' \in \{a, b\}$. From lemma 3, we can legally travel from $v'$ to $w'$. Combining these paths yields a legal path from $v$ to $w$, as desired.

It remains, then, to construct $S$, $\{a, b\}$, and $T$ satisfying conditions (i) and (ii). Consider all possible ways to partition the vertices in $G$ into three nonempty sets $S$, $\{a, b\}$, $T$ such that $S$ and $T$ are not adjacent. If there is no such partition, then consider any distinct vertices $x, y$. There is no valid partition for $(a, b) = (x, y)$, so for any other two vertices $v_1, v_2$ there must exist $v_1 \rightsquigarrow v_2$ not passing through $x$ or $y$. It follows that the desired result holds trivially, because if we remove *any* vertex $x$ then the graph remains 2-connected.

Hence, we may assume that a partition $S \cup \{a, b\} \cup T$ exists where $S$ and $T$ are not adjacent. Find such a partition with $|S|$ minimal. Let $G_1$ be the graph obtained from $G$ by removing the vertices in $T$ and then adding the edge $\overline{ab}$ (if $a$ and $b$ are not already adjacent in $G$). We wish to prove that $G_1$ is 3-connected (that is, it remains connected even if we remove any two vertices). If not, then we may partition the vertices of $G_1$ into three nonempty sets $S_1, \{a_1, b_1\}, T_1$ such that $S_1$ and $T_1$ are not adjacent in $G_1$.

Notice that neither $(a, b)$ nor $(b, a)$ can lie in $S_1 \times T_1$, because $a$ and $b$ are adjacent in $G_1$. Hence, we may assume without loss of generality that $S_1$ contains neither $a$ nor $b$. In this case, $S_1 \subseteq S$, so that $S_1$ and $T$ are not adjacent in $G$. Also, by the construction of $S_1$ and $T_1$, $S_1$ and $T_1$ are not adjacent in $G_1$; hence, $S_1$ and $T_1$ are not adjacent in $G$ either. We now partition the vertices in $G$ into the three sets $S_1$, $\{a_1, b_1\}$, $T_1 \cup T$. We showed above that no vertex in $S_1$ is adjacent in $G$ to any vertex in $T_1 \cup T$. However, $|S_1| = |S| - |T_1| < |S|$, contradicting the minimal definition of $S$.

Thus, there indeed exist $S$, $\{a, b\}$, $T$ satisfying conditions (i) and (ii). This completes the proof.

## Problem 22

The perpendicular bisectors of sides $\overline{AB}$ and $\overline{BC}$ of nonequilateral triangle $ABC$ meet lines $BC$ and $AB$ at $A_1$ and $C_1$, respectively. Let the bisectors of angles $A_1AC$ and $C_1CA$ meet at $B'$, and define $C'$ and $A'$ analogously. Prove that the points $A', B', C'$ lie on a line passing through the circumcenter of triangle $ABC$.

**Solution.** Let $I$ be the incenter and $O$ the circumcenter of triangle $ABC$. We will show that $A', B', C'$ lie on the line $OI$.

We prove that $B'$ lies on the line $OI$; similar proofs show that $A'$ and $C'$ do as well. Let $P, Q, R, S$ be the second intersections of the lines $CB', AB', CI, AI$, respectively, with the circumcircle of triangle $ABC$. Note that

$$\angle RCA + \angle ABC + \angle CAQ$$

$$= \frac{1}{2}\angle ACB + \angle ABC + \frac{1}{2}(\angle CAB - \angle A_1 AB)$$

$$= \frac{1}{2}\angle ACB + \angle ABC + \frac{1}{2}(\angle CAB - \angle ABC) = \frac{\pi}{2}.$$

Thus, the points $R$ and $Q$ are diametrically opposed on the circumcircle of triangle $ABC$. Similarly, the points $P$ and $S$ are diametrically opposed. By Pascal's Theorem applied to the cyclic hexagon $AQRCPS$, the points $O, I, B'$ lie on a line.

## Problem 23

Is it possible to select 102 17-element subsets of a 102-element set, such that the intersection of any two of the subsets has at most 3 elements?

**Solution.** The answer is "yes." More generally, suppose that $p$ is a prime congruent to 2 modulo 3. We show that it is possible to select $p(p+1)/3$ $p$-element subsets of a $p(p+1)/3$-element set, such that the intersection of any two of the subsets has at most 3 elements. Setting $p = 17$ yields the claim.

Let $\mathcal{P}$ be *the projective plane of order $p$* (which this solution refers to as "the projective plane," for short), defined as follows. Let $\mathcal{A}$ be the ordered triples $(a, b, c)$ of integers modulo $p$, and define the equivalence relation $\sim$ by $(a, b, c) \sim (d, e, f)$ if and only if $(a, b, c) = (d\kappa, e\kappa, f\kappa)$ for some $\kappa$. Then let $\mathcal{P} = (\mathcal{A} - \{(0, 0, 0)\}) / \sim$. We let $[a, b, c] \in \mathcal{P}$ denote the equivalence class containing $(a, b, c)$, and we call it a *point* of $\mathcal{P}$. Because $\mathcal{A} - \{(0, 0, 0)\}$ contains $p^3 - 1$ elements, and each equivalence class under $\sim$ contains $p - 1$ elements, we find that

$$|\mathcal{P}| = (p^3 - 1)/(p - 1) = p^2 + p + 1.$$

Given $q \in \mathcal{P}$, we may write $q = [\alpha, \beta, \gamma]$ and consider the solutions $[x, y, z]$ to

$$\alpha x + \beta y + \gamma z \equiv 0 \pmod{p}.$$

The set of these solutions is called a *line* in the projective plane; it is easy to check that this line is well-defined regardless of how we write $q = [\alpha, \beta, \gamma]$, and that $(x, y, z)$ satisfies the above equation if and only if every triple in $[x, y, z]$ does. We let $[[\alpha, \beta, \gamma]]$ denote the above line.

It is easy to check that $\mathcal{P}$ is in one-to-one correspondence with $\mathcal{P}^*$, the set of lines in the projective plane, via the correspondence $[\alpha, \beta, \gamma] \longleftrightarrow [[\alpha, \beta, \gamma]]$. It is also easy to check that any two distinct points lie on exactly one line, and that any two distinct lines intersect at exactly one point. Furthermore, any line contains exactly $p+1$ points. (The projective plane is not an invention of this solution, but a standard object in algebraic geometry; the properties described up to this point are also well known.)

Define $\varphi : \mathcal{P} \to \mathcal{P}$ by $\varphi([a, b, c]) = [b, c, a]$. Given a point $q \in \mathcal{P}$, we say we *rotate* it to obtain $q'$ if $q' = \varphi(q)$. Similarly, given a subset $T \subseteq \mathcal{P}$, we say we rotate it to obtain $T'$ if $T' = \varphi(T)$.

Given a point $q \neq [1, 1, 1]$ in the projective plane, we rotate it once and then a second time to obtain two additional points. Together, these three points form a *triplet*. We will show below that (i) the corresponding triplet actually contains three points. Observe that any two triplets obtained in this manner are either identical or disjoint. Because there are $p^2 + p$ points in $\mathcal{P} - \{[1, 1, 1]\}$, it follows that there are $p(p+1)/3$ distinct triplets. Let $\mathcal{S}$ be the set of these triplets.

Given a line $\ell = [[\alpha, \beta, \gamma]] \neq [[1, 1, 1]]$ in the projective plane, it is easy to show that rotating it once and then a second time yields the lines $[[\beta, \gamma, \alpha]]$ and $[[\gamma, \alpha, \beta]]$. The points $q \neq [1, 1, 1]$ on $[[\alpha, \beta, \gamma]]$, $[[\beta, \gamma, \alpha]]$, and $[[\gamma, \alpha, \beta]]$ can be partitioned into triplets. More specifically, we will show below that (ii) there are exactly $3p$ such points $q \neq [1, 1, 1]$. Hence, these points can be partitioned into exactly $p$ distinct triplets; let $T_\ell$ be the set of such triplets.

Take any two lines $\ell_1, \ell_1' \neq [[1, 1, 1]]$, and suppose that $|T_{\ell_1} \cap T_{\ell_1'}| > 3$. We claim that $\ell_1$ and $\ell_1'$ are rotations of each other. Suppose otherwise for sake of contradiction. Let $\ell_2, \ell_3$ be the rotations of $\ell_1$, and let $\ell_2', \ell_3'$ be the rotations of $\ell_1'$. We are given that $T_{\ell_1}$ and $T_{\ell_1'}$ share more than three triplets; that is, $\ell_1 \cup \ell_2 \cup \ell_3$ intersects $\ell_1' \cup \ell_2' \cup \ell_3'$ in more than 9 points. Because $\ell_1$ and $\ell_1'$ are not rotations of each other, each $\ell_i$ is distinct from all the $\ell_j'$. Hence, $\ell_i \cap \ell_j'$ contains exactly one point for each $i$ and $j$. It follows that $\ell_1 \cup \ell_2 \cup \ell_3$ and $\ell_1' \cup \ell_2' \cup \ell_3'$ consists of at most $3 \cdot 3 = 9$ points, a contradiction. Hence, our assumption as wrong, and $|T_{\ell_1} \cap T_{\ell_1'}| > 3$ only if $\ell_1$ and $\ell_1'$ are rotations of each other.

Just as there are $p(p + 1)$ points in $\mathcal{P} - \{[1, 1, 1]\}$, there are $p(p + 1)$ lines in $\mathcal{P}^* - \{[[1, 1, 1]]\}$. We can partition these into $p(p + 1)/3$ triples $(\ell_1, \ell_2, \ell_3)$, where the lines in each triple are rotations of each other. Now, pick one line $\ell$ from each triple and take the corresponding set $T_\ell$ of triplets. From the previous paragraph, any two of these $p(p + 1)/3$ sets intersect in at most 3 triplets.

Hence, we have found a set $\mathcal{S}$ of $p(p + 1)/3$ elements (namely, the triplets of $\mathcal{P}$), along with $p(p+1)/3$ subsets of $\mathcal{S}$ (namely, the appropriate $T_\ell$) such that no two of these subsets have four elements in common. This completes the proof.

Well, not quite. We have yet to prove that (i) if $q \neq [1, 1, 1]$, then the triplet $\{q, \varphi(q), \varphi^2(q)\}$ contains three distinct points, and (ii) if $\ell_1 = [[\alpha, \beta, \gamma]] \neq [[1, 1, 1]]$, then there are $3p$ points $q \neq [1, 1, 1]$ on $[[\alpha, \beta, \gamma]] \cup [[\beta, \gamma, \alpha]] \cup [[\gamma, \alpha, \beta]]$.

To prove (i), we first show that $x^3 \equiv 1 \pmod{p}$ only if $x \equiv 1 \pmod{p}$. Because 3 is coprime to $p - 1$, we can write $1 = 3r + (p - 1)s$. We are given that $x^3 \equiv 1 \pmod{p}$, and by Fermat's Little Theorem we also have $x^{p-1} \equiv 1 \pmod{p}$. Hence,

$$x = x^{3r+(p-1)s} = \left(x^3\right)^r \left(x^{p-1}\right)^s \equiv 1^r \cdot 1^s \equiv 1 \pmod{p}.$$

(Alternatively, let $g$ be a primitive element modulo $p$, and write $x = g^m$

for some nonnegative integer $m$. Then

$$1 \equiv (g^m)^3 = g^{3m} \quad (\text{mod } p),$$

implying that $p - 1$ divides $3m$. Because $p - 1$ is relatively prime to 3, we must have $(p - 1) \mid m$. Writing $m = (p - 1)n$, we have

$$x \equiv g^m \equiv (g^{p-1})^n \equiv 1 \, (\text{mod } p).)$$

Now, if $q = [a, b, c] \neq [1, 1, 1]$, then suppose (for sake of contradiction) that $[a, b, c] = [b, c, a]$. There exists $\kappa$ such that $(a, b, c) = (b\kappa, c\kappa, a\kappa)$. Thus,

$$ab^{-1} \equiv bc^{-1} \equiv ca^{-1} \quad (\text{mod } p),$$

because all three quantities are congruent to $\kappa$ modulo $p$. Hence,

$$(ab^{-1})^3 \equiv (ab^{-1})(bc^{-1})(ca^{-1}) \equiv 1 \, (\text{mod } p).$$

From this and the result in the last paragraph, we conclude that $ab^{-1} \equiv 1 \,(\text{mod } p)$. Therefore, $a \equiv b \,(\text{mod } p)$, and similarly $b \equiv c \,(\text{mod } p)$ — implying that $[a, b, c] = [1, 1, 1]$, a contradiction.

Next, we prove (ii). Let $\ell_1 = [[\alpha, \beta, \gamma]] \neq [[1, 1, 1]]$, $\ell_2 = [[\beta, \gamma, \alpha]]$, and $\ell_3 = [[\gamma, \alpha, \beta]]$. Because $[[\alpha, \beta, \gamma]] \neq [[1, 1, 1]]$, we know (from a proof similar to that in the previous paragraph) that $\ell_1, \ell_2, \ell_3$ are pairwise distinct. Hence, any two of these lines intersect at exactly one point. We consider two cases: $\ell_1$ and $\ell_2$ intersect at $[1, 1, 1]$, or they intersect elsewhere.

If $[1, 1, 1]$ lies on $\ell_1$ and $\ell_2$, then it lies on $\ell_3$ as well. Each line contains $p+1$ points in total and hence $p$ points distinct from $[1, 1, 1]$. Counting over all three lines, we find $3p$ points distinct from $[1, 1, 1]$; these points must be distinct from each other, because any two of the lines $\ell_i, \ell_j$ intersect at only $[1, 1, 1]$.

If instead $q_0 = \ell_1 \cap \ell_2$ is not equal to $[1, 1, 1]$, then $[1, 1, 1]$ cannot lie on any of the lines $\ell_1, \ell_2, \ell_3$. We have $\varphi(q_0) = \ell_2 \cap \ell_3$ and $\varphi^2(q_0) = \ell_3 \cap \ell_1$; because $q_0 \neq [1, 1, 1]$, the three intersection points $q_0, \varphi(q_0), \varphi^2(q_0)$ are pairwise distinct. Now, each of the three lines $\ell_1, \ell_2, \ell_3$ contains $p + 1$ points (all distinct from $[1, 1, 1]$), for a total of $3p+3$ points. However, we count each of $q_0, \varphi(q_0), \varphi^2(q_0)$ twice in this manner, so in fact we have $(3p + 3) - 3 = 3p$ points on $\ell_1 \cup \ell_2 \cup \ell_3 - \{[1, 1, 1]\}$, as desired. *This completes the proof.*

# 3

# 2001 National Contests: Problems

## 3.1 Belarus

### Problem 1

The problem committee of a mathematical olympiad prepares some variants of the contest. Each variant contains four problems, chosen from a shortlist of $n$ problems, and any two variants have at most one problem in common.

(a) If $n = 14$, determine the largest possible number of variants the problem committee can prepare.

(b) Find the smallest value of $n$ such that it is possible to prepare ten variants of the contest.

### Problem 2

Let $x_1$, $x_2$, and $x_3$ be real numbers in $[-1, 1]$, and let $y_1$, $y_2$, and $y_3$ be real numbers in $[0, 1)$. Find the maximum possible value of the expression

$$\frac{1 - x_1}{1 - x_2 y_3} \cdot \frac{1 - x_2}{1 - x_3 y_1} \cdot \frac{1 - x_3}{1 - x_1 y_2}.$$

### Problem 3

Let $ABCD$ be a convex quadrilateral circumscribed about a circle. Lines $AB$ and $DC$ intersect at $E$, and $B$ and $C$ lie on $\overline{AE}$ and $\overline{DE}$, respectively; lines $DA$ and $CB$ intersect at $F$, and $A$ and $B$ lie on $\overline{DF}$ and $\overline{CF}$, respectively. Let $I_1$, $I_2$, and $I_3$ be the incenters of triangles $AFB$, $BEC$, and $ABC$, respectively. Line $I_1 I_3$ intersects lines $EA$ and $ED$ at $K$ and $L$, respectively, and line $I_2 I_3$ intersects lines $FC$ and $FD$ at $M$ and $N$, respectively. Prove that $EK = EL$ if and only if $FM = FN$.

## Problem 4

On the Cartesian coordinate plane, the graph of the parabola $y = x^2$ is drawn. Three distinct points $A$, $B$, and $C$ are marked on the graph with $A$ lying between $B$ and $C$. Point $N$ is marked on $\overline{BC}$ so that $\overline{AN}$ is parallel to the $y$-axis. Let $K_1$ and $K_2$ be the areas of triangles $ABN$ and $ACN$, respectively. Express $AN$ in terms of $K_1$ and $K_2$.

## Problem 5

Prove that for every positive integer $n$ and every positive real $a$,

$$a^n + \frac{1}{a^n} - 2 \geq n^2 \left( a + \frac{1}{a} - 2 \right).$$

## Problem 6

Three distinct points $A$, $B$, and $N$ are marked on the line $\ell$, with $B$ lying between $A$ and $N$. For an arbitrary angle $\alpha \in (0, \frac{\pi}{2})$, points $C$ and $D$ are marked in the plane on the same side of $\ell$ such that $N$, $C$, and $D$ are collinear; $\angle NAD = \angle NBC = \alpha$; and $A$, $B$, $C$, and $D$ are concyclic. Find the locus of the intersection points of the diagonals of $ABCD$ as $\alpha$ varies between 0 and $\frac{\pi}{2}$.

## Problem 7

In the increasing sequence of positive integers $a_1, a_2, \ldots$, the number $a_k$ is said to be *funny* if it can be represented as the sum of some other terms (not necessarily distinct) of the sequence.

(a) Prove that all but finitely terms of the sequence are funny.

(b) Does the result in (a) always hold if the terms of the sequence can be any positive rational numbers?

## Problem 8

Let $n$ be a positive integer. Each square of a $(2n - 1) \times (2n - 1)$ square board contains an arrow, either pointing up, down, to the left, or to the right. A beetle sits in one of the cells. Each year it creeps from one square in the direction of the arrow in that square, either reaching another square or leaving the board. Each time the beetle moves, the arrow in the square it leaves turns $\pi/2$ clockwise. Prove that the beetle leaves the board in at most $2^{3n-1}(n - 1)! - 4$ years after it first moves.

## Problem 9

The convex quadrilateral $ABCD$ is inscribed in the circle $S_1$. Let $O$ be the intersection of $\overline{AC}$ and $\overline{BD}$. Circle $S_2$ passes through $D$ and $O$, intersecting $\overline{AD}$ and $\overline{CD}$ at $M$ and $N$, respectively. Lines $OM$ and $AB$ intersect at $R$, lines $ON$ and $BC$ intersect at $T$, and $R$ and $T$ lie on the same side of line $BD$ as $A$. Prove that $O$, $R$, $T$, and $B$ are concyclic.

## Problem 10

There are $n$ aborigines on an island. Any two of them are either friends or enemies. One day, the chieftain orders that all citizens (including himself) make and wear a necklace with zero or more stones so that (i) given a pair of friends, there exists a color such that each has a stone of that color; (ii) given a pair of enemies, there does *not* exist a color such that each has a stone of that color.

(a) Prove that the aborigines can carry out the chieftain's order.

(b) What is the minimum number of colors of stones required for the aborigines to carry out the chieftain's order?

## 3.2 Bulgaria

### Problem 1

Diagonals $\overline{AC}$ and $\overline{BD}$ of a cyclic quadrilateral $ABCD$ intersect at point $E$. Prove that if $\angle BAD = \pi/3$ and $AE = 3CE$, then the sum of some two sides of the quadrilateral equals the sum of the other two.

### Problem 2

Find the least positive integer $n$ such that it is possible for a set of $n$ people to have the following properties: (i) among any four of the $n$ people, some two are not friends with each other; given any $k \geq 1$ of the $n$ people among whom there is no pair of friends, there exists three people among the remaining $n - k$ people such that every two of the three are friends. (If a person $A$ is a friend of a person $B$, then $B$ is a friend of $A$ as well.)

### Problem 3

Let $ABC$ be a right triangle with hypotenuse $\overline{AB}$. A point $D$ distinct from $A$ and $C$ is chosen on $\overset{\frown}{AC}$ such that the line through the incenter of triangle $ABC$ parallel to the internal bisector of angle $ADB$ is tangent to the incircle of triangle $BCD$. Prove that $AD = BD$.

### Problem 4

Find all triples of positive integers $(a, b, c)$ such that $a^3 + b^3 + c^3$ is divisible by $a^2 b$, $b^2 c$, and $c^2 a$.

### Problem 5

Consider the sequence $\{a_n\}$ such that $a_0 = 4$, $a_1 = 22$, and $a_n - 6a_{n-1} + a_{n-2} = 0$ for $n \geq 2$. Prove that there exist sequences $\{x_n\}$ and $\{y_n\}$ of positive integers such that

$$a_n = \frac{y_n^2 + 7}{x_n - y_n}$$

for any $n \geq 0$.

### Problem 6

Let $I$ be the incenter and $k$ be the incircle of nonisosceles triangle $ABC$. Let $k$ intersect $\overline{BC}$, $\overline{CA}$, and $\overline{AB}$ at $A_1$, $B_1$, and $C_1$, respectively. Let

$\overline{AA_1}$ intersect $k$ again at $A_2$, and define $B_2$ and $C_2$ similarly. Finally, choose $A_3$ and $B_3$ on $\overline{B_1C_1}$ and $\overline{A_1C_1}$, respectively, such that $\overline{A_1A_3}$ and $\overline{B_1B_3}$ are angle bisectors in triangle $A_1B_1C_1$. Prove that (a) $\overline{A_2A_3}$ bisects angle $B_1A_2C_1$; (b) if the circumcircles of triangles $A_1A_2A_3$ and $B_1B_2B_3$ intersect at $P$ and $Q$, then $I$ lies on $\overleftrightarrow{PQ}$.

## Problem 7

Given a permutation $(a_1, a_2, \ldots, a_n)$ of the numbers $1, 2, \ldots, n$, one may interchange any two consecutive "blocks" — that is, one may transform

$$(a_1, \ldots, a_i, \underbrace{a_{i+1}, \ldots, a_{i+p}}_{A}, \underbrace{a_{i+p+1}, \ldots, a_{i+q}}_{B}, a_{i+q+1}, \ldots, a_n)$$

into

$$(a_1, \ldots, a_i, \underbrace{a_{i+p+1}, \ldots, a_{i+q}}_{B}, \underbrace{a_{i+1}, \ldots, a_{i+p}}_{A}, a_{i+q+1}, \ldots, a_n)$$

by interchanging the "blocks" $A$ and $B$. Find the least number of such changes which are needed to transform $(n, n-1, \ldots, 1)$ into $(1, 2, \ldots, n)$.

## Problem 8

Let $n \geq 2$ be a fixed integer. At any lattice point $(i, j)$ we write the unique integer $k \in \{0, 1, \ldots, n-1\}$ such that $i + j \equiv k \pmod{n}$. Find all pairs $a, b$ of positive integers such that the rectangle with vertices $(0, 0)$, $(a, 0)$, $(a, b)$, and $(0, b)$ has the following properties: (i) the numbers $0, 1, \ldots, n-1$ appear in its interior an equal number of times; (ii) the numbers $0, 1, \ldots, n-1$ appear on its boundary an equal number of times.

## Problem 9

Find all real numbers $t$ for which there exist real numbers $x, y, z$ such that

$$3x^2 + 3xz + z^2 = 1,$$
$$3y^2 + 3yz + z^2 = 4,$$
$$x^2 - xy + y^2 = t.$$

## Problem 10

Let $p$ be a prime number congruent to 3 modulo 4, and consider the equation

$$(p+2)x^2 - (p+1)y^2 + px + (p+2)y = 1.$$

Prove that this equation has infinitely many solutions in positive integers, and show that if $(x, y) = (x_0, y_0)$ is a solution of the equation in positive integers, then $p \mid x_0$.

## 3.3 Canada

### Problem 1

Let $ABC$ be a triangle with $AC > AB$. Let $P$ be the intersection point of the perpendicular bisector of $\overline{BC}$ and the internal angle bisector of angle $CAB$. Let $X$ and $Y$ be the feet of the perpendiculars from $P$ to lines $AB$ and $AC$, respectively. Let $Z$ be the intersection point of lines $XY$ and $BC$. Determine the value of $\frac{BZ}{ZC}$.

### Problem 2

Let $n$ be a positive integer. Nancy is given a matrix in which each entry is a positive integer. She is permitted to make either of the following two moves:

(i) select a row and multiply each entry in this row by $n$;

(ii) select a column and subtract $n$ from each entry in this column.

Find all possible values of $n$ for which given any matrix, it is possible for Nancy to perform a finite sequence of moves to obtain a matrix in which each entry is 0.

### Problem 3

Let $P_0$, $P_1$, and $P_2$ be three points on a circle with radius 1, where $P_1 P_2 = t < 2$. Define the sequence of points $P_3, P_4, \ldots$ recursively by letting $P_i$ be the circumcenter of triangle $P_{i-1} P_{i-2} P_{i-3}$ for each integer $i \geq 3$.

(a) Prove that the points $P_1, P_5, P_9, P_{13}, \ldots$ are collinear.

(b) Let $x = P_1 P_{1001}$ and $y = P_{1001} P_{2001}$. Prove that $\sqrt[500]{x/y}$ depends only on $t$, not on the position of $P_0$, and determine all values of $t$ for which $\sqrt[500]{x/y}$ is an integer.

## 3.4  China

### Problem 1

Let $a$ be a fixed real number with $\sqrt{2} < a < 2$. Let $ABCD$ be a quadrilateral inscribed in a circle $\omega$ of radius one, such that the circumcenter of
quadrilateral $ABCD$ lies in the quadrilateral's interior. Lines $\ell_A$, $\ell_B$, $\ell_C$,
$\ell_D$ are tangent to $\omega$ at $A$, $B$, $C$, $D$, respectively. Let lines $\ell_A$ and $\ell_B$, $\ell_B$
and $\ell_C$, $\ell_C$ and $\ell_D$, $\ell_D$ and $\ell_A$ intersect at $A'$, $B'$, $C'$, $D'$, respectively.
Determine the minimum value of

$$\frac{[A'B'C'D']}{[ABCD]}$$

among all such quadrilaterals $ABCD$ for which the longest and shortest
sides of the quadrilateral have length $a$ and $\sqrt{4-a^2}$, respectively.

### Problem 2

Determine the smallest positive integer $m$ such that for any $m$-element
subsets $W$ of $X = \{1, 2, \ldots, 2001\}$, there are two elements $u$ and $v$ (not
necessarily distinct) in $W$ with $u + v = 2^n$ for some positive integer $n$.

### Problem 3

Two triangle are said to be *of the same type* if they are both acute triangles,
both right triangles, or both obtuse triangles. Let $n$ be a positive integer
and let $\mathcal{P}$ be a $n$-sided regular polygon. Exactly one magpie sits at each
vertex of $\mathcal{P}$. A hunter passes by, and the magpies fly away. When they
return, exactly one magpie lands on each vertex of $\mathcal{P}$, not necessarily in its
original position. Find all $n$ for which there must exist three magpies with
the following property: the triangle formed by the vertices the magpies
originally sit at, and the triangle formed by the vertices they return to after
the hunter passes by, are of the same type.

### Problem 4

We are given three integers $a$, $b$, $c$ such that $a$, $b$, $c$, $a+b-c$, $a+c-b$,
$b+c-a$, and $a+b+c$ are seven distinct primes. Let $d$ be the difference
between the largest and smallest of these seven primes. Suppose that
$800 \in \{a+b, b+c, c+a\}$. Determine the maximum possible value of $d$.

## Problem 5

Let $P_1 P_2 \ldots P_{24}$ be a regular 24-sided polygon inscribed in a circle $\omega$ with circumference 24. Determine the number of ways to choose sets of eight distinct vertices $\{P_{i_1}, P_{i_2}, \ldots, P_{i_8}\}$ such that none of the arcs $P_{i_j} P_{i_k}$ has length 3 or 8.

## Problem 6

Let $a = 2001$. Consider the set $A$ of all pairs of positive integers $(m, n)$ such that

(i) $m < 2a$;

(ii) $2am - m^2 + n^2$ is divisible by $2n$;

(iii) $n^2 - m^2 + 2mn \leq 2a(n - m)$.

For $(m, n) \in A$, let

$$f(m, n) = \frac{2am - m^2 - mn}{n}.$$

Determine the maximum and minimum values of $f$, respectively.

## Problem 7

For each integer $k > 1$, find the smallest integer $m$ greater than 1 with the following property: there exists a polynomial $f(x)$ with integer coefficients such that $f(x) - 1$ has at least 1 integer root and $f(x) - m$ has exactly $k$ distinct integer roots.

## Problem 8

Given positive integers $k$, $m$, $n$ such that $k \leq m \leq n$, express

$$\sum_{i=0}^{n} (-1)^i \frac{1}{n + k + i} \cdot \frac{(m + n + i)!}{i!(n - i)!(m + i)!}$$

in closed form.

## Problem 9

Let $a$ be a positive integer with $a \geq 2$, and let $N_a$ be the number of positive integers $k$ such that

$$k_1^2 + k_2^2 + \cdots + k_n^2 = k,$$

where $k_1 k_2 \ldots k_n$ is the base $a$ representation of $k$. Prove that:

(a) $N_a$ is odd;

(b) for any positive integer $M$, there is some $a$ for which $N_a \geq M$.

## Problem 10

Let $n$ be a positive integer, and define

$$M = \{(x, y) \mid x, y \in \mathbb{N}, 1 \leq x, y \leq n\}.$$

Determine the number of functions $f$ defined on $M$ such that

(i) $f(x, y)$ is a nonnegative integer for any $(x, y) \in M$;

(ii) for $1 \leq x \leq n$, $\sum_{y=1}^{n} f(x, y) = n - 1$;

(iii) if $f(x_1, y_1) f(x_2, y_2) > 0$, then $(x_1 - x_2)(y_1 - y_2) \geq 0$.

## 3.5   Czech and Slovak Republics

### Problem 1

Find all triples $a, b, c$ of real numbers for which a real number $x$ satisfies

$$\sqrt{2x^2 + ax + b} > x - c$$

if and only if $x \leq 0$ or $x > 1$.

### Problem 2

In a certain language there are $n$ letters. A sequence of letters is called a *word* if and only if between any pair of identical letters, there is no other pair of equal letters. Prove that there exists a word of maximum possible length, and find the number of words which have that length.

### Problem 3

Let $n \geq 1$ be an integer, and let $a_1, a_2, \ldots, a_n$ be positive integers. Let $f : \mathbb{Z} \to \mathbb{R}$ be a function such that $f(x) = 1$ for each integer $x < 0$ and

$$f(x) = 1 - f(x - a_1)f(x - a_2) \cdots f(x - a_n)$$

for each integer $x \geq 0$. Show that there exist positive integers $s$ and $t$ such that $f(x + t) = f(x)$ for any integer $x > s$.

## 3.6   Hungary

### Problem 1

Let $x$, $y$, and $z$ be positive real numbers smaller than 4. Prove that among
the numbers

$$\frac{1}{x} + \frac{1}{4-y}, \quad \frac{1}{y} + \frac{1}{4-z}, \quad \frac{1}{z} + \frac{1}{4-x},$$

there is at least one which is greater than or equal to 1.

### Problem 2

Find all integers $x$, $y$, and $z$ such that $5x^2 - 14y^2 = 11z^2$.

### Problem 3

Find all triangles $ABC$ for which it is true that the median from $A$ and
the altitude from $A$ are reflections of each other across the internal angle
bisector from $A$.

### Problem 4

Let $m$ and $n$ be integers such that $1 \le m \le n$. Prove that $m$ is a divisor
of

$$n \sum_{k=0}^{m-1} (-1)^k \binom{n}{k}.$$

### Problem 5

Find all real numbers $c$ with the following property: Given any triangle,
one can find two points $A$ and $B$ on its perimeter so that they divide the
perimeter in two parts of equal length and so that $AB$ is at most $c$ times
the perimeter.

### Problem 6

The circles $k_1$ and $k_2$ and the point $P$ lie in a plane. There exists a line $\ell$
and points $A_1, A_2, B_1, B_2, C_1, C_2$ with the following properties: $\ell$ passes
through $P$ and intersects $k_i$ at $A_i$ and $B_i$ for $i = 1, 2$; $C_i$ lies on $k_i$ for
$i = 1, 2$; and $A_1C_1 = B_1C_1 = A_2C_2 = B_2C_2$. Describe how to construct
such a line and such points given only $k_1$, $k_2$, and $P$.

## Problem 7

Let $k$ and $m$ be positive integers, and let $a_1, a_2, \ldots, a_k$ and $b_1, b_2, \ldots, b_m$ be distinct integers greater than 1. Each $a_i$ is the product of an even number of primes, not necessarily distinct, while each $b_i$ is the product of an odd number of primes, again not necessarily distinct. How many ways can we choose several of the $k + m$ given numbers such that each $b_i$ has an even number of divisors among the chosen numbers?

## 3.7   India

### Problem 1

Every vertex of the unit squares on an $m \times n$ chessboard is colored either blue, green, or red, such that all the vertices on the boundary of the board are colored red. We say that a unit square of the board is *properly colored* if exactly one pair of adjacent vertices of the square are the same color. Show that the number of properly colored squares is even.

### Problem 2

Let $ABCD$ be a rectangle, and let $\Gamma$ be an arc of a circle passing through $A$ and $C$. Let $\Gamma_1$ be a circle which is tangent to lines $CD$ and $DA$ as well as tangent to $\Gamma$. Similarly, let $\Gamma_2$ be a circle lying completely inside rectangle $ABCD$ which is tangent to lines $AB$ and $BC$ as well as tangent to $\Gamma$. Suppose that $\Gamma_1$ and $\Gamma_2$ both lie completely in the closed region bounded by rectangle $ABCD$. Let $r_1$ and $r_2$ be the radii of $\Gamma_1$ and $\Gamma_2$, respectively, and let $r$ be the inradius of triangle $ABC$.

(a) Prove that $r_1 + r_2 = 2r$.

(b) Show that one of the common internal tangents to $\Gamma_1$ and $\Gamma_2$ is parallel to $\overline{AC}$ and has length $|AB - BC|$.

### Problem 3

Let $a_1$, $a_2$, ... be a strictly increasing sequence of positive integers such that $\gcd(a_m, a_n) = a_{\gcd(m,n)}$ for all positive integers $m$ and $n$. Let $k$ be the least positive integer for which there exist positive integers $r < k$ and $s > k$ such that $a_k^2 = a_r a_s$. Prove that $r$ divides $k$ and that $k$ divides $s$.

### Problem 4

Let $a \geq 3$ be a real number and $p(x)$ be a polynomial of degree $n$ with real coefficients. Prove that

$$\max_{0 \leq j \leq n+1} \{|a^j - p(j)|\} \geq 1.$$

## 3.8   Iran

### Problem 1

Let $\alpha$ be a real number between 1 and 2, exclusive. Prove that $\alpha$ has a unique representation as an infinite product

$$\alpha = \prod_{k=1}^{\infty} \left( 1 + \frac{1}{n_k} \right),$$

where each $n_k$ is a natural number and $n_k^2 \leq n_{k+1}$ for all $k \geq 1$.

### Problem 2

We flip a fair coin repeatedly until encountering three consecutive flips of the form (i) two tails followed by heads, or (ii) heads, followed by tails, followed by heads. Which sequence, (i) or (ii), is more likely to occur first?

### Problem 3

Suppose that $x$, $y$, and $z$ are natural numbers such that $xy = z^2 + 1$. Prove that there exist integers $a$, $b$, $c$, and $d$ such that $x = a^2 + b^2$, $y = c^2 + d^2$, and $z = ac + bd$.

### Problem 4

Let $ACE$ be a triangle, $B$ be a point on $\overline{AC}$, and $D$ be a point on $\overline{AE}$. Let $F$ be the intersection of $\overline{CD}$ and $\overline{BE}$. If $AB + BF = AD + DF$, prove that $AC + CF = AE + EF$.

### Problem 5

Suppose that $a_1, a_2, \ldots$ is a sequence of natural numbers such that for all natural numbers $m$ and $n$, $\gcd(a_m, a_n) = a_{\gcd(m,n)}$. Prove that there exists a sequence $b_1, b_2, \ldots$ of natural numbers such that $a_n = \prod_{d \mid n} b_d$ for all integers $n \geq 1$.

### Problem 6

Let a *generalized diagonal* in an $n \times n$ matrix be a set of entries which contains exactly one element from each row and one element from each column. Let $A$ be an $n \times n$ matrix filled with 0s and 1s which contains exactly one generalized diagonal whose entries are all 1. Prove that it is possible to permute the rows and columns of $A$ to obtain an *upper-triangular matrix*, a matrix $(b_{ij})_{1\leq i,j\leq n}$ such that $b_{ij} = 0$ whenever $1 \leq j < i \leq n$.

## Problem 7

Let $O$ and $H$ be the circumcenter and orthocenter, respectively, of triangle $ABC$. The *nine-point circle* of triangle $ABC$ is the circle passing through the midpoints of the sides, the feet of the altitudes, and the midpoints of $\overline{AH}$, $\overline{BH}$, and $\overline{CH}$. Let $N$ be the center of this circle, and let $N'$ be the point such that

$$\angle N'BA = \angle NBC \qquad \text{and} \qquad \angle N'AB = \angle NAC.$$

Let the perpendicular bisector of $\overline{OA}$ intersect line $BC$ at $A'$, and define $B'$ and $C'$ similarly. Prove that $A'$, $B'$, and $C'$ lie on a line $\ell$ which is perpendicular to line $ON'$.

## Problem 8

Let $n = 2^m + 1$ for some positive integer $m$. Let $f_1, f_2, \ldots, f_n : [0,1] \to [0,1]$ be increasing functions. Suppose that for $i = 1, 2, \ldots, n$, $f_i(0) = 0$ and

$$|f_i(x) - f_i(y)| \leq |x - y|$$

for all $x, y \in [0,1]$. Prove that there exist distinct integers $i$ and $j$ between 1 and $n$, inclusive, such that

$$|f_i(x) - f_j(x)| \leq \frac{1}{m}$$

for all $x \in [0,1]$.

## Problem 9

In triangle $ABC$, let $I$ be the incenter and let $I_a$ be the excenter opposite $A$. Suppose that $\overline{II_a}$ meets $\overline{BC}$ and the circumcircle of triangle $ABC$ at $A'$ and $M$, respectively. Let $N$ be the midpoint of arc $MBA$ of the circumcircle of triangle $ABC$. Let lines $NI$ and $NI_a$ intersect the circumcircle of triangle $ABC$ again at $S$ and $T$, respectively. Prove that $S$, $T$, and $A'$ are collinear.

## Problem 10

The set of *n-variable formulas* is a subset of the functions of $n$ variables $x_1, \ldots, x_n$, and it is defined recursively as follows: the formulas $x_1, \ldots, x_n$ are $n$-variable formulas, as is any formula of the form

$$(x_1, \ldots, x_n) \mapsto \max \{ f_1(x_1, \ldots, x_n), \ldots, f_k(x_1, \ldots, x_n) \}$$

or

$$(x_1, \ldots, x_n) \mapsto \min \{ f_1(x_1, \ldots, x_n), \ldots, f_k(x_1, \ldots, x_n) \},$$

where each $f_i$ is an $n$-variable formula. For example,

$$\max \big( x_2, x_3, \min(x_1, \max(x_4, x_5)) \big)$$

is a 5-variable formula. Suppose that $P$ and $Q$ are two $n$-variable formulas such that

$$P(x_1, \ldots, x_n) = Q(x_1, \ldots, x_n) \qquad (*)$$

for all $x_1, \ldots, x_n \in \{0, 1\}$. Prove that $(*)$ also holds for all $x_1, \ldots, x_n \in \mathbb{R}$.

## 3.9   Japan

### Problem 1

Each square of an $m \times n$ chessboard is painted black or white. Each black square is adjacent to an odd number of black squares. Prove that the number of black squares is even. (Two squares are adjacent if they are different and share a common edge.)

### Problem 2

Find all positive integers $n$ such that

$$n = \prod_{k=0}^{m} (a_k + 1),$$

where $\overline{a_m a_{m-1} \ldots a_0}$ is the decimal representation of $n$ — that is, where $a_0, a_1, \ldots, a_m$ is the unique sequence of integers in $\{0, 1, \ldots, 9\}$ such that $n = \sum_{k=0}^{m} a_k 10^k$ and $a_m \neq 0$.

### Problem 3

Three real numbers $a, b, c \geq 0$ satisfy the inequalities $a^2 \leq b^2 + c^2$, $b^2 \leq c^2 + a^2$, and $c^2 \leq a^2 + b^2$. Prove that

$$(a + b + c)(a^2 + b^2 + c^2)(a^3 + b^3 + c^3) \geq 4(a^6 + b^6 + c^6),$$

and determine when equality holds.

### Problem 4

Let $p$ be a prime number and $m$ be a positive integer. Show that there exists a positive integer $n$ such that there exist $m$ consecutive zeroes in the decimal representation of $p^n$.

### Problem 5

Two triangles $ABC$ and $PQR$ satisfy the following properties: $A$ and $P$ are the midpoints of $\overline{QR}$ and $\overline{BC}$, respectively, and lines $QR$ and $BC$ are the internal angle bisectors of angles $BAC$ and $QPR$, respectively. Prove that $AB + AC = PQ + PR$.

## 3.10 Korea

### Problem 1

Given an odd prime $p$, find all functions $f : \mathbb{Z} \to \mathbb{Z}$ satisfying the following two conditions:

(i) $f(m) = f(n)$ for all $m, n \in \mathbb{Z}$ such that $m \equiv n \pmod{p}$;

(ii) $f(mn) = f(m)f(n)$ for all $m, n \in \mathbb{Z}$.

### Problem 2

Let $P$ be a point inside convex quadrilateral $O_1O_2O_3O_4$, where we write $O_0 = O_4$ and $O_5 = O_1$. For each $i = 1, 2, 3, 4$, consider the lines $\ell$ that pass through $P$ and meet the rays $O_iO_{i-1}$ and $O_iO_{i+1}$ at distinct points $A_i(\ell)$ and $B_i(\ell)$. Let $f_i(\ell) = PA_i(\ell) \cdot PB_i(\ell)$. Among all such lines $\ell$, let $m_i$ be a line for which $f_i$ is the minimum. Show that if $m_1 = m_3$ and $m_2 = m_4$, then the quadrilateral $O_1O_2O_3O_4$ is a parallelogram.

### Problem 3

Let $x_1, x_2, \ldots, x_n$ and $y_1, y_2, \ldots, y_n$ be real numbers satisfying $\sum_{i=1}^{n} x_i^2 = \sum_{i=1}^{n} y_i^2 = 1$. Show that

$$(x_1y_2 - x_2y_1)^2 \leq 2 \left| 1 - \sum_{i=1}^{n} x_iy_i \right|,$$

and determine when equality holds.

### Problem 4

Given positive integers $n$ and $N$, let $\mathcal{P}_n$ be the set of all polynomials $f(x) = a_0 + a_1x + \cdots + a_nx^n$ with integer coefficients satisfying the following two conditions:

(i) $|a_j| \leq N$ for $j = 0, 1, \ldots, n$;

(ii) at most two of $a_0, a_1, \ldots, a_n$ equal $N$.

Find the number of elements in the set $\{f(2N) \mid f(x) \in \mathcal{P}_n\}$.

### Problem 5

In acute triangle $ABC$, $\angle ABC < \pi/4$. Point $D$ lies on $\overline{BC}$ so that the angle bisector of angle $ADB$ contains $O$, the circumcenter of triangle

$ABC$. Let $\omega$ be the circumcircle of triangle $AOC$. Let $P$ be the point of intersection of the two tangent lines to $\omega$ at $A$ and $C$. Let $Q$ be the point of intersection of lines $AD$ and $CO$, and let $X$ be the point of intersection of line $PQ$ and the tangent line to $\omega$ at $O$. Show that $XO = XD$.

## Problem 6

Let $n \geq 5$ be a positive integer, and let $a_1, b_1, a_2, b_2, \ldots, a_n, b_n$ be integers satisfying the following two conditions:

(i) the pairs $(a_i, b_i)$ are all distinct for $i = 1, 2, \ldots, n$;

(ii) $|a_i b_{i+1} - a_{i+1} b_i| = 1$ for $i = 1, 2, \ldots, n$, where $(a_{n+1}, b_{n+1}) = (a_1, b_1)$.

Show that there exist $i, j$ with $1 \leq i, j \leq n$ such that $1 < |i - j| < n - 1$ and $|a_i b_j - a_j b_i| = 1$.

## 3.11   Poland

### Problem 1

Let $n \geq 2$ be an integer. Show that

$$\sum_{k=1}^{n} k x_k \leq \binom{n}{2} + \sum_{k=1}^{n} x_k^k$$

for all nonnegative reals $x_1, x_2, \ldots, x_n$.

### Problem 2

Let $P$ be a point inside a regular tetrahedron whose edges have length 1. Show that the sum of the distances from $P$ to the vertices of the tetrahedron is at most 3.

### Problem 3

The sequence $x_1, x_2, x_3, \ldots$ is defined recursively by $x_1 = a$, $x_2 = b$, and $x_{n+2} = x_{n+1} + x_n$ for $n = 1, 2, \ldots$, where $a$ and $b$ are real numbers. Call a number $c$ a *repeated value* if $x_k = x_\ell = c$ for some two distinct positive integers $k$ and $\ell$. Prove that one can choose the initial terms $a$ and $b$ so that there are more than 2000 repeated values in the sequence $x_1, x_2, \ldots$, but that it is impossible to choose $a$ and $b$ so that there are infinitely many repeated values.

### Problem 4

The integers $a$ and $b$ have the property that for every nonnegative integer $n$, the number $2^n a + b$ is a perfect square. Show that $a = 0$.

### Problem 5

Let $ABCD$ be a parallelogram, and let $K$ and $L$ be points lying on $\overline{BC}$ and $\overline{CD}$, respectively, such that $BK \cdot AD = DL \cdot AB$. Let $\overline{DK}$ and $\overline{BL}$ intersect at $P$. Show that $\angle DAP = \angle BAC$.

### Problem 6

Let $n_1 < n_2 < \cdots < n_{2000} < 10^{100}$ be positive integers. Prove that one can find two nonempty disjoint subsets $A$ and $B$ of $\{n_1, n_2, \ldots, n_{2000}\}$ such that $|A| = |B|$, $\sum_{x \in A} x = \sum_{x \in B} x$, and $\sum_{x \in A} x^2 = \sum_{x \in B} x^2$.

## 3.12  Romania

### Problem 1

Determine the ordered systems $(x, y, z)$ of positive rational numbers for which $x + \frac{1}{y}$, $y + \frac{1}{z}$, and $z + \frac{1}{x}$ are integers.

### Problem 2

Let $m$ and $k$ be positive integers such that $k < m$, and let $M$ be a set with $m$ elements. Let $p$ be an integer such that there exist subsets $A_1, A_2, \ldots, A_p$ of $M$ for which $A_i \cap A_j$ has at most $k$ elements for each pair of distinct numbers $i, j \in \{1, 2, \ldots, p\}$. Prove that the maximum possible value of $p$ is

$$p_{max} = \binom{m}{0} + \binom{m}{1} + \binom{m}{2} + \cdots + \binom{m}{k+1}.$$

### Problem 3

Let $n \geq 2$ be an even integer, and let $a$ and $b$ be real numbers such that $b^n = 3a + 1$. Show that the polynomial $p(x) = (x^2 + x + 1)^n - x^n - a$ is divisible by $q(x) = x^3 + x^2 + x + b$ if and only if $b = 1$.

### Problem 4

Show that if $a$, $b$, and $c$ are complex numbers such that

$$(a + b)(a + c) = b,$$
$$(b + c)(b + a) = c,$$
$$(c + a)(c + b) = a,$$

then $a$, $b$, and $c$ are real numbers.

### Problem 5

(a) Let $f, g \colon \mathbb{Z} \to \mathbb{Z}$ be injective maps. Show that the function $h \colon \mathbb{Z} \to \mathbb{Z}$, defined by $h(x) = f(x)g(x)$ for all $x \in \mathbb{Z}$, cannot be surjective.

(b) Let $f \colon \mathbb{Z} \to \mathbb{Z}$ be a surjective map. Show that there exist surjective functions $g, h \colon \mathbb{Z} \to \mathbb{Z}$ such that $f(x) = g(x)h(x)$ for all $x \in \mathbb{Z}$.

## Problem 6

Three schools each have 200 students. Every student has at least one friend in each school. (If student $a$ is a friend of student $b$, then $b$ is a friend of $a$; also, for the purposes of this problem, no student is a friend of himself.) There exists a set $E$ of 300 students (chosen from among the 600 students at the three schools) with the following property: for any school $S$ and any two students $x, y \in E$ who are not in the school $S$, $x$ and $y$ do not have the same number of friends in $S$. Show that one can find three students, one in each school, such that any two are friends with each other.

## Problem 7

The vertices of square $ABCD$ lie outside a circle centered at $M$. Let $\overline{AA'}, \overline{BB'}, \overline{CC'}, \overline{DD'}$ be tangents to the circle. We assume that these segments can be arranged to be the four consecutive sides of a quadrilateral $p$ in which some circle is inscribed. Prove that $p$ has an axis of symmetry.

## Problem 8

Find the least number $n$ with the following property: given any $n$ rays in three-dimensional space sharing a common endpoint, the angle between some two of these rays is acute.

## Problem 9

Let $f(x) = a_0 + a_1 x + \cdots + a_m x^m$, with $m \geq 2$ and $a_m \neq 0$, be a polynomial with integer coefficients. Let $n$ be a positive integer, and suppose that:

(i) $a_2, a_3, \ldots a_m$ are divisible by all the prime factors of $n$;

(ii) $a_1$ and $n$ are relatively prime.

Prove that for any positive integer $k$, there exists a positive integer $c$ such that $f(c)$ is divisible by $n^k$.

## Problem 10

Find all pairs $(m, n)$ of positive integers, with $m, n \geq 2$, such that $a^n - 1$ is divisible by $m$ for each $a \in \{1, 2, \ldots, n\}$.

## Problem 11

Prove that there is no function $f : (0, \infty) \to (0, \infty)$ such that

$$f(x + y) \geq f(x) + yf(f(x))$$

for all $x, y \in (0, \infty)$.

## Problem 12

Let $P$ be a convex polyhedron with vertices $V_1$, $V_2$, ..., $V_p$. Two vertices $V_i$ and $V_j$ are called *neighbors* if they are distinct and belong to the same face of the polyhedron. The $p$ sequences $(v_i(n))_{n \geq 0}$, for $i = 1, 2, \ldots, p$, are defined recursively as follows: the $v_i(0)$ are chosen arbitrarily; and for $n \geq 0$, $v_i(n + 1)$ is the arithmetic mean of the numbers $v_j(n)$ for all $j$ such that $V_i$ and $V_j$ are neighbors. Suppose that $v_i(n)$ is an integer for all $1 \leq i \leq p$ and $n \in \mathbb{N}$. Prove that there exist $N \in \mathbb{N}$ and $k \in \mathbb{Z}$ such that $v_i(n) = k$ for all $n \geq N$ and $i = 1, 2, \ldots, p$.

## 3.13 Russia

### Problem 1

Peter and Alex play a game starting with an ordered pair of integers $(a, b)$. On each turn, the current player increases or decreases either $a$ or $b$: Peter by 1, and Alex by 1 or 3. Alex wins if at some point in the game the roots of $x^2 + ax + b$ are integers. Is it true that given any initial values $a$ and $b$, Alex can guarantee that he wins?

### Problem 2

Let $M$ and $N$ be points on sides $\overline{AB}$ and $\overline{BC}$, respectively, of parallelogram $ABCD$ such that $AM = NC$. Let $Q$ be the intersection of $\overline{AN}$ and $\overline{CM}$. Prove that $\overline{DQ}$ is an angle bisector of angle $CDA$.

### Problem 3

A target consists of an equilateral triangle broken into 100 equilateral triangles of unit side length by three sets of parallel lines. A sniper shoots at the target repeatedly as follows: he aims at one of the small triangles and then hits either that triangle or one of the small triangles which shares a side with it. He may choose to stop shooting at any time. What is the greatest number of triangles that he can be sure to hit exactly five times?

### Problem 4

Two points are selected inside a convex pentagon. Prove that it is possible to select four of the pentagon's vertices so that the quadrilateral they form contains both points.

### Problem 5

Does there exist a positive integer such that the product of its proper divisors ends with exactly 2001 zeroes?

### Problem 6

A circle is tangent to rays $OA$ and $OB$ at $A$ and $B$, respectively. Let $K$ be a point on minor arc $AB$ of this circle. Let $L$ be a point on line $OB$ such that $\overline{OA} \parallel \overline{KL}$. Let $M$ be the intersection (distinct from $K$) of line $AK$ and the circumcircle $\omega$ of triangle $KLB$. Prove that line $OM$ is tangent to $\omega$.

## Problem 7

Let $a_1, a_2, \ldots, a_{10^6}$ be nonzero integers between 1 and 9, inclusive. Prove that at most 100 of the numbers $\overline{a_1 a_2 \ldots a_k}$ ($1 \leq k \leq 10^6$) are perfect squares. (Here, $\overline{a_1 a_2 \ldots a_k}$ denotes the decimal number with the $k$ digits $a_1, a_2, \ldots, a_k$.)

## Problem 8

The lengths of the sides of an $n$-gon equal $a_1, a_2, \ldots, a_n$. If $f$ is a quadratic such that

$$f(a_k) = f\left(\left(\sum_{i=1}^{n} a_i\right) - a_k\right)$$

for $k = 1$, prove that this equality holds for $k = 2, 3, \ldots, n$ as well.

## Problem 9

Given any point $K$ in the interior of diagonal $\overline{AC}$ of parallelogram $ABCD$, construct the line $\ell_K$ as follows. Let $s_1$ be the circle tangent to lines $AB$ and $AD$ such that of $s_1$'s two intersection points with $\overline{AC}$, $K$ is the point farther from $A$. Similarly, let $s_2$ be the circle tangent to lines $CB$ and $CD$ such that of $s_2$'s two intersection points with $\overline{CA}$, $K$ is the point farther from $C$. Then let $\ell_K$ be the line connecting the centers of $s_1$ and $s_2$. Prove that as $K$ varies along $\overline{AC}$, all the lines $\ell_K$ are parallel to each other.

## Problem 10

Describe all possible ways to color each positive integer in one of three colors such that any positive integers $a$, $b$, $c$ (not necessarily distinct) which satisfy $2000(a + b) = c$ are colored either in one color or in three different colors.

## Problem 11

Three sets of ten parallel lines are drawn. Find the greatest possible number of triangles whose sides lie along the lines but whose interiors do not intersect any of the lines.

## Problem 12

Let $a$, $b$, and $c$ be integers such that $b \neq c$. If $ax^2 + bx + c$ and $(c-b)x^2 + (c-a)x + (a+b)$ have a common root, prove that $a+b+2c$ is divisible by 3.

## Problem 13

Let $ABC$ be a triangle with $AC \neq AB$, and select point $B_1$ on ray $AC$ such that $AB = AB_1$. Let $\omega$ be the circle passing through $C$, $B_1$, and the foot of the internal bisector of angle $CAB$. Let $\omega$ intersect the circumcircle of triangle $ABC$ again at $Q$. Prove that $\overline{AC}$ is parallel to the tangent to $\omega$ at $Q$.

## Problem 14

We call a set of squares in a checkerboard plane *rook-connected* if it is possible to travel between any two squares in the set by moving finitely many times like a rook, without visiting any square outside of the set. (One moves "like a rook" by moving between two distinct—not necessarily adjacent—squares which lie in the same row or column.) Prove that any rook-connected set of 100 squares can be partitioned into fifty pairs of squares, such that the two squares in each pair lie in the same row or column.

## Problem 15

At each of one thousand distinct points on a circle are written two positive integers. The sum of the numbers at each point $P$ is divisible by the product of the numbers on the point which is the clockwise neighbor of $P$. What is the maximum possible value of the greatest of the 2000 numbers?

## Problem 16

Find all primes $p$ and $q$ such that $p + q = (p - q)^3$.

## Problem 17

The monic polynomial $f(x)$ with real coefficients has exactly two distinct real roots. Suppose that $f(f(x))$ has exactly three distinct real roots. Is it possible that $f(f(f(x)))$ has exactly seven distinct real roots?

## Problem 18

Let $\overline{AD}$ be the internal angle bisector of $A$ in triangle $BAC$, with $D$ on $\overline{BC}$. Let $M$ and $N$ be points on the circumcircles of triangles $ADB$ and $ADC$, respectively, so that $\overline{MN}$ is tangent to these two circles. Prove that line $MN$ is tangent to the circle passing through the midpoints of $\overline{BD}$, $\overline{CD}$, and $\overline{MN}$.

## Problem 19

In tetrahedron $A_1A_2A_3A_4$, let $\ell_k$ be the line connecting $A_k$ with the incenter of the opposite face. If $\ell_1$ and $\ell_2$ intersect, prove that $\ell_3$ and $\ell_4$ intersect.

## Problem 20

An infinite set $S$ of points on the plane has the property that no $1 \times 1$ square of the plane contains infinitely many points from $S$. Prove that there exist two points $A$ and $B$ from $S$ such that $\min\{XA, XB\} \geq .999AB$ for any other point $X$ in $S$.

## Problem 21

Prove that from any set of 117 pairwise distinct three-digit numbers, it is possible to select 4 pairwise disjoint subsets such that the sums of the numbers in each subset are equal.

## Problem 22

The numbers from 1 to 999999 are divided into two groups. For each such number $n$, if the square closest to $n$ is odd, then $n$ is placed in the first group; otherwise, $n$ is placed in the second group. The sum of the numbers in each group is computed. Which group yields the larger sum?

## Problem 23

Two polynomials $P(x) = x^4 + ax^3 + bx^2 + cx + d$ and $Q(x) = x^2 + px + q$ take negative values on some common real interval $I$ of length greater than 2, and outside of $I$ they take on nonnegative values. Prove that $P(x_0) < Q(x_0)$ for some real number $x_0$.

## Problem 24

The point $K$ is selected inside parallelogram $ABCD$ such that the midpoint of $\overline{AD}$ is equidistant from $K$ and $C$ and such that the midpoint of $\overline{CD}$ is equidistant from $K$ and $A$. Let $N$ be the midpoint of $\overline{BK}$. Prove that $\angle NAK = \angle NCK$.

## Problem 25

We are given a 2000-sided polygon in which no three diagonals are concurrent. Each diagonal is colored in one of 999 colors. Prove that there exists a triangle whose sides lie entirely on diagonals of one color. (The triangle's vertices need not be vertices of the 2000-sided polygon.)

## Problem 26

Jury lays 2001 coins, each worth 1, 2, or 3 kopecks, in a row. Between any two $k$-kopeck coins lie at least $k$ coins for $k = 1, 2, 3$. For which $n$ is it possible that Jury lays down exactly $n$ 3-kopeck coins?

## Problem 27

A company of $2n + 1$ people has the property that for each group of $n$ people, there is a person among the other $n + 1$ who knows everybody in that group. Prove that some person in the company knows everybody else. (If a person $A$ knows a person $B$, then $B$ knows $A$ as well.)

## Problem 28

Side $\overline{AC}$ is the longest of the three sides in triangle $ABC$. Let $N$ be a point on $\overline{AC}$. Let the perpendicular bisector of $\overline{AN}$ intersect line $AB$ at $K$, and let the perpendicular bisector of $\overline{CN}$ intersect line $BC$ at $M$. Prove that the circumcenter of triangle $ABC$ lies on the circumcircle of triangle $KBM$.

## Problem 29

Find all odd positive integers $n$ greater than 1 such that for any distinct prime divisors $a$ and $b$ of $n$, the number $a + b - 1$ is also a divisor of $n$.

## Problem 30

Each of the subsets $A_1, A_2, \ldots, A_{100}$ of a line is the union of 100 pairwise disjoint closed intervals. Prove that the intersection of these 100 sets is the union of no more than 9901 disjoint closed intervals. (A closed interval is a single point or a segment.)

## Problem 31

Two circles are internally tangent at a point $N$, and a point $K$ different from $N$ is chosen on the smaller circle. A line tangent to the smaller circle at $K$ intersects the larger circle at $A$ and $B$. Let $M$ be the midpoint of the arc $AB$ of the larger circle not containing $N$. Prove that the circumradius of triangle $BMK$ is constant as $K$ varies along the smaller circle (regardless of which arc $MN$ point $B$ lies on).

## Problem 32

In a country, two-way roads connect some cities in pairs such that given two cities $A$ and $B$, there exists a unique path from $A$ to $B$ which does not pass through the same city twice. It is known that exactly 100 cities in the country lie at the end of exactly one road. Prove that it is possible to construct 50 new two-way roads so that if any single road were closed, it would still be possible to travel from any city to any other.

## Problem 33

The polynomial $P(x) = x^3 + ax^2 + bx + c$ has three distinct real roots. The polynomial $P(Q(x))$, where $Q(x) = x^2 + x + 2001$, has no real roots. Prove that $P(2001) > \frac{1}{64}$.

## Problem 34

Each number $1, 2, \ldots, n^2$ is written once in an $n \times n$ grid such that each square contains one number. Given any two squares in the grid, a vector is drawn from the center of the square containing the larger number to the center of the other square. If the sums of the numbers in each row or column of the grid are equal, prove that the sum of the drawn vectors is zero.

## Problem 35

Distinct points $A_1$, $B_1$, $C_1$ are selected inside triangle $ABC$ on the altitudes from $A$, $B$, and $C$, respectively. If $[ABC_1]+[BCA_1]+[CAB_1] = [ABC]$, prove that the circumcircle of triangle $A_1B_1C_1$ passes through the orthocenter $H$ of triangle $ABC$.

## Problem 36

We are given a set of 100 stones with total weight $2S$. Call an integer $k$ *average* if it is possible to select $k$ of the 100 stones whose total weight equals $S$. What is the maximum possible number of integers which are average?

## Problem 37

Two finite sets $S_1$ and $S_2$ of convex polygons in the plane are given with the following properties: (i) given any polygon from $S_1$ and any polygon from $S_2$, the two polygons have a common point; (ii) each of the two sets contains a pair of disjoint polygons. Prove that there exists a line which intersects all the polygons in both sets.

## Problem 38

In a contest consisting of $N$ problems, the jury defines the difficulty of each problem by assigning it a positive integral number of points. (The same number of points may be assigned to different problems.) Any participant who answers the problem correctly receives that number of points for that problem; any other participants receive 0 points. After the participants submitted their answers, the jury realizes that given any ordering of the participants (where ties are not permitted), it could have defined the problems' difficulty levels to make that ordering coincide with the participants' ranking according to their total scores. Determine, in terms of $N$, the maximum number of participants for which such a scenario to occur.

## Problem 39

The monic quadratics $f$ and $g$ take negative values on disjoint nonempty intervals of the real numbers, and the four endpoints of these intervals are also distinct. Prove that there exist positive numbers $\alpha$ and $\beta$ such that

$$\alpha f(x) + \beta g(x) > 0$$

for all real numbers $x$.

## Problem 40

Let $a$ and $b$ be distinct positive integers such that $ab(a+b)$ is divisible by $a^2 + ab + b^2$. Prove that $|a - b| > \sqrt[3]{ab}$.

## Problem 41

In a country of 2001 cities, some cities are connected in pairs by two-way roads. We call two cities which are connected by a road *adjacent*. Each city is adjacent to at least one other city, and no city is adjacent to every other city. A set $D$ of cities is called *dominating* if any city not included in $D$ is adjacent to some city in $D$. It is known that any dominating set contains at least $k$ cities. Prove that the country can be divided into $2001 - k$ republics such that no two cities in any single republic are adjacent.

## Problem 42

Let $SABC$ be a tetrahedron. The circumcircle of $ABC$ is a great circle of a sphere $\omega$, and $\omega$ intersects $\overline{SA}$, $\overline{SB}$, and $\overline{SC}$ again at $A_1$, $B_1$, and $C_1$, respectively. The planes tangent to $\omega$ at $A_1$, $B_1$, and $C_1$ intersect at a point $O$. Prove that $O$ is the circumcenter of tetrahedron $SA_1B_1C_1$.

## 3.14   Taiwan

### Problem 1

Let $O$ be the excenter of triangle $ABC$ opposite $A$. Let $M$ be the midpoint of $\overline{AC}$, and let $P$ be the intersection point of $\overline{MO}$ and $\overline{BC}$. Prove that $AB = BP$ if $\angle BAC = 2\angle ACB$.

### Problem 2

Let $n \geq 3$ be an integer, and let $A$ be a set of $n$ distinct integers. Let the minimal and maximal elements of $A$ be $m$ and $M$, respectively. Suppose that there exists a polynomial $p$ with integer coefficients such that (i) $m < p(a) < M$ for all $a \in A$, and (ii) $p(m) < p(a)$ for all $a \in A - \{m, M\}$. Show that $n \leq 5$, and prove that there exist integers $b$ and $c$ such that each element of $A$ is a solution to the equation $p(x) + x^2 + bx + c = 0$.

### Problem 3

Let $n \geq 3$ be an integer and let $A_1, A_2, \ldots, A_n$ be $n$ distinct subsets of $S = \{1, 2, \ldots, n\}$. Show that there exists an element $x \in S$ such that the $n$ subsets $A_1 \setminus \{x\}$, $A_2 \setminus \{x\}$, $\ldots$, $A_n \setminus \{x\}$ are also distinct.

### Problem 4

Let $\Gamma$ be the circumcircle of a fixed triangle $ABC$. Suppose that $M$ and $N$ are the midpoints of arcs $BC$ and $CA$, respectively, and let $X$ be any point on arc $AB$. (Here, arc $AB$ refers to the arc not containing $C$; analogous statements hold for arcs $BC$ and $CA$.) Let $O_1$ and $O_2$ be the incenters of triangles $XAC$ and $XBC$, respectively. Let $\Gamma$ and the circumcircle of triangle $XO_1O_2$ intersect at $Q$. Prove that $\triangle QNO_1 \sim \triangle QMO_2$, and determine the locus of $Q$.

### Problem 5

Let $x, y$ be distinct real numbers, and let $f : \mathbb{N} \to \mathbb{R}$ be defined by $f(n) = \sum_{k=0}^{n-1} y^k x^{n-1-k}$ for all $n \in \mathbb{N}$. Suppose that $f(m)$, $f(m+1)$, $f(m+2)$, and $f(m+3)$ are integers for some positive integer $m$. Prove that $f(n)$ is an integer for all $n \in \mathbb{N}$.

## Problem 6

We are given $n$ stones $A_1, A_2 \ldots, A_n$ labelled with distinct real numbers. We may *compare* two stones by asking what the order of their corresponding numbers are. We are given that the numbers on $A_1, A_2, \ldots, A_{n-1}$ are increasing in that order; the $n$ orderings of the numbers on $A_1, A_2, \ldots, A_n$ which satisfy this condition are assumed to be equally likely. Based on this information, an algorithm is created that minimizes the expected number of comparisons needed to determine the order of the numbers on $A_1, A_2, \ldots, A_n$. What is this expected number?

## 3.15   United States of America

### Problem 1

Each of eight boxes contains six balls. Each ball has been colored with one of $n$ colors, such that no two balls in the same box are the same color, and no two colors occur together in more than one box. Determine, with justification, the smallest integer $n$ for which this is possible.

### Problem 2

Let $ABC$ be a triangle and let $\omega$ be its incircle. Denote by $D_1$ and $E_1$ the points where $\omega$ is tangent to sides $\overline{BC}$ and $\overline{AC}$, respectively. Denote by $D_2$ and $E_2$ the points on sides $\overline{BC}$ and $\overline{AC}$, respectively, such that $CD_2 = BD_1$ and $CE_2 = AE_1$, and denote by $P$ the point of intersection of $\overline{AD_2}$ and $\overline{BE_2}$. Circle $\omega$ intersects $\overline{AD_2}$ at two points, the closer of which to the vertex $A$ is denoted by $Q$. Prove that $AQ = D_2P$.

### Problem 3

Let $a, b$, and $c$ be nonnegative real numbers such that

$$a^2 + b^2 + c^2 + abc = 4.$$

Prove that

$$0 \le ab + bc + ca - abc \le 2.$$

### Problem 4

Let $P$ be a point in the plane of triangle $ABC$ such that there exists an obtuse triangle whose sides are congruent to $\overline{PA}$, $\overline{PB}$, and $\overline{PC}$. Assume that in this triangle the obtuse angle opposes the side congruent to $\overline{PA}$. Prove that angle $BAC$ is acute.

### Problem 5

Let $S$ be a set of integers (not necessarily positive) such that

(a) there exist $a, b \in S$ with $\gcd(a, b) = \gcd(a - 2, b - 2) = 1$;

(b) if $x$ and $y$ are elements of $S$ (possibly equal), then $x^2 - y$ also belongs to $S$.

Prove that $S$ is the set of all integers.

## Problem 6

Each point in the plane is assigned a real number such that, for any triangle, the number at the center of its inscribed circle is equal to the arithmetic mean of the three numbers at its vertices. Prove that all points in the plane are assigned the same number.

# 3.16 Vietnam

## Problem 1

The sequence of integers $a_0, a_1, \ldots$ is defined recursively by the initial condition $a_0 = 1$ and the recursive relation $a_n = a_{n-1} + a_{\lfloor n/3 \rfloor}$ for all integers $n \geq 1$. (Here, $\lfloor x \rfloor$ denotes the greatest integer less than or equal to $x$.) Prove that for every prime number $p \leq 13$, there exists an infinite number of natural numbers $k$ such that $a_k$ is divisible by $p$.

## Problem 2

In the plane, two circles intersect at $A$ and $B$, and a common tangent intersects the circles at $P$ and $Q$. Let the tangents at $P$ and $Q$ to the circumcircle of triangle $APQ$ intersect at $S$, and let $H$ be the reflection of $B$ across line $PQ$. Prove that the points $A$, $S$, and $H$ are collinear.

## Problem 3

A club has 42 members. Among each group of 31 members, there is at least one pair of participants — one male, one female — who know each other. (Person A knows person B if and only if person B knows person A.) Prove that there exist 12 distinct males $a_1, \ldots, a_{12}$ and 12 distinct females $b_1, \ldots, b_{12}$ such that $a_i$ knows $b_i$ for all $i$.

## Problem 4

The positive real numbers $a$, $b$, and $c$ satisfy the condition $21ab + 2bc + 8ca \leq 12$. Find the least possible value of the expression $\frac{1}{a} + \frac{2}{b} + \frac{3}{c}$.

## Problem 5

Let $n > 1$ be an integer, and let $T$ be the set of points $(x, y, z)$ in three-dimensional space such that $x$, $y$, and $z$ are integers between 1 and $n$, inclusive. We color the points in $T$ so that if $x_0 \leq x_1$, $y_0 \leq y_1$, and $z_0 \leq z_1$, then $(x_0, y_0, z_0)$ and $(x_1, y_1, z_1)$ are either equal or not both colored. At most how many points in $T$ can be colored?

## Problem 6

Let $a_1, a_2, \ldots$ be a sequence of positive integers satisfying the condition $0 < a_{n+1} - a_n \leq 2001$ for all integers $n \geq 1$. Prove that there exist an infinite number of ordered pairs $(p, q)$ of distinct positive integers such that $a_p$ is a divisor of $a_q$.

# 4

# 2001 Regional Contests: Problems

## 4.1 Asian Pacific Mathematical Olympiad

### Problem 1

For each positive integer $n$, let $S(n)$ be the sum of digits in the decimal representation of $n$. Any positive integer obtained by removing several (at least one) digits from the right-hand end of the decimal representation of $n$ is called a *stump* of $n$. Let $T(n)$ be the sum of all stumps of $n$. Prove that $n = S(n) + 9T(n)$.

### Problem 2

Find the largest positive integer $N$ so that the number of integers in the set $\{1, 2, \ldots, N\}$ which are divisible by 3 is equal to the number of integers which are divisible by either 5 or 7 (or both).

### Problem 3

Let two congruent regular $n$-sided ($n \geq 3$) polygonal regions $S$ and $T$ be located in the plane such that their intersection is a $2n$-sided polygonal region $P$. The sides of $S$ are colored red and the sides of $T$ are colored blue. Prove that the sum of the lengths of the blue sides of $P$ is equal to the sum of the lengths of its red sides.

### Problem 4

A point in the Cartesian coordinate plane is called a *mixed point* if one of its coordinates is rational and the other one is irrational. Find all polynomials with real coefficients such that their graphs do not contain any mixed point.

## Problem 5

Find the greatest integer $n$, such that there are $n + 4$ points $A, B, C, D$, $X_1, \ldots, X_n$ in the plane with the following properties: the lengths $AB$ and $CD$ are distinct; and for each $i = 1, 2, \ldots, n$, triangles $ABX_i$ and $CDX_i$ are congruent (although not necessarily in that order).

## 4.2 Austrian-Polish Mathematics Competition

### Problem 1

Let $k$ be a fixed positive integer. Consider the sequence defined recursively by $a_0 = 1$ and

$$a_{n+1} = a_n + \lfloor \sqrt[k]{a_n} \rfloor$$

for $n = 0, 1, \ldots$. (Here, $\lfloor x \rfloor$ denotes the greatest integer less than or equal to $x$.) For each $k$, find the set $A_k$ consisting of all integers in the sequence $\sqrt[k]{a_0}, \sqrt[k]{a_1}, \ldots$.

### Problem 2

Consider the set $A$ of all positive integers $n$ with the following properties: the decimal expansion contains no 0, and the sum of the (decimal) digits of $n$ divides $n$.

(a) Prove that there exist infinitely many elements in $A$ with the following property: the digits that appear in the decimal expansion of $A$ appear the same number of times.

(b) Show that for each positive integer $k$, there exists an element in $A$ with exactly $k$ digits.

### Problem 3

We are given a right prism with a regular octagon for its base, whose edges all have length 1. The points $M_1, M_2, \ldots, M_{10}$ are the centers of the faces of the prism. Let $P$ be a point inside the prism, and let $P_i$ denote the second intersection of line $M_i P$ with the surface of the prism. Suppose that the interior of each face contains exactly one of $P_1, P_2, \ldots, P_{10}$. Prove that

$$\sum_{i=1}^{10} \frac{M_i P}{M_i P_i} = 5.$$

### Problem 4

Let $n > 10$ be a positive integer and let $A$ be a set containing $2n$ elements. The family $\{A_i \mid i = 1, 2, \ldots, m\}$ of subsets of the set $A$ is called *suitable* if:

- for each $i = 1, 2, \ldots, m$, the set $A_i$ contains $n$ elements;
- for all $1 \leq i < j < k \leq m$, the set $A_i \cap A_j \cap A_k$ contains at most one element.

For each $n$, determine the largest $m$ for which there exists a suitable family of $m$ sets.

## 4.3  Balkan Mathematical Olympiad

### Problem 1

Let $n$ be a positive integer. Show that if $a$ and $b$ are integers greater than 1 such that $2^n - 1 = ab$, then $ab - (a - b) - 1 = k \cdot 2^{2m}$ for some odd integer $k$ and some positive integer $m$.

### Problem 2

Prove that if a convex pentagon satisfies the following conditions, then it is a regular pentagon:

(i) all the interior angles of the pentagon are congruent;

(ii) the lengths of the sides of the pentagon are rational numbers.

### Problem 3

A $3 \times 3 \times 3$ cube is divided into 27 congruent $1 \times 1 \times 1$ cells. One of these cells is empty, and the others are filled with unit cubes labelled $1, 2, \ldots, 26$ in some order. An *admissible move* consists of moving a unit cube which shares a face with the empty cell into the empty cell. Does there always exist — for any initial empty cell and any labelling of the 26 cubes — a finite sequence of admissible moves after which each unit cube labelled with $k$ is in the cell originally containing the unit cube labelled with $27 - k$, for each $k = 1, 2, \ldots, 26$?

## 4.4   Baltic Mathematics Competition

### Problem 1

Let 2001 given points on a circle be colored either red or green. In one
*step* all points are recolored simultaneously in the following way: If before
the recoloring, both neighbors of a point $P$ have the same color as $P$, then
the color of $P$ remains unchanged; otherwise, the color of $P$ is changed.
Starting with an initial coloring $F_1$, we obtain the colorings $F_2, F_3, \ldots$
after several steps. Prove that there is a number $n_0 \le 1000$ such that
$F_{n_0} = F_{n_0+2}$. Is this assertion also true if 1000 is replaced by 999?

### Problem 2

In a triangle $ABC$, the bisector of angle $BAC$ meets $\overline{BC}$ at $D$. Suppose
that $BD \cdot CD = AD^2$ and $\angle ADB = \pi/4$. Determine the angles of
triangle $ABC$.

### Problem 3

Let $a_0, a_1, \ldots$ be a sequence of positive real numbers satisfying

$$i \cdot a_i^2 \ge (i+1) \cdot a_{i-1} a_{i+1}$$

for $i = 1, 2, \ldots$. Furthermore, let $x$ and $y$ be positive reals, and let
$b_i = x a_i + y a_{i-1}$ for $i = 1, 2, \ldots$. Prove that

$$i \cdot b_i^2 > (i+1) \cdot b_{i-1} b_{i+1}$$

for all integers $i \ge 2$.

### Problem 4

Let $a$ be an odd integer. Prove that $a^{2^n} + 2^{2^n}$ and $a^{2^m} + 2^{2^m}$ are relatively
prime for all positive integers $n$ and $m$ with $n \ne m$.

## 4.5 Czech-Slovak-Polish Match

### Problem 1

Let $n \geq 2$ be an integer. Show that

$$(a_1^3 + 1)(a_2^3 + 1) \cdots (a_n^3 + 1) \geq (a_1^2 a_2 + 1)(a_2^2 a_3 + 1) \cdots (a_n^2 a_1 + 1)$$

for all positive numbers $a_1, a_2, \ldots, a_n$.

### Problem 2

In triangle $ABC$, angles $CAB$ and $ABC$ are acute. Isosceles triangles $ACD$ and $BCE$ with bases $\overline{AC}$ and $\overline{BC}$, respectively, are constructed externally to triangle $ABC$ such that $\angle ADC = \angle ABC$ and $\angle BEC = \angle BAC$. Let $O$ be the circumcenter of triangle $ABC$. Prove that $DO + OE = AB + BC + CA$ if and only if $\angle ACB = \pi/2$.

### Problem 3

Let $n$ and $k$ be positive integers satisfying $\frac{1}{2}n < k \leq \frac{2}{3}n$. Find the smallest number of pieces that can be placed on an $n \times n$ chessboard so that no column or row of the chessboard contains $k$ adjacent unoccupied squares.

### Problem 4

Two distinct points $A$ and $B$ are given in the plane. Consider all triangles $ABC$ with the following property: There exist points $D$ and $E$ in the interior of $\overline{BC}$ and $\overline{CA}$, respectively, such that

(i) $\dfrac{BD}{BC} = \dfrac{CE}{CA} = \dfrac{1}{3}$;

(ii) the points $A$, $B$, $D$, and $E$ are concyclic.

Find the locus of the intersection of lines $AD$ and $BE$ for all such triangles $ABC$.

### Problem 5

Find all functions $f : \mathbb{R} \to \mathbb{R}$ satisfying the equation

$$f(x^2 + y) + f(f(x) - y) = 2f(f(x)) + 2y^2$$

for all $x, y \in \mathbb{R}$.

## Problem 6

We color 2000 lattice points of three-dimensional space red and another 2000 lattice points blue. Among the segments with one red endpoint and one blue endpoint, suppose that no two have a common interior point. Consider the smallest right parallelepiped with edges parallel to the coordinate axes which contains all the lattice points we have colored. Show that this parallelepiped contains at least $5 \cdot 10^5$ lattice points, and give an example of a coloring in which this parallelepiped contains at most $8 \cdot 10^6$ lattice points.

# 4.6   St. Petersburg City
## Mathematical Olympiad (Russia)

### Problem 1

In the parliament of the country Alternativia, for any two deputies there exists a third who is acquainted with exactly one of the two. Each deputy belongs to one of two parties. Each day the president (not a member of the parliament) selects a group of deputies and orders them to change parties, at which time each deputy acquainted with at least one member of the group also changes parties. Prove that the president can arrange that at some point, every deputy belongs to a single party.

### Problem 2

Do there exist distinct numbers $x, y, z \in [0, \pi/2]$ such that the six numbers $\sin x, \sin y, \sin z, \cos x, \cos y, \cos z$ can be divided into three pairs with equal sum?

### Problem 3

A country has 2000 cities and a complete lack of roads. Show that it is possible to join pairs of cities by (two-way) roads so that for $n = 1, \ldots, 1000$, there are exactly two cities where exactly $n$ roads meet.

### Problem 4

The points $A_1, B_1, C_1$ are the midpoints of sides $\overline{BC}, \overline{CA}, \overline{AB}$ of acute triangle $ABC$. On lines $B_1C_1$ and $A_1B_1$ are chosen points $E$ and $F$ such that line $BE$ bisects angle $AEB_1$ and line $BF$ bisects angle $CFB_1$. Prove that $\angle BAE = \angle BCF$.

### Problem 5

For all positive integers $m > n$, prove that

$$\operatorname{lcm}(m, n) + \operatorname{lcm}(m+1, n+1) > \frac{2mn}{\sqrt{m-n}}.$$

## Problem 6

Acute triangle $ABC$ has incenter $I$ and orthocenter $H$. The point $M$ is the midpoint of minor arc $AC$ of the circumcircle of $ABC$. Given that $MI = MH$, find $\angle ABC$.

## Problem 7

Find all functions $f : \mathbb{Z} \to \mathbb{Z}$ such that

$$f\big(x + y + f(y)\big) = f(x) + 2y$$

for all integers $x, y$.

## Problem 8

From a $20 \times 20$ grid are removed 20 rectangles of sizes $1 \times 20, 1 \times 19$, $\ldots, 1 \times 1$, where the sides of the rectangle lie along gridlines. Prove that at least 85 $1 \times 2$ rectangles can be removed from the remainder.

## Problem 9

In a $10 \times 10$ table are written natural numbers not exceeding 10. Any two numbers that appear in adjacent or diagonally adjacent spaces of the table are relatively prime. Prove that some number appears in the table at least 17 times.

## Problem 10

The bisectors of angles $A$ and $B$ of convex quadrilateral $ABCD$ meet at $P$, and the bisectors of angles $C$ and $D$ meet at $Q$. Suppose that $P \neq Q$ and that line $PQ$ passes through the midpoint of side $\overline{AB}$. Prove that $\angle ABC = \angle BAD$ or $\angle ABC + \angle BAD = \pi$.

## Problem 11

Do there exist quadratic polynomials $f$ and $g$ with leading coefficients 1, such that for any integer $n$, $f(n)g(n)$ is an integer but none of $f(n)$, $g(n)$, and $f(n) + g(n)$ are integers?

## Problem 12

Ten points, labelled 1 to 10, are chosen in the plane. Permutations of $\{1, \ldots, 10\}$ are obtained as follows: for each rectangular coordinate system in which the ten points have distinct first coordinates, the labels of the points are listed in increasing order of the first coordinates of the points. Over all sequences of 10 labelled points, what is the maximum number of permutations of $\{1, \ldots, 10\}$ obtained in this fashion?

## Problem 13

A natural number is written on a chalkboard. Two players take turns, each turn consisting of replacing the number $n$ with either $n-1$ or $\lfloor(n+1)/2\rfloor$. The first player to write the number 1 wins. If the starting number is 1000000, which player wins with correct play?

## Problem 14

The altitudes of triangle $ABC$ meet at $H$. Point $K$ is chosen such that the circumcircles of $BHK$ and $CHK$ are tangent to line $BC$. Point $D$ is the foot of the altitude from $B$. Prove that $A$ is equidistant from lines $KB$ and $KD$.

## Problem 15

Let $m, n, k$ be positive integers with $n > 1$. Show that $\sigma(n)^k \neq n^m$, where $\sigma(n)$ is the sum of the positive integers dividing $n$.

## Problem 16

At a chess club, players may play against each other or against the computer. Yesterday there were $n$ players at the club. Each player played at most $n$ games, and every pair of players that did not play each other played at most $n$ games in total. Prove that at most $n(n+1)/2$ games were played.

## Problem 17

Show that there exist infinitely many positive integers $n$ such that the largest prime divisor of $n^4 + 1$ is greater than $2n$.

## Problem 18

In the interior of acute triangle $ABC$ is chosen a point $M$ such that $\angle AMC + \angle ABC = \pi$. Line $AM$ meets side $\overline{BC}$ at $D$, and line $CM$ meets side $\overline{AB}$ at $E$. Show that the circumcircle of triangle $BDE$ passes through some fixed point different from $B$, independent of the choice of $M$.

## Problem 19

Given are real numbers $x_1, \ldots, x_{10}$ in the interval $[0, \pi/2]$ such that $\sin^2 x_1 + \sin^2 x_2 + \cdots + \sin^2 x_{10} = 1$. Prove that

$$3(\sin x_1 + \cdots + \sin x_{10}) \leq \cos x_1 + \cdots + \cos x_{10}.$$

## Problem 20

The convex 2000-gon $\mathcal{M}$ satisfies the following property: the maximum distance between two vertices is equal to 1. It is known that among all convex 2000-gons with this same property, $\mathcal{M}$ has maximal area. Prove that some two diagonals of $\mathcal{M}$ are perpendicular.

## Problem 21

Let $a, b$ be integers greater than 1. The sequence $x_1, x_2, \ldots$ is defined by the initial conditions $x_0 = 0, x_1 = 1$ and the recursion

$$x_{2n} = ax_{2n-1} - x_{2n-2}, \qquad x_{2n+1} = bx_{2n} - x_{2n-1},$$

for $n \geq 1$. Prove that for any natural numbers $m$ and $n$, the product $x_{n+m}x_{n+m-1} \cdots x_{n+1}$ is divisible by $x_m x_{m-1}$.

# Glossary

**Abel summation**   For an integer $n > 0$ and reals $a_1, a_2, \ldots, a_n$ and $b_1, b_2, \ldots, b_n$,

$$\sum_{i=1}^{n} a_i b_i = b_n \sum_{i=1}^{n} a_i + \sum_{i=1}^{n-1} \left( (b_i - b_{i+1}) \sum_{j=1}^{i} a_j \right).$$

**Angle bisector theorem**   If $D$ is the intersection of either angle bisector of angle $ABC$ with line $AC$, then $BA/BC = DA/DC$.

**Arithmetic mean-geometric mean (AM-GM) inequality**   If $a_1, a_2, \ldots,$ $a_n$ are $n$ nonnegative numbers, then their **arithmetic mean** is defined as $\frac{1}{n} \sum_{i=1}^{n} a_i$ and their **geometric mean** is defined as $(a_1 a_2 \cdots a_n)^{\frac{1}{n}}$. The arithmetic mean-geometric mean inequality states that

$$\frac{1}{n} \sum_{i=1}^{n} a_i \geq (a_1 a_2 \cdots a_n)^{\frac{1}{n}}$$

with equality if and only if $a_1 = a_2 = \cdots = a_n$. The inequality is a special case of the **power mean inequality**.

**Arithmetic mean-harmonic mean (AM-HM) inequality**   If $a_1, a_2, \ldots,$ $a_n$ are $n$ positive numbers, then their **arithmetic mean** is defined as

$$\frac{1}{n} \sum_{i=1}^{n} a_i$$

and their **harmonic mean** is defined as

$$\frac{1}{\frac{1}{n} \sum_{i=1}^{n} \frac{1}{a_i}}.$$

259

The arithmetic mean-harmonic mean inequality states that

$$\frac{1}{n}\sum_{i=1}^{n} a_i \geq \frac{1}{\frac{1}{n}\sum_{i=1}^{n}\frac{1}{a_i}}$$

with equality if and only if $a_1 = a_2 = \cdots = a_n$. Like the arithmetic mean-geometric mean inequality, this inequality is a special case of the **power mean inequality**.

**Bernoulli's inequality**   For $x > -1$ and $a > 1$,

$$(1+x)^a \geq 1 + ax,$$

with equality when $x = 0$.

**Binomial coefficient**

$$\binom{n}{k} = \frac{n!}{k!(n-k)!},$$

the coefficient of $x^k$ in the expansion of $(x+1)^n$.

**Binomial theorem**

$$(x+y)^n = \sum_{k=0}^{n}\binom{n}{k}x^{n-k}y^k.$$

**Brianchon's theorem**   If hexagon $ABCDEF$ is circumscribed about a conic in the projective plane such that $A \neq D$, $B \neq E$, and $C \neq F$, then lines $AD$, $BE$, and $CF$ concur. (If they lie on a conic in the affine plane, then these lines either concur or are parallel.) This theorem is the dual to **Pascal's theorem**.

**Brocard angle**   See **Brocard points**.

**Brocard points**   Given a triangle $ABC$, there exists a unique point $P$ such that $\angle ABP = \angle BCP = \angle CAP$ and a unique point $Q$ such that $\angle BAQ = \angle CBQ = \angle ACQ$. The points $P$ and $Q$ are the Brocard points of triangle $ABC$. Moreover, $\angle ABP$ and $\angle BAQ$ are equal; their value $\phi$ is the Brocard angle of triangle $ABC$.

**Cauchy–Schwarz inequality**   For any real numbers $a_1, a_2, \ldots, a_n$, and $b_1, b_2, \ldots, b_n$

$$\sum_{i=1}^{n} a_i^2 \cdot \sum_{i=1}^{n} b_i^2 \geq \left(\sum_{i=1}^{n} a_i b_i\right)^2,$$

with equality if and only if $a_i$ and $b_i$ are proportional, $i = 1, 2, \ldots, n$.

**Centrally symmetric**   A geometric figure is centrally symmetric (centrosymmetric) about a point $O$ if, whenever $P$ is in the figure and $O$ is the midpoint of a segment $PQ$, then $Q$ is also in the figure.

**Centroid of a triangle**   Point of intersection of the medians.

**Centroid of a tetrahedron**   Point of the intersection of the segments connecting the midpoints of the opposite edges, which is the same as the point of intersection of the segments connecting each vertex with the centroid of the opposite face.

**Ceva's theorem and its trigonometric form**   Let $AD, BE, CF$ be three cevians of triangle $ABC$. The following are equivalent:

(i) $AD, BE, CF$ are concurrent;

(ii) $\dfrac{AF}{FB} \cdot \dfrac{BD}{DC} \cdot \dfrac{CE}{EA} = 1$;

(iii) $\dfrac{\sin \angle ABE}{\sin \angle EBC} \cdot \dfrac{\sin \angle BCF}{\sin \angle FCA} \cdot \dfrac{\sin \angle CAD}{\sin \angle DAB} = 1.$

**Cevian**   A cevian of a triangle is any segment joining a vertex to a point on the opposite side.

**Chinese remainder theorem**   Let $k$ be a positive integer.   Given integers $a_1, a_2, \ldots, a_k$ and pairwise relatively prime positive integers $n_1, n_2, \ldots, n_k$, there exists a unique integer $a$ such that $0 \le a < \prod_{i=1}^{k} n_i$ and $a \equiv a_i \pmod{n_i}$ for $i = 1, 2, \ldots, k$.

**Circumcenter**   Center of the circumscribed circle or sphere.

**Circumcircle**   Circumscribed circle.

**Complex numbers in planar geometry**   If we introduce a Cartesian coordinate system in the Euclidean plane, we can assign a complex number to each point in the plane by assigning $\alpha + \beta i$ to the point $(\alpha, \beta)$ for all reals $\alpha$ and $\beta$. Suppose that $A, B, \ldots, F$ are points and $a, b, \ldots, f$ are the corresponding complex numbers. Then:

- $a + (c - b)$ corresponds to the translation of $A$ under the vector $\overrightarrow{BC}$;
- given an angle $\theta$, $b + e^{i\theta}(a - b)$ corresponds to the image of $A$ under a rotation through $\theta$ about $B$;
- given a real scalar $\lambda$, $b + \lambda(a - b)$ corresponds to the image of $A$ under a homothety of ratio $\lambda$ centered at $B$;

- the absolute value of $a - b$ equals $AB$;

- the argument of $(c - b)/(a - c)$ equals $\angle ABC$ (directed and modulo $2\pi$).

Using these facts, one can translate much of the language of geometry in the Euclidean plane into language about complex numbers.

**Congruence**   For integers $a$, $b$, and $n$ with $n \geq 1$, $a \equiv b \pmod{n}$ (or "$a$ is congruent to $b$ modulo $n$") means that $a - b$ is divisible by $n$.

**Concave Up (Down) Function**   A function $f(x)$ is concave up (down) on $[a, b]$ if $(x, f(x))$ lies under (above) the line connecting $(a_1, f(a_1))$ and $(b_1, f(b_1))$ for all

$$a \leq a_1 < x < b_1 \leq b.$$

A function $g(x)$ is concave up (down) on the Euclidean plane if it is concave up (down) on each line in the plane, where we identify the line naturally with $\mathbb{R}$.

Concave up and down functions are also called **convex** and **concave**, respectively.

**Convex hull**   Given a nonempty set $S$ of points in Euclidean space, the convex hull of $S$ is the intersection of all convex sets containing $S$. This convex hull is itself a convex set.

**Convex set**   A set $S$ of points in Euclidean space is convex if for all pairs of distinct points $A, B \in S$, the segment $\overline{AB}$ is a subset of $S$.

**Cyclic polygon**   Polygon that can be inscribed in a circle.

**de Moivre's formula**   For any angle $\alpha$ and for any integer $n$,

$$(\cos \alpha + i \sin \alpha)^n = \cos n\alpha + i \sin n\alpha.$$

**Derangement**   A derangement of $n$ items $a_1, \ldots, a_n$ is a permutation $(b_1, b_2, \ldots, b_n)$ of these items such that $b_i \neq a_i$ for all $i$. According to a formula of Euler's, there are exactly

$$n! - \frac{n!}{1!} + \frac{n!}{2!} - \frac{n!}{3!} + \cdots + (-1)^n \frac{n!}{n!}$$

derangements of $n$ items.

**Desargues' theorem**   Two triangles have corresponding vertices joined by lines which are concurrent or parallel if and only if the intersections of corresponding sides are collinear.

**Directed angles**   A directed angle contains information about both the angle's measure and the angle's orientation (clockwise or counterclockwise). If two directed angles sum to zero, then they have the same angle measure but opposite orientations. One often takes directed angles modulo $\pi$ or $2\pi$. Some important features of directed angles modulo $\pi$ follow:

- If $A, B, C, D$ are points such that $\angle ABC$ and $\angle ABD$ are well-defined, then $\angle ABC = \angle ABD$ if and only if $B, C, D$ are collinear.

- If $A, B, C, D$ are points such that $\angle ABC$ and $\angle ADC$ are well-defined, then $\angle ABC = \angle ADC$ if and only if $A, B, C, D$ are concyclic.

- Because $2(\theta) = 2(\pi/2 + \theta)$, but $\theta \neq \pi/2 + \theta$, one cannot divide directed angles by 2. For example, if $\angle ABC = 2\angle ADC$, $D$ lies either on the internal angle bisector of angle $ABC$, or on the *external* angle bisector of angle $ABC$ — we cannot write $\angle ADC = \frac{1}{2}\angle ABC$ to determine which line $D$ lies on.

These features show that using directed angles modulo $\pi$ allows one to deal with multiple possible configurations of a geometry problem at once, but at the expense of possibly losing important information about a configuration.

**Dirichlet's Theorem**   A set $S$ of primes is said to have *Dirichlet density* if

$$\lim_{s \to 1} \frac{\sum_{p \in S} p^{-s}}{\ln(s-1)^{-1}}$$

exists, where $\ln x$ denotes the natural logarithm of $x$. If the limit exists, we call it the Dirichlet density of $S$ and denote it by $d(S)$.

There are infinitely many primes in any arithmetic sequence of integers for which the common difference is relatively prime to the terms. In other words, if $a$ and $m$ are relatively prime positive integers, then there are infinitely many primes $p$ such that $p \equiv a \pmod{m}$. More precisely, let $S(a; m)$ denote the set of all such primes. Then $d(S(a; m)) = 1/\phi(m)$, where $\phi$ is the Euler function.

**Euler's formula (for planar graphs)**   If $F$, $V$, and $E$ are the number of faces, vertices, and edges, respectively, of a planar graph, then $F + V - E = 2$. This is a special case of an invariant of topological surfaces called the Euler characteristic.

**Euler's formula (in planar geometry)**   Let $O$ and $I$ be the circumcenter and incenter, respectively, of a triangle with circumradius $R$ and inradius

$r$. Then

$$OI^2 = R^2 - 2rR.$$

**Euler line**   The orthocenter, centroid and circumcenter of any triangle are collinear. The centroid divides the distance from the orthocenter to the circumcenter in the ratio of 2 : 1. The line on which these three points lie is called the Euler line of the triangle.

**Euler's theorem**   Given relatively prime integers $a$ and $m$ with $m \geq 1$, $a^{\phi(m)} \equiv a \pmod{m}$, where $\phi(m)$ is the number of positive integers less than or equal to $m$ and relatively prime to $m$. Euler's theorem is a generalization of Fermat's little theorem.

**Euler (Totient) Function**   Let $n$ be a positive integer. The Euler (totient) function $\phi(n)$ is defined to be the number of positive integers less than or equal to $n$ that are relatively prime to $n$. The following are three fundamental properties of this function:

- $\phi(nm) = \phi(n)\phi(m)$ for relatively prime positive integers $m$ and $n$;
- if $n = p_1^{a_1} p_2^{a_2} \cdots p_k^{a_k}$ is a prime factorization of $n$ (with distinct primes $p_i$), then

$$\phi(n) = n \left(1 - \frac{1}{p_1}\right) \left(1 - \frac{1}{p_2}\right) \cdots \left(1 - \frac{1}{p_k}\right);$$

- $\displaystyle\sum_{d|n} \phi(d) = n.$

**Excircles or escribed circles**   Given a triangle $ABC$, there are four circles tangent to the lines $AB, BC, CA$. One is the inscribed circle, which lies in the interior of the triangle. One lies on the opposite side of line $BC$ from $A$, and is called the excircle (escribed circle) opposite $A$, and similarly for the other two sides. The excenter opposite $A$ is the center of the excircle opposite $A$; it lies on the internal angle bisector of $A$ and the external angle bisectors of $B$ and $C$.

**Excenters**   See **excircles**.

**Exradii**   The radii of the three excircles of a triangle.

**Fermat number**   A number of the form $2^{2^n}$ for some positive integer $n$.

**Fermat's little theorem**   If $p$ is prime, then $a^p \equiv a \pmod{p}$ for all integers $a$.

**Feuerbach circle**   The feet of the three altitudes of any triangle, the midpoints of the three sides, and the midpoints of segments from the three vertices to the orthocenter, all lie on the same circle, the Feuerbach circle or the **nine-point circle** of the triangle. Let $R$ be the circumradius of the triangle. The nine-point circle of the triangle has radius $R/2$ and is centered at the midpoint of the segment joining the orthocenter and the circumcenter of the triangle.

**Feuerbach's theorem**   The nine-point circle of a triangle is tangent to the incircle and to the three excircles of the triangle.

**Fibonacci Numbers**   The sequence $\{F_n\}_{n=0}^{\infty}$ defined recursively by

$$F_0 = 0, \quad F_1 = 1, \quad F_{n+1} = F_n + F_{n-1}$$

for all $n \geq 1$ is called the Fibonacci sequence. It can be shown that

$$F_n = \frac{1}{\sqrt{5}} \left[ \left( \frac{1 + \sqrt{5}}{2} \right)^n - \left( \frac{1 - \sqrt{5}}{2} \right)^n \right]$$

for all $n \geq 0$.

**Fibonacci sequence**   The sequence $F_0, F_1, \ldots$ defined recursively by $F_0 = 0$, $F_1 = 1$, and $F_{n+2} = F_{n+1} + F_n$ for all $n \geq 0$.

**Generating function**   If $a_0, a_1, a_2, \ldots$ is a sequence of numbers, then the generating function for the sequence is the infinite series

$$a_0 + a_1 x + a_2 x^2 + \cdots .$$

If $f$ is a function such that

$$f(x) = a_0 + a_1 x + a_2 x^2 + \cdots ,$$

then we also refer to $f$ as the generating function for the sequence.

**Graph**   A graph is a collection of *vertices* and *edges*, where the edges are distinct unordered pairs of distinct vertices. We say that the two vertices in one of these unordered pairs are *adjacent* and connected by that edge. The *degree* of a vertex is the number of edges which contain it. A *path* is a sequence of vertices $v_1, v_2, \ldots, v_n$ such that $v_i$ is adjacent to $v_{i+1}$ for each $i$. A graph is called *connected* if for every two vertices $v$ and $w$, there exists a path from $v$ to $w$. A *cycle* of the graph is an ordered collection of vertices $v_1, v_2, \ldots, v_n$ such that $v_1 = v_n$ and such that the $(v_i, v_{i+1})$ are distinct edges. A connected graph which contains no cycles is called a *tree*, and every tree contains at least two *leaves*, vertices with degree 1.

**Harmonic conjugates**   Let $A$, $C$, $B$, $D$ be four points on a line in that order. If the points $C$ and $D$ divide $AB$ internally and externally in the same ratio, (i.e., $AC : CB = AD : DB$), then the points $C$ and $D$ are said to be harmonic conjugates of each other with respect to the points $A$ and $B$, and $AB$ is said to be **harmonically divided** by the points $C$ and $D$. If $C$ and $D$ are harmonic with respect to $A$ and $B$, then $A$ and $B$ are harmonic with respect to $C$ and $D$.

**Harmonic range**   The four points $A$, $B$, $C$, $D$ are referred to as a harmonic range, denoted by $(ABCD)$, if $C$ and $D$ are harmonic conjugates with respect to $A$ and $B$.

**Helly's theorem**   If $n > d$ and $C_1, \ldots, C_n$ are convex subsets of $\mathbb{R}^d$, each $d + 1$ of which have nonempty intersection, then there is a point in common to all the sets.

**Heron's Formula**   The area of a triangle $ABC$ with side lengths $a$, $b$, $c$ is equal to

$$[ABC] = \sqrt{s(s - a)(s - b)(s - c)},$$

where $s = (a + b + c)/2$ is the semiperimeter of the triangle.

Further, suppose that $P, Q, R$ are the points of tangency of the incircle with sides $\overline{AB}, \overline{BC}, \overline{CA}$, respectively. Let $AP = x$, $BQ = y$, $CR = z$. Then $AR = x$, $BP = y$, $CQ = z$, and

$$x = s - a, \; y = s - b, \; y = s - c, \; x + y + z = s.$$

Then, Heron's formula becomes

$$[ABC] = \sqrt{xyz(x + y + z)}.$$

**Hölder's inequality**   Let $w_1, \ldots, w_n$ be positive real numbers whose sum is 1. For any positive real numbers $a_{ij}$,

$$\prod_{i=1}^{n} \left( \sum_{j=1}^{m} a_{ij} \right)^{w_i} \geq \sum_{j=1}^{m} \prod_{i=1}^{n} a_{ij}^{w_i}.$$

**Homothety**   A homothety (central similarity) is a transformation that fixes one point $O$ (its center) and maps each point $P$ to a point $P'$ for which $O, P, P'$ are collinear and the ratio $OP : OP' = k$ is constant ($k$

can be either positive or negative), where $k$ is called the **magnitude** of the homothety.

**Homothetic triangles**   Two triangles $ABC$ and $DEF$ are homothetic if they have parallel sides. Suppose that $AB \parallel DE$, $BC \parallel EF$, and $CA \parallel FD$. Then lines $AD$, $BE$, and $CF$ concur at a point $X$, as given by a special case of Desargues' theorem. Furthermore, some homothety centered at $X$ maps triangle $ABC$ onto triangle $DEF$.

**Incenter**   Center of inscribed circle.

**Incircle**   Inscribed circle.

**Inversion of Center $O$ and Radius $r$**   Given a point $O$ in the plane and a real number $r > 0$, the inversion through $O$ with radius $r$ maps every point $P \neq O$ to the point $P'$ on the ray $\overrightarrow{OP}$ such that $OP \cdot OP' = r^2$. We also refer to this map as inversion through $\omega$, the circle with center $O$ and radius $r$. Key properties of inversion are:

1. Lines passing through $O$ invert to themselves (though the individual points on the line are not all fixed).

2. Lines not passing through $O$ invert to circles through $O$, and vice versa.

3. Circles not passing through $O$ invert to other circles not through $O$.

4. A circle other than $\omega$ inverts to itself (as a whole, not point-by-point) if and only if it is orthogonal to $\omega$, that is, it intersects $\omega$ and the tangents to the circle and to $\omega$ at either intersection point are perpendicular.

5. **Inversive Distance Formula**: If $A'$ and $B'$ are the images of $A$ and $B$, respectively, under the inversion, then

$$A'B' = \frac{r^2 \cdot AB}{OA \cdot OB}.$$

**Isogonal conjugate**   Let $ABC$ be a triangle and let $P$ be a point in the plane which does not lie on any of the lines $AB$, $BC$, and $CA$. There exists a unique point $Q$ in the plane such that $\angle ABP = \angle QBC$, $\angle BCP = \angle QCA$, and $\angle CAP = \angle QAB$, where the angles in these equations are directed modulo $\pi$. We call $Q$ the isogonal conjugate of $P$. With this definition, we see that $P$ is also the isogonal conjugate of $Q$.

**Jensen's inequality**   If $f$ is concave up on an interval $[a, b]$ and $\lambda_1$, $\lambda_2$, ..., $\lambda_n$ are nonnegative numbers with sum equal to 1, then

$$\lambda_1 f(x_1) + \lambda_2 f(x_2) + \cdots + \lambda_n f(x_n) \geq f(\lambda_1 x_1 + \lambda_2 x_2 + \cdots + \lambda_n x_n)$$

for any $x_1, x_2, \ldots, x_n$ in the interval $[a, b]$. If the function is concave down, the inequality is reversed.

**Kite**   A quadrilateral with its sides forming two pairs of congruent adjacent sides. A kite is symmetric with respect to one of its diagonals. (If it is symmetric about both diagonals, it is a rhombus.) The two diagonals of a kite are perpendicular to each other. For example, if $ABCD$ is a quadrilateral with $AB = AD$ and $CB = CD$, then quadrilateral $ABCD$ is a kite and is symmetric about diagonal $\overline{AC}$.

**Kummer's Theorem**   Given nonnegative integers $a$ and $b$ and a prime $p$, $p^t \mid \binom{a+b}{a}$ if and only if $t$ is less than or equal to the number of carries in the addition $a + b$ in base $p$.

**Lattice point**   In the Cartesian plane, the lattice points are the points $(x, y)$ for which $x$ and $y$ are both integers.

**Lagrange's Interpolation Formula**   Let $x_0, x_1, \ldots, x_n$ be distinct real numbers, and let $y_0, y_1, \ldots, y_n$ be arbitrary real numbers. Then there exists a unique polynomial $P(x)$ of degree at most $n$ such that $P(x_i) = y_i$, $i = 0, 1, \ldots, n$. This polynomial is given by

$$P(x) = \sum_{i=0}^{n} y_i \frac{(x - x_0) \cdots (x - x_{i-1})(x - x_{i+1}) \cdots (x - x_n)}{(x_i - x_0) \cdots (x_i - x_{i-1})(x_i - x_{i+1}) \cdots (x_i - x_n)}.$$

**Law of cosines**   In a triangle $ABC$,

$$CA^2 = AB^2 + BC^2 - 2AB \cdot BC \cos \angle ABC,$$

and analogous equations hold for $AB^2$ and $BC^2$.

**Law of quadratic reciprocity**   If $p, q$ are distinct odd primes, then

$$\left( \frac{p}{q} \right) \left( \frac{q}{p} \right) = (-1)^{\frac{(p-1)(q-1)}{4}},$$

where $\left( \frac{p}{q} \right)$ and $\left( \frac{q}{p} \right)$ are **Legendre symbols**.

**Law of sines**   In a triangle $ABC$ with circumradius equal to $R$ one has

$$\frac{\sin A}{BC} = \frac{\sin B}{AC} = \frac{\sin C}{AB} = 2R.$$

**Legendre symbol** If $m$ is an integer and $n$ is a positive prime, then the Legendre symbol $\left(\frac{m}{n}\right)$ is defined to equal 0 if $n \mid m$, 1 if $m$ is a nonzero quadratic residue modulo $n$, and $-1$ if $m$ is a quadratic nonresidue modulo $n$ (i.e. not a quadratic residue modulo $n$).

**Lucas Numbers** The sequence $\{L_n\}_{n=0}^{\infty}$ defined recursively by

$$L_0 = 2, \quad L_1 = 1, \quad L_{n+1} = L_n + L_{n-1}$$

for all $n \geq 1$ is the Lucas sequence. It can be shown that

$$L_n = \left(\frac{1 + \sqrt{5}}{2}\right)^n + \left(\frac{1 - \sqrt{5}}{2}\right)^n$$

for all $n \geq 0$.

**Lucas's theorem** Let $p$ be a prime; let $a$ and $b$ be two positive integers such that

$$a = a_k p^k + a_{k-1} p^{k-1} + \cdots + a_1 p + a_0,$$

$$b = b_k p^k + b_{k-1} p^{k-1} + \cdots + b_1 p + b_0,$$

where $0 \leq a_i, b_i < p$ are integers for $i = 0, 1, \ldots, k$. Then

$$\binom{a}{b} \equiv \binom{a_k}{b_k}\binom{a_{k-1}}{b_{k-1}} \cdots \binom{a_1}{b_1}\binom{a_0}{b_0} \pmod{p}.$$

**Matrix** A matrix is a rectangular array of objects. A matrix $A$ with $m$ rows and $n$ columns is an $m \times n$ matrix. The object in the $i$th row and $j$th column of matrix $A$ is denoted $a_{i,j}$. If a matrix has the same number of rows as it has columns, then the matrix is called a square matrix. In a square $n \times n$ matrix $A$, the **main diagonal** consists of the elements $a_{1,1}, a_{2,2}, \ldots, a_{n,n}$.

**Menelaus' theorem** Given a triangle $ABC$, let $F$, $G$, $H$ be points on lines $BC$, $CA$, $AB$, respectively. Then $F$, $G$, $H$ are collinear if and only if, using directed lengths,

$$\frac{AH}{HB} \cdot \frac{BF}{FC} \cdot \frac{CG}{GA} = -1.$$

**Minkowski's inequality** Given a positive integer $n$, a real number $r \geq 1$, and positive reals $a_1, a_2, \ldots, a_n$ and $b_1, b_2, \ldots, b_n$, we have

$$\left(\sum_{i=1}^{n}(a_n + b_n)^r\right)^{1/r} \leq \left(\sum_{i=1}^{n}a_i^r\right)^{1/r} + \left(\sum_{i=1}^{n}b_i^r\right)^{1/r}.$$

**Multiset**   Informally, a multiset is a set in which an element may appear more than once. For instance, $\{1, 2, 3, 2\}$ and $\{2, 2, 2, 3, 1\}$ are distinct multisets.

**Nine point circle**   See **Feuerbach circle**.

**Orbit**   Suppose that $S$ is a collection of functions on a set $T$, such that $S$ is closed under composition and each $f \in S$ has an inverse. $T$ can be partitioned into its orbits under $S$, sets of elements such that $a$ and $b$ are in the same set if and only if $f(a) = b$ for some $f \in S$.

**Order**   Given a nonzero element $g$ of a finite field, there exists a smallest positive integer $d$, named the order of $g$, such that $g^d = 1$.

**Orthocenter of a triangle**   Point of intersection of the altitudes.

**Pascal's theorem**   If $ABCDEF$ is a hexagon inscribed in a conic in the projective plane, such that each pair of opposite sides intersects at most one point, then the three intersection points formed in this manner are collinear. (If the hexagon is inscribed in a conic in the affine plane, then either the above result holds, or else each pair of opposite sides is parallel.) This theorem the dual to **Brianchon's theorem**.

**Pell's equations**   If $D$ is a prime congruent to 3 modulo 4, then the Diophantine equation

$$x^2 - Dy^2 = 1$$

in $x$ and $y$ is known as a Pell's equation. This equation has infinitely many integer solutions in $x$ and $y$.

**Phi function**   For all positive integers $n$, $\phi(n)$ is defined to be the number of integers in $\{1, 2, \ldots, n\}$ which are relatively prime to $n$. This function is multiplicative — that is, $\phi(ab) = \phi(a)\phi(b)$ for all $a, b$ relatively prime.

**Periodic function**   $f(x)$ is periodic with period $T > 0$ if

$$f(x + T) = f(x)$$

for all $x$.

**Permutation**   Let $S$ be a set. A permutation of $S$ is a one-to-one function $\pi : S \to S$ that maps $S$ onto $S$. If $S = \{x_1, x_2, \ldots, x_n\}$ is a finite set, then we may denote a permutation $\pi$ of $S$ by $\{y_1, y_2, \ldots, y_n\}$, where $y_k = \pi(x_k)$.

**Pick's theorem**   Given a non self-intersecting polygon $\mathcal{P}$ in the coordinate plane whose vertices are at lattice points, let $B$ denote the number of lattice points on its boundary and let $I$ denote the number of lattice points in its interior. The area of $\mathcal{P}$ is given by the formula $I + \frac{1}{2}B - 1$.

**Pigeonhole principle**   If $n$ objects are distributed among $k < n$ boxes, some box contains at least two objects.

**Pole-polar transformation**   Let $C$ be a circle with center $O$ and radius $R$. The pole-polar transformation with respect to $C$ maps points different from $O$ to lines, and lines that do not pass through $O$ to points. If $P \neq O$ is a point then the **polar** of $P$ is the line $p'$ that is perpendicular to ray $\overrightarrow{OP}$ and satisfies

$$d(O, P)d(O, p') = R^2,$$

where $d(A, B)$ denote the distance between the objects $A$ and $B$. If $q$ is a line that does not pass through $O$, then the **pole** of $q$ is the point $Q'$ that has polar $q$. The pole-polar transformation with respect to the circle $C$ is also called **reciprocation** in the circle $C$.

**Polynomial in $x$ of degree $n$**   Function of the form $f(x) = \sum_{k=0}^{n} a_k x^k$.

**Power of a Point Theorem**   Given a fixed point $P$ and a fixed circle $\omega$, draw a line through $P$ which intersects the circle at $X$ and $Y$. The power of the point $P$ with respect to $\omega$ is defined to be the signed product $PX \cdot PY$. The power of a point theorem states that this quantity is a constant; i.e., does not depend on the line through $P$. It does not matter whether $P$ is in, on, or outside $\omega$.

**Power mean inequality**   Let $a_1, a_2, \ldots, a_n$ be any positive numbers for which $a_1 + a_2 + \cdots + a_n = 1$. For positive numbers $x_1, x_2, \ldots, x_n$ we define

$$M_{-\infty} = \min\{x_1, x_2, \ldots, x_k\},$$
$$M_{\infty} = \max\{x_1, x_2, \ldots, x_k\},$$
$$M_0 = x_1^{a_1} x_2^{a_2} \cdots x_n^{a_n},$$
$$M_t = (a_1 x_1^t + a_2 x_2^t + \cdots + a_k x_k^t)^{1/t},$$

where $t$ is a nonzero real number. Then

$$M_{-\infty} \leq M_s \leq M_t \leq M_{\infty}$$

for $s \leq t$.

**Primitive element**   For each prime $p$, a field $F$ with $p$ elements contains an element $g$, called a primitive element of $F$, with the following property: for any nonzero element $h$ of $F$, there exists an integer $k$ such that $g^k = h$.

**Projective plane**   Let $\mathbb{K}$ be a field. The projective plane over $\mathbb{K}$ is the set of equivalence classes of $\mathbb{K}^3 - \{(0,0,0)\}$, under equivalence by scalar multiplication (that is, where $(a,b,c)$ and $(d,e,f)$ are equivalent if and only if $(a,b,c) = (d\kappa, e\kappa, f\kappa)$ for some $\kappa \in \mathbb{K}$). The elements of $\mathbb{K}$ are called *points*, and the equivalence class containing $(a,b,c)$ is often denoted $[a,b,c]$ or $[a:b:c]$. Also, given $(\alpha, \beta, \gamma) \in \mathbb{K}^3 - \{(0,0,0)\}$, the set of solutions $[x,y,z]$ to

$$\alpha x + \beta y + \gamma z = 0$$

is called a *line* in the projective plane over $\mathbb{K}$. Any two distinct points (resp. lines) are said to "intersect in" or "lie on" a unique line (resp. point).

**Ptolemy's theorem**   In a convex cyclic quadrilateral $ABCD$,

$$AC \cdot BD = AB \cdot CD + AD \cdot BC.$$

**Quadratic residue**   The integer $m$ is a quadratic residue modulo $n$ if $m \equiv k^2 \pmod{n}$ for some $k$. Otherwise, $m$ is a quadratic nonresidue.

**Radical axis**   Let $\omega_1$ and $\omega_2$ be two non-concentric circles. The locus of all points of equal power with respect to these circles is called the radical axis of $\omega_1$ and $\omega_2$.

**Radical axis theorem**   Let $\omega_1, \omega_2, \omega_3$ be three circles whose centers are not collinear. There is exactly one point whose powers with respect to the three circles are all equal. This point is called the **radical center** of $\omega_1, \omega_2, \omega_3$.

**Root of an equation**   Solution to the equation.

**Root of unity**   Solution to the equation $z^n - 1 = 0$.

**RMS-AM (Root Mean Square-Arithmetic Mean) Inequality**   For positive real numbers $x_1, x_2, \ldots, x_n$,

$$\sqrt{\frac{x_1^2 + x_2^2 + \cdots + x_k^2}{n}} \geq \frac{x_1 + x_2 + \cdots + x_k}{n}.$$

This is a special case of Jensen's Inequality with $f(x) = x^2$, and it is also a special case of the Power Mean Inequality.

**Sigma function**   For all positive integers $n$, $\sigma(n)$ is defined to be the sum of all positive integer divisors of $n$. This function is multiplicative — that is, $\sigma(ab) = \sigma(a)\sigma(b)$ for all $a, b$ relatively prime.

**Simson line**   For any point $P$ on the circumcircle of $\triangle ABC$, the feet of the perpendiculars from $P$ to the sides of $\triangle ABC$ all lie on a line called the Simson line of $P$ with respect to $\triangle ABC$.

**Solid triangle inequality**   Given four points $A, B, C, P$ in three-dimensional space which are not coplanar, we have

$$\angle APB + \angle BPC > \angle APC.$$

**Stewart's theorem**   In a triangle $ABC$ with cevian $\overline{AD}$, write $a = BC$, $b = CA$, $c = AB$, $m = BD$, $n = DC$, and $d = AD$. Then

$$d^2a + man = c^2n + b^2m.$$

This formula can be used to express the lengths of the altitudes and angle bisectors of a triangle in terms of its side lengths.

**Thue-Morse sequence**   The sequence $t_0, t_1, \ldots$, defined by $t_0 = 0$ and the recursive relations $t_{2k} = t_k$, $t_{2k+1} = 1 - t_{2k}$ for $k \geq 1$. The binary representation of $n$ contains an odd number of 1's if and only if $t_n$ is odd.

**Transformation**   A transformation of the plane is a mapping of the plane onto itself such that every point $P$ is mapped to a unique image $P'$ and every point $Q'$ has a unique prototype (preimage, inverse image, counterimage) $Q$.

- A **reflection across a line** (in the plane) is a transformation which takes every point in the plane to its mirror image, with the line as mirror. A **rotation** is a transformation in which the entire plane is rotated about a fixed point in the plane.

- A **similarity** is a transformation that preserves ratios of distances. If $P'$ and $Q'$ are the respective images of points $P$ and $Q$ under a similarity $\mathbf{T}$, then the ratio $P'Q'/PQ$ depends only on $\mathbf{T}$. This ratio is the **similitude** of $\mathbf{T}$. A **dilation** is a direction-preserving similarity, i.e., a similarity that takes each line into a parallel line.

- The **product $\mathbf{T_2 T_1}$ of two transformations** is transformation defined by $\mathbf{T_2 T_1} = \mathbf{T_2} \circ \mathbf{T_1}$, where $\circ$ defines function composition. A **spiral similarity** is the product of a rotation and a dilation, or vice versa.

**Triangular number**   A number of the form $n(n+1)/2$, where $n$ is some positive integer.

**Trigonometric identities**

$$\sin^2 x + \cos^2 x = 1,$$
$$1 + \cot^2 x = \csc^2 x,$$
$$\tan^2 x + 1 = \sec^2 x;$$

*addition and subtraction formulas:*

$$\sin(a \pm b) = \sin a \cos b \pm \cos a \sin b,$$
$$\cos(a \pm b) = \cos a \cos b \mp \sin a \sin b,$$
$$\tan(a \pm b) = \frac{\tan a \pm \tan b}{1 \mp \tan a \tan b};$$

*double-angle formulas:*

$$\sin 2a = 2 \sin a \cos a$$
$$= \frac{2 \tan a}{1 + \tan^2 a},$$

$$\cos 2a = 2 \cos^2 a - 1 = 1 - 2 \sin^2 a$$
$$= \frac{1 - \tan^2 a}{1 + \tan^2 a},$$

$$\tan 2a = \frac{2 \tan a}{1 - \tan^2 a};$$

*triple-angle formulas:*

$$\sin 3a = 3 \sin a - 4 \sin^3 a,$$
$$\cos 3a = 4 \cos^3 a - 3 \cos a,$$
$$\tan 3a = \frac{3 \tan a - \tan^3 a}{1 - 3 \tan^2 a};$$

*half-angle formulas:*

$$\sin^2 \frac{a}{2} = \frac{1 - \cos a}{2},$$
$$\cos^2 \frac{a}{2} = \frac{1 + \cos a}{2};$$

*sum-to-product formulas:*

$$\sin a + \sin b = 2 \sin \frac{a+b}{2} \cos \frac{a-b}{2},$$

$$\cos a + \cos b = 2 \cos \frac{a+b}{2} \cos \frac{a-b}{2},$$

$$\tan a + \tan b = \frac{\sin(a+b)}{\cos a \cos b};$$

*difference-to-product formulas:*

$$\sin a - \sin b = 2 \sin \frac{a-b}{2} \cos \frac{a+b}{2},$$

$$\cos a - \cos b = -2 \sin \frac{a-b}{2} \sin \frac{a+b}{2},$$

$$\tan a - \tan b = \frac{\sin(a-b)}{\cos a \cos b};$$

*product-to-sum formulas:*

$$2 \sin a \cos b = \sin(a+b) + \sin(a-b),$$

$$2 \cos a \cos b = \cos(a+b) + \cos(a-b),$$

$$2 \sin a \sin b = -\cos(a+b) + \cos(a-b).$$

**Wilson's theorem**   If $n > 1$ be a positive integer, then

$$(n-1)! \equiv -1 \pmod{n}$$

if and only if $n$ is prime.

**Zeckendorf representation**   Let $F_0, F_1, \ldots$ be the Fibonacci numbers $1, 2, \ldots$. Each nonnegative integer $n$ can be written uniquely as a sum of nonconsecutive positive Fibonacci numbers; that is, each nonnegative integer $n$ can be written uniquely in the form

$$n = \sum_{k=0}^{\infty} \alpha_k F_k,$$

where $\alpha_k \in \{0, 1\}$ and $(\alpha_k, \alpha_{k+1}) \neq (1, 1)$ for each $k$. This expression for $n$ is called its Zeckendorf representation.

# Classification of Problems

## Algebra

| | |
|---|---|
| Belarus | 00-8 |
| Bulgaria | 01-9 |
| Belarus | 00-8 |
| Bulgaria | 01-10 |
| China | 01-7, 8 |
| India | 01-3 |
| Iran | 00-2; 01-1, 10 |
| Italy | 00-3 |
| Japan | 01-2 |
| Korea | 01-1, 4 |
| Mongolia | 00-1 |
| Poland | 00-1, 3, 6; 01-3 |
| Romania | 00-1; 01-3, 4 |
| Russia | 00-32; 01-12, 17, 22, 33 |
| Taiwan | 01-2 |
| Asian Pacific | 00-1, 2; 01-4 |
| St. Petersburg | 01-2, 11, 21 |

## Algebra and Combinatorics

| | |
|---|---|
| China | 00-2, 4 |
| Poland | 01-6 |
| Romania | 01-12 |
| Russia | 01-1, 36 |

| | |
|---|---|
| Mediterranean | 00-1 |
| St. Petersburg | 00-9 |

## Functional Equations

| | |
|---|---|
| Belarus | 00-4 |
| Czech and Slovak | 01-3 |
| Estonia | 00-4 |
| India | 00-4 |
| Iran | 00-5, 6 |
| Korea | 00-2 |
| Mongolia | 00-3 |
| Romania | 01-5, 11 |
| Russia | 00-7, 40 |
| Taiwan | 00-4 |
| Turkey | 00-9 |
| Vietnam | 00-5 |
| Czech-Slovak-Polish | 01-5 |
| St. Petersburg | 01-7 |

## Combinatorics

*Also see "Algebra and Combinatorics" under Algebra and "Combinatorics and Number Theory" under Number Theory.*

| | |
|---|---|
| Belarus | 00-2, 5, 7, 9; 01-1, 8, 10 |
| Bulgaria | 00-2, 12; 01-2, 7 |
| Canada | 00-1 |
| China | 00-3, 6; 01-3, 5, 10 |
| Czech and Slovak | 01-2 |
| Iran | 00-7; 01-2, 6 |
| Japan | 01-1 |
| Korea | 00-5 |
| Mongolia | 00-6 |
| Poland | 00-5 |
| Romania | 00-2, 10; 01-2, 6 |
| Russia | 00-6, 12, 15, 24, 29, 31, 37, 39, 46; 01-3, 14, 26, 34, 38 |

*(Combinatorics, continued)*

| | |
|---|---|
| Taiwan | 00-3; 01-3, 6 |
| Turkey | 00-3 |
| United States | 00-3, 4; 01-1 |
| Vietnam | 01-3 |
| Asian Pacific | 00-5 |
| Austrian-Polish | 00-2; 01-4 |
| Balkan | 00-2 |
| Baltic | 01-1 |
| Czech-Slovak-Polish | 01-3 |
| St. Petersburg | 00-2, 6, 11, 13, 16, 23; 01-1, 8, 13, 16 |

## Combinatorial Geometry

| | |
|---|---|
| Bulgaria | 00-8; 01-8 |
| Czech and Slovak | 00-4 |
| Iran | 00-4 |
| Israel | 00-4 |
| Japan | 00-4 |
| Romania | 00-9; 01-8 |
| Russia | 00-9, 18, 23, 26, 33; |
| | 01-4, 11, 20, 25, 30, 37 |
| United States | 01-6 |
| Vietnam | 01-5 |
| Czech-Slovak-Polish | 01-6 |
| Austrian-Polish | 00-3, 5 |
| St. Petersburg | 00-15; 01-12 |
| Vietnam | 01-5 |

## Graph Theory

| | |
|---|---|
| Hungary | 00-4 |
| India | 00-3 |
| Russia | 00-3, 36, 43; 01-27, 41 |
| St. Petersburg | 00-4, 10, 21; 01-3 |

# Geometry

*Also see "Combinatorial Geometry" under* Combinatorics *and "Geometric Inequalities" under* Inequalities.

| | |
|---|---|
| Belarus | 00-1, 3, 11; 01-3, 4, 6, 9 |
| Bulgaria | 00-1, 3, 5, 7, 9, 11; 01-1, 3, 6 |
| Canada | 00-2; 01-1, 3 |
| China | 00-1; 01-1 |
| Czech and Slovak | 00-2, 3 |
| Estonia | 00-3, 5 |
| Hungary | 00-3, 6; 01-3, 5, 6 |
| India | 00-1; 01-1, 2 |
| Iran | 00-1; 01-4, 7, 9 |
| Iran | 00-3, 8, 9 |
| Israel | 00-3 |
| Italy | 00-1 |
| Japan | 01-5 |
| Korea | 00-3; 01-2, 5 |
| Mongolia | 00-2, 4 |
| Poland | 00-2; 01-5 |
| Romania | 00-5, 8; 01-7 |
| Russia | 00-2, 5, 13, 17, 20, 21, 27, 30, 35, 38, 42, 45; 01-2, 6, 8, 9, 13, 18, 19, 24, 28, 31, 35, 42 |
| Taiwan | 00-1; 01-1, 4 |
| Turkey | 00-2, 6, 7, 8 |
| United Kingdom | 00-1 |
| United States | 00-2, 5; 01-2, 4 |
| Vietnam | 00-1, 2, 3; 01-2 |
| Asian Pacific | 00-3; 01-3, 5 |
| Austrian-Polish | 00-4 |
| Balkan | 00-1; 01-2, 3 |
| Baltic | 01-2 |
| Czech-Slovak-Polish | 01-2 |
| Mediterranean | 00-2 |
| St. Petersburg | 00-1, 8, 12, 17, 19, 22; 01-4, 6, 10, 14, 18, 20 |

# Inequalities

| | |
|---|---|
| Belarus | 00-10; 01-2, 5 |
| Canada | 00-3 |
| China | 01-6 |
| Czech and Slovak | 00-1; 01-1 |
| Estonia | 00-1 |
| Hungary | 00-5; 01-1 |
| India | 01-4 |
| Iran | 00-10; 01-8 |
| Japan | 01-3 |
| Korea | 00-6; 01-3 |
| Poland | 01-1 |
| Romania | 00-3 |
| Russia | 00-4, 10, 19, 41, 44; 01-23, 39 |
| United Kingdom | 00-2 |
| United States | 00-1, 6; 01-3 |
| Vietnam | 01-4 |
| Asian Pacific | 00-4 |
| Austrian-Polish | 00-6 |
| Baltic | 01-3 |
| Czech-Slovak-Polish | 01-1, 4 |
| St. Petersburg | 00-5, 14, 20; 01-5, 19 |

## Geometric Inequalities

| | |
|---|---|
| Israel | 00-2 |
| Japan | 00-2 |
| Poland | 00-4; 01-2 |
| Turkey | 00-4 |
| Austrian-Polish | 01-3 |
| Mediterranean | 00-3 |

# Number Theory

| | |
|---|---|
| Belarus | 00-6 |
| Bulgaria | 00-4, 6; 01-4, 10 |

*(Number Theory, continued)*

| China | 00-5; 01-4, 9 |
|---|---|
| Estonia | 00-2 |
| Hungary | 00-1, 2; 01-2, 4 |
| India | 00-2 |
| Iran | 01-3, 5 |
| Israel | 00-1 |
| Japan | 01-4 |
| Korea | 00-1, 4 |
| Mongolia | 00-5 |
| Poland | 01-4 |
| Romania | 00-4, 6, 7; 01-1, 9, 10 |
| Russia | 00-11, 16, 22, 25, 28, 34; 01-5, 7, 16, 29, 40 |
| Taiwan | 00-2; 01-5 |
| Asian Pacific | 01-1, 2 |
| Austrian-Polish | 00-1; 01-1, 2 |
| Balkan | 00-3, 01-1 |
| Baltic | 01-4 |
| St. Petersburg | 00-3, 18; 01-15, 17 |

## Combinatorics and Number Theory

| Belarus | 01-7 |
|---|---|
| Bulgaria | 00-10; 01-5 |
| Canada | 01-2 |
| China | 01-2 |
| Estonia | 00-6 |
| Hungary | 00-7; 01-7 |
| Italy | 00-2 |
| Japan | 00-1, 3 |
| Korea | 01-6 |
| Russia | 00-1, 8, 14; 01-10, 15, 21, 32 |
| Turkey | 00-1, 5 |
| United Kingdom | 00-3 |
| United States | 01-5 |
| Vietnam | 00-4; 01-1, 6 |
| St. Petersburg | 00-7; 01-9 |